A Colour Atlas of Salmonid Diseases

David W. Bruno • Patricia A. Noguera
Trygve T. Poppe

A Colour Atlas of Salmonid Diseases

Second Edition

Springer

David W. Bruno
Marine Scotland Science
Aberdeen, UK

Patricia A. Noguera
Marine Scotland Science
Aberdeen, UK

Trygve T. Poppe
Norwegian School of Veterinary Science
Oslo, Norway

ISBN 978-94-007-2009-1 ISBN 978-94-007-2010-7 (eBook)
DOI 10.1007/978-94-007-2010-7
Springer Dordrecht Heidelberg New York London

Library of Congress Control Number: 2013947087

Printed on acid-free paper

Springer is part of Springer Science+Business Media (www.springer.com)

Preface

Understanding disease is a challenging and complex process and different from diagnosing pathogen involvement. Sophisticated molecular techniques are becoming available for a range of infectious diseases and represent valuable tools; however, for disease diagnosis they generally focus on confirming or ruling out the presence of an aetiological agent. The fish itself is a complicated biological system, and the disease process the result of interactions between the agent, the host and the environment. An understanding of the normal structure and function is important so that deviations from normality that occur during disease are recognised. A good pathologist should therefore possess knowledge in diverse disciplines such as biology, genetics and physiology, parasitology, microbiology and immunology, to mention a few.

Within aquaculture, fish farming has experienced a tremendous growth and is still developing rapidly with salmonid production being a relevant component of this expansion, if not in terms of tonnage, certainly in terms of its economic value. As with all intensification in animal production, new diseases and diverse problems arise and represent diagnostic challenges, economic losses and, importantly, animal welfare issues. In this scenario and in spite of all the new techniques available, morphological pathology will remain the gold standard and cannot be replaced in the initial stages of diagnostic work. In addition, fish are increasingly being used as experimental animals, and therefore knowledge of basic fish pathology remains important for scientists.

It is our hope and intention that this book will increase the awareness and significance of diseases and animal welfare of this iconic group of fish—the salmonids. We anticipate that fish farmers, fish veterinarians and practitioners, hobbyists, anglers, policy makers and regulators will find this book useful, and even for those not working with disease as their primary focus, we believe that by understanding the pathological processes this can make their job more rewarding. For the authors, fish pathology has been a life-long experience and passion, something we hope a new generation of professionals will give continuity. The pathogen diversity and new manifestations of disease makes it a very dynamic discipline and an all-embracing art where there will always be something new to learn.

'Make an early start in your career to become a good pathologist'

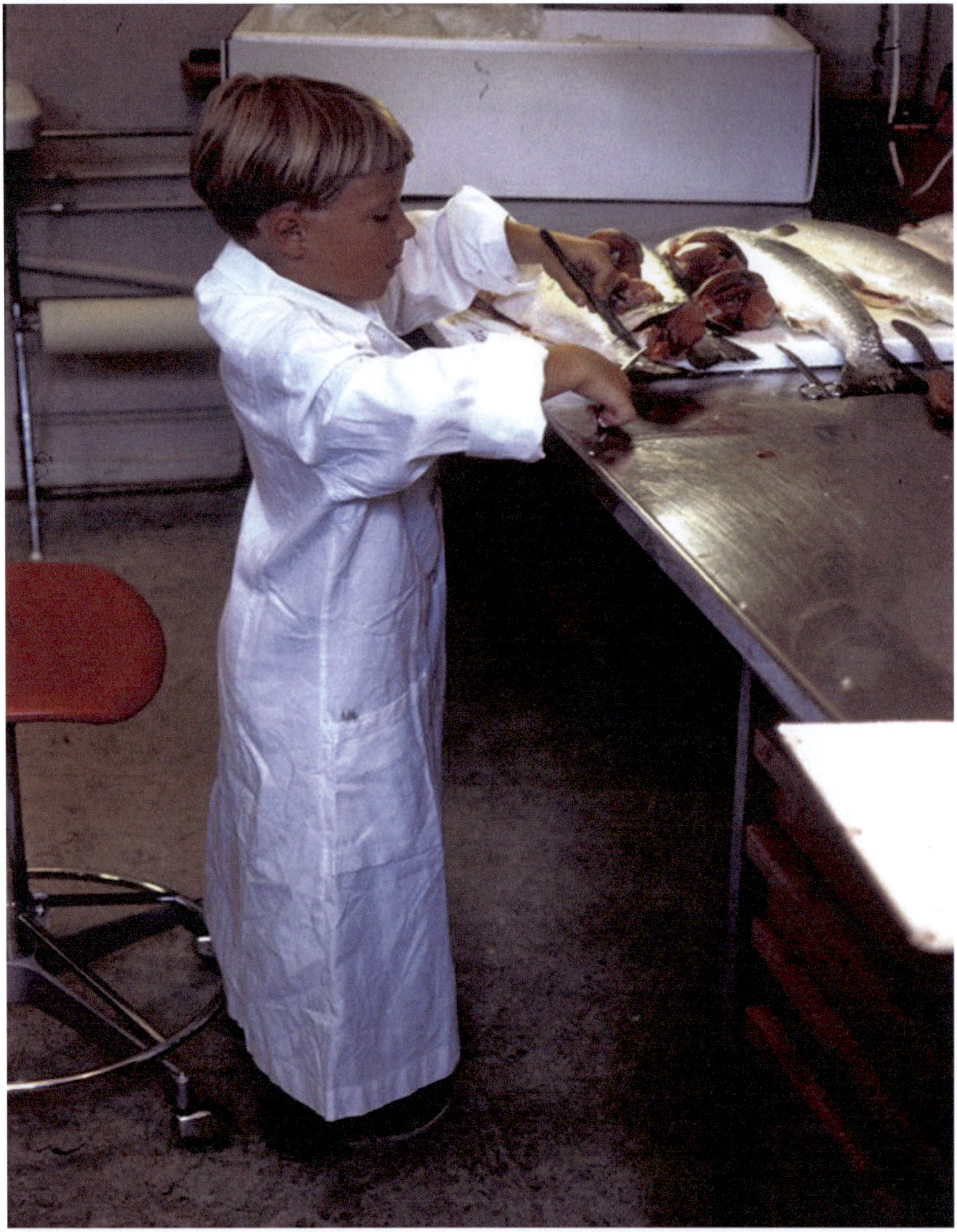

David Bruno, Patricia Noguera, Trygve Poppe

Acknowledgements

The authors would like to thank the Marine Scotland Science and the Norwegian School of Veterinary Science for their support. In addition we are grateful to Tor Atle Mo who advised on the recent taxonomic changes within the Protist and Metazoa and to Graham Bruno who has tirelessly dealt with our IT enquiries. We also appreciate the help of Mark Fordyce and Nichola McManus from Marine Scotland for cutting and staining additional material. We are also grateful to many colleagues who have generously donated images or wax blocks from which sections were prepared for this book, and these are acknowledged below.

Alaska Department of Fish and Game	6.2, 9.55
Amanda Walker	10.4
Andrea Warwick	6.11
Brian Shaw	11.1
Carla Conway	5.29, 5.30
Charles McGurk	10.14
Daniel Pendry	6.56
David Cox	8.22
Dave Groman	4.2 (b), 6.44
Diane Elliott	8.12
EirikWelde	12.5
Erik Sterud	7.1
Gael Kurath	5.26
Heike Schmidt	12.7
Iva Dykova	8.19
Jayde Ferguson	8.5, 12.8 (b), 12.16
John McArdle	4.3 (d)
Lars Lonnstrom	6.16, 6.17
Lars Thomas Poppe	1.1 (b)
Mamoru Yoshimizu	5.15
Mansour El-Matbouli	9.9
Mar Marcos-Lopez	8.1, 11.2, 11.4, 11.5
Marshall Halliday	7.2
Michail Belov	1.1 (k)
Mike Kent	5.20, 5.21, 5.46 (a)
Perttu Koski	10.24
Sandra Bravo	4.3 (a), 9.51, 9.52
SatuViljamaa-Dirks	6.52, 6.53, 6.54
Scott Foot	9.13
Simon Jones	9.3
Sonia Duguid	7.15, 10.59
Sonia Mumford	7.13, 8.14, 8.23, 8.29, 9.6, 9.32, 10.11, 10.22, 10.33

(continued)

(continued)

Stephen Atkinson	9.1 (a)
Stephen Feist	8.4, 8.20, 9.2, 9.17
Torsten Boutrup	5.24, 5.25
Thomas Wahli	8.7
Wendy Sealey / Charlie Smith	10.23

'Thank you Arko for keeping our samples cool and guarding them well'

Contents

1 Introduction . . . 1
Further Reading . . . 4

2 Functional Anatomy . . . 5
2.1 Respiratory System (including the operculum and pseudobranch) . . . 6
2.2 Circulatory System (Cardiovascular and Secondary Circulation) . . . 8
2.2.1 The Cardiovascular System . . . 8
2.2.2 Secondary Circulation . . . 11
2.3 Integument System . . . 11
2.4 Musculoskeletal System . . . 12
2.4.1 The Skeletal System . . . 12
2.4.2 Fins . . . 13
2.5 Excretory System . . . 14
2.5.1 Kidney . . . 14
2.6 Digestive System . . . 15
2.6.1 Liver and Gall Bladder . . . 18
2.6.2 Pancreas . . . 19
2.6.3 The Swim or Air Bladder . . . 19
2.7 Sensory System . . . 20
2.7.1 The Mechanosensory Lateral Line . . . 20
2.7.2 Visual . . . 20
2.8 Nervous System . . . 21
2.8.1 Brain . . . 21
2.8.2 The Sympathetic and Parasympathetic Ganglions and Nerves . . . 22
2.9 Endocrine System . . . 22
2.10 Reproductive System . . . 24
2.10.1 Ovary . . . 24
2.10.2 Testes . . . 25
Further Reading . . . 26

3 Post-mortem Examination and the Recognition of Tissue Abnormalities . . . 27
3.1 Introduction . . . 27
3.2 Sample Size and Euthanasia . . . 28
3.3 Necropsy Procedure . . . 28
3.3.1 External Examination . . . 28
3.3.2 Internal Examination . . . 30

4 Pathology and Disease Diagnosis . . . 33
4.1 Inflammation and Proliferation . . . 33
4.2 Circulation Disorders . . . 44
4.3 Cell Injury and Death (Necrosis and Apoptosis) . . . 46
4.4 Pigments and Mineralization . . . 48

4.5 Neoplasia . . . 48
4.6 Artifacts . . . 50
Further Reading . . . 51

5 Viral Diseases . . . 53
5.1 Infectious Pancreatic Necrosis . . . 53
5.2 Infectious Salmon Anaemia . . . 56
5.3 *Oncorhynchus masou* Virus . . . 58
5.4 Piscine Reovirus (Heart and Skeletal Muscle Inflammation, HSMI) . . . 59
5.5 Salmon Leukaemia Virus . . . 60
5.6 Viral Haemorrhagic Septicaemia . . . 60
5.7 Infectious Haematopoietic Necrosis . . . 62
5.8 Hirame Rhabdovirus . . . 64
5.9 Salmonid Alphavirus (Pancreas and Sleeping Disease) . . . 64
5.10 Piscine Myocarditis Virus (Cardiomyopathy Syndrome) . . . 67
5.11 Erythrocytic Inclusion Body Syndrome . . . 69
Further Reading . . . 70

6 Bacterial Diseases . . . 73
6.1 *Aeromonas hydrophila* . . . 74
6.2 *Aeromonas salmonicida* subsp. *salmonicida* . . . 75
6.3 *Listonella anguillarum* . . . 76
6.4 *Aliivibrio* (*Vibrio*) *salmonicida* . . . 78
6.5 *Moritella viscosa* . . . 79
6.6 *Yersinia ruckeri* . . . 80
6.7 *Pseudomonas fluorescens* . . . 81
6.8 *Flavobacterium psychrophilum* . . . 82
6.9 *Flavobacterium columnare* . . . 84
6.10 *Tenacibaculum maritimum* . . . 85
6.11 *Hafnia alvei* . . . 86
6.12 *Chryseobacterium* spp. . . . 86
6.13 *Pasteurella skyensis* . . . 86
6.14 *Francisella noatunensis* subsp. *noatunensis* . . . 87
6.15 *Renibacterium salmoninarum* . . . 87
6.16 *Carnobacterium maltaromaticum* . . . 89
6.17 *Streptococcus phocae* . . . 90
6.18 *Mycobacterium* spp. . . . 90
6.19 *Nocardia* sp. . . . 91
6.20 *Piscirickettsia salmonis* . . . 91
6.21 Epitheliocystis . . . 93
6.22 *Candidatus arthromitus* . . . 94
6.23 Red Mark Syndrome = Cold Water Strawberry Disease . . . 94
Further Reading . . . 96

7 Fungal and Related Oomycete Infections . . . 99
7.1 *Saprolegnia* spp. . . . 99
7.2 *Isaria farinosa* (formally *Paecilomyces farinosus*) . . . 101
7.3 *Phialophora* sp. . . . 103
7.4 *Phoma herbarum* . . . 103
7.5 *Exophiala* spp. . . . 104
Further Reading . . . 105

8 Protists . . . 107
8.1 Amoebozoa . . . 107
8.1.1 *Paramoeba perurans* . . . 107
8.2 Ciliophora . . . 108
8.2.1 *Capriniana piscium* . . . 108
8.2.2 *Chilodonella piscicola* . . . 110
8.2.3 *Ichthyophthirius multifiliis* . . . 110
8.2.4 *Trichodina truttae* . . . 111
8.2.5 *Scyphidia* (*Riboscyphidia*, *Ambiphyra*) . . . 112
8.3 Mesomycetozoea . . . 112
8.3.1 *Ichthyophonus hoferi* . . . 112
8.3.2 *Sphaerothecum destruens* . . . 113
8.3.3 *Dermocystidium salmonis* . . . 114
8.4 Microsporidia . . . 115
8.4.1 *Kabatana takedai* . . . 115
8.4.2 *Loma salmonae* . . . 116
8.4.3 *Candidatus* Paranucleospora theridon (=*Desmozoon lepeophtheiri*) . . . 116
8.5 Sarcomastigophora . . . 117
8.5.1 *Cryptobia salmositica* . . . 117
8.5.2 *Ichthyobodo* spp. . . . 117
8.5.3 *Spironucleus* spp. . . . 118
Further Reading . . . 120

9 Metazoa . . . 123
9.1 Myxozoa . . . 123
9.1.1 *Ceratomyxa shasta* . . . 123
9.1.2 *Chloromyxum* spp. . . . 124
9.1.3 *Kudoa thyrsites* . . . 125
9.1.4 *Henneguya* spp. . . . 126
9.1.5 *Myxidium* spp. . . . 126
9.1.6 *Myxobolus cerebralis* . . . 127
9.1.7 *Parvicapsula* spp. . . . 128
9.1.8 *Sphaerospora truttae* . . . 130
9.1.9 *Tetracapsuloides bryosalmonae* . . . 130
9.2 Nematoda . . . 132
9.2.1 *Philonema oncorhynchi* . . . 132
9.2.2 *Cystidicola farionis* . . . 133
9.2.3 *Pseudoterranova decipiens* . . . 133
9.2.4 *Eustrongylides* sp. . . . 133
9.2.5 *Anisakis simplex* . . . 133
9.3 Cestoda . . . 135
9.3.1 *Eubothrium* spp. . . . 136
9.3.2 *Diphyllobothrium* spp. . . . 137
9.4 Trematoda . . . 137
9.4.1 *Sanguinicola* spp. . . . 137
9.4.2 *Cryptocotyle lingua* . . . 138
9.4.3 *Diplostomum spathaceum* . . . 138
9.4.4 *Phyllodistomum umblae* . . . 139
9.4.5 *Apatemon gracilis* . . . 139
9.4.6 *Cotylurus* spp. . . . 140
9.4.7 *Ichthyocotylurus erraticus* . . . 140
9.4.8 *Stephanostomum tenue* . . . 141

9.5 Monogenea . . . 141
9.5.1 *Dactylogyrus* sp. . . . 141
9.5.2 *Gyrodactylus salaris* . . . 141
9.5.3 *Discocotyle sagittata* . . . 142
9.6 Acanthocephala . . . 142
9.6.1 *Acanthocephalus* spp. . . . 143
9.6.2 *Pomphorhynchus* spp. . . . 143
9.7 Maxillopoda . . . 143
9.7.1 *Lepeophtheirus salmonis* . . . 143
9.7.2 *Caligus elongatus* . . . 145
9.7.3 *Caligus rogercresseyii* . . . 145
9.7.4 *Argulus coregoni, A. foliaceus* and *A. japonicus* . . . 146
9.7.5 *Salmincola* spp. . . . 146
9.8 Bilvalvia . . . 146
9.8.1 Glochidiosis . . . 147
9.9 Annelida . . . 148
9.9.1 Leeches . . . 148
Further Reading . . . 148

10 Production Diseases and Other Disorders . . . 151
10.1 Environment . . . 151
10.1.1 Water Quality and Husbandry . . . 151
10.1.2 Jellyfish . . . 152
10.1.3 Phytoplankton and Algal Blooms . . . 153
10.1.4 Gas Bubble Disease . . . 153
10.1.5 Nephrocalcinosis . . . 154
10.2 Vaccination Side-Effects . . . 155
10.3 Dietary Imbalance . . . 156
10.3.1 Inadequate Digestion of Pellet . . . 157
10.3.2 Hepatic Lipidosis . . . 157
10.3.3 Steatitis . . . 158
10.3.4 Heart Fat Infiltration . . . 158
10.3.5 Soybean-Induced Enteritis . . . 158
10.3.6 Vitamin Deficiencies . . . 158
10.3.7 Mineral Deficiencies . . . 162
10.4 Diseases Associated with the Eye . . . 162
10.5 Developmental Disorders . . . 164
10.5.1 Deformities Among Eggs and Fry . . . 164
10.5.2 Yolk Sac Constrictions . . . 164
10.5.3 Soft Egg Disease . . . 164
10.5.4 Blue Sac Disease . . . 164
10.5.5 White Spot Disease . . . 164
10.6 Abnormalities of the Opercula . . . 166
10.7 Jaw and Head Deformities . . . 166
10.7.1 Mandibular Ankylosis . . . 167
10.7.2 Pug Head . . . 167
10.7.3 Microstomia . . . 167
10.7.4 Double Mouth Deformity . . . 168
10.8 Deformities of the Vertebral Column . . . 168
10.9 Congenital Malformations . . . 169
10.10 Cardiac Abnormalities . . . 170
10.10.1 Cardiomegaly . . . 171
10.10.2 Absent Septum Transversum (Ectopia Cordis) . . . 171

10.10.3 Situs Inversus Cordis 172
10.10.4 Skewed Ventricle 172
10.10.5 Arteriosclerosis 172
10.10.6 Hypoplasia of the Outer Compact Myocardium 173
10.10.7 Aneurysms 173
10.10.8 Sub-endocardial Fibrosis 173
10.10.9 Ventricular Cysts 174
10.11 Predators 174
10.12 Lesions Associated with Angling 176
Further Reading 176

11 Idiopathic Diseases 179
11.1 Ulcerative Dermal Necrosis 179
11.2 Puffy Skin 181
11.3 Polycystic Syndrome 181
11.4 White Eye Syndrome 182
11.5 Haemorrhagic Smolt Syndrome 183
11.6 Swim Bladder Stress Syndrome 183
Further Reading 185

12 Neoplasia 187
12.1 Papilloma 187
12.2 Chondroma 187
12.3 Pigment Cells 189
12.3.1 Melanoma 189
12.3.2 Iridophoroma 189
12.4 Leiomyosarcoma 190
12.5 Fibrosarcoma 190
12.6 Lymphoma and Lymphosarcoma 191
12.7 Hepatoma 192
12.8 Nephroblastoma 192
12.9 Adenocarcinoma 192
12.10 Haemangioma 192
Further Reading 193

Glossary 195

Index 205

1 Introduction

Abstract

Global production of Atlantic salmon and rainbow trout continues to increase, however, despite advances in prophylaxis and vaccines, disease outbreaks are one of the major limiting factors for the production of farmed fish worldwide. Infectious agents can increase rapidly in susceptible stocks, especially where the general health status of the population is poor. Therefore, the maintenance of health and the accurate diagnosis of infection are of major importance. This book (A Colour Atlas of Salmonid Diseases) represents our current knowledge of infectious and non-infectious diseases affecting salmonids from the view of a diagnostic pathologist. The reader can familiarise themselves with the wide range of conditions that we consider as 'abnormal' thus providing an invaluable guide to those involved in diagnosing fish diseases. Curiosity and an open mind combined with broad knowledge and experience remain important for the fish pathologist.

Keywords

Aquaculture • Infectious disease • Salmon • Trout • Profitability

Aquaculture has been practiced for many thousands of years and history provides numerous examples where farming has emerged to supplement traditional net or line fishing from the sea, for example ancient fish farming in China. World production of fish has not kept pace with human population growth and today the decline or stagnation of capture fisheries has been compensated by the rapid increase of aquaculture production throughout the world. Species such as carp were farmed in Roman and medieval times and remain part of European culture and heritage, now with sea bream and sea bass in the southern European regions.

Salmonid faming largely occurs within latitudes 40–70° in the northern hemisphere and 40–50° in the southern hemisphere. This covers Norway, Scotland, Ireland, the Faroe Islands, Canada, and the North Eastern seaboard of the USA in the north and Chile, Australia (Tasmania) in the south. Minor production also occurs in France and Spain and in New Zealand, Peru and Argentina. By the twenty-first century farmed Atlantic salmon, rainbow trout and to a lesser extent, Arctic char and brown trout, experienced a tremendous growth in cold water aquaculture and was expected to continue to increase by ~15 %, reaching a global production of ~1.8 million tonnes by 2012. For many years Norway has led the world in Atlantic salmon and rainbow trout production with just over a million tonnes in 2011 with a projection of an additional 10 % increase. In Scotland, 2011 marked the salmon sectors 40th anniversary of its first commercial farm, currently producing ~154,000 tones and projecting a growth of 4 % per annum over the subsequent 10 years, and widening their horizons to include a 50 % increase in sea water reared rainbow trout. In the southern hemisphere and since the emergence of Chile as a strong salmon producer back in the 1980s, the region has contributed to the global salmon market with peaks in production similar to those of Norway.

The commercial rearing of salmon starts in fresh water with hatcheries and juvenile growing units and subsequently fish are transferred to sea water as they become 'smolts'. Growers are held in net pens (cages) or in land based tanks. Rainbow trout typically remain in fresh water throughout their life, although increasingly, production in sea water cages are being utilised. Fish reach the market size around 15–18 months later. The major species from the family are shown in Fig. 1.1

D.W. Bruno et al., *A Colour Atlas of Salmonid Diseases*,
DOI 10.1007/978-94-007-2010-7_1, © Springer Science+Business Media B.V. 2013

Subfamily: Salmoninae	
Genus: *Salmo* – Atlantic salmon, trout	
S. salar (Linnaeus)	Atlantic salmon [a]
S. marmoratus (Cuvier)	Marble trout
S. obtusirostris (Heckel)	Adriatic trout
S. trutta (Linnaeus)	Brown trout [b]
Genus: *Oncorhynchus* – Pacific salmon, trout	
O. gorbuschca (Walbaum)	Pink, humpback salmon [c]
O. keta (Walbaum)	Chum, dog salmon
O. kisutch (Walbaum)	Coho, silver salmon
O. masou (Brevoort)	Masou, yamame salmon
O. nerka (Walbaum)	Sockeye, kokanee salmon
O. rhodurus (Jordan & McGregor)	Amago salmon
O. tshawytscha (Walbaum)	Chinook, king salmon
O. mykiss (Walbaum)	Rainbow, steelhead trout [d]
O. clarki (Richardson)	Cutthroat trout
Genus: *Salvelinus* - char, trout	
S. alpinus (Linnaeus)	Arctic char [e]
S. confluentus (Suckley)	Bull trout
S. leucomaenisleucomaenis (Pallas)	Whitespotted char
S. leucomaenis pluvius (Hilgendorf)	Japanese char
S. fontinalis (Mitchell)	Brook trout, speckled trout
S. malma (Walbaum)	Dolly Varden trout
S. namaycush (Walbaum)	Lake trout [f]
Genus: *Hucho*	
H. hucho (Linnaeus)	Danube salmon, huchen
H. perryi (Pallas)	Sakhalin / Japanese taimen
H. taimen (Pallas)	Siberian taimen [g]
Genus: *Brachymystax*	
B. lenok (*Pallas*)	Lenok, manchurian trout [h]
Subfamily: Thymallinae	
Genus: *Thymallus* - grayling	
T. arcticus (Berg)	Arctic grayling
T. thymallus (Linaeus)	Grayling [i]
Subfamily: Coregoninae	
Genus: *Coregonus*	
C. albula (Linnaeus)	Vendace
C. lavaretus (Linnaeus)	European whitefish [j]
C. nasus (Pallus)	Broad whitefish
Genus: *Stenodus*	
S. leucichthys nelma (Pallas)	Nelma, inconnu, sheefish [k]

Fig. 1.1 Major groups within the family Salmonidae

Disease outbreaks have always been one of the major obstacles to profitable fish farming worldwide and directly linked to fish survival. A build-up of infectious agents can occur in susceptible stocks, especially where the general health status of the population is poor and occurrence of disease can reduce profit dramatically. The success of health management in controlling the spread of infectious diseases can be illustrated by reference to the devastating impact of the viral disease infectious salmon anaemia (ISA). The infection was first reported in Norway in 1984 among sea-farmed Atlantic salmon resulting in significant loses and economic costs. In 1998 Scottish salmon were reported with ISA, and in 2007 this virus also impacted on the fast-growing Chilean aquaculture and resulted in the temporary collapse of their Atlantic salmon production. Specific control measures including health certification, segregation of year classes, fallowing of sites and the disinfection of water effluent from slaughtering facilities, were introduced in several countries. The practice of pumping sea water into tanks of pre-smolts to facilitate smolting was also ceased. These measures reduced the impact of ISA and other infectious diseases to a level where outbreaks declined significantly. In addition, similar policies have resulted in an overall improvement in the health of farmed fish in other countries with the additional benefit that there has been a marked reduction in the use of antibiotics to control bacterial infections.

Despite all these actions, losses from all possible causes during sea water production can reach 20 % and therefore the health of farmed fish is a major concern as impaired health or any disease state is not acceptable, neither from a welfare point of view nor economically sustainable.

Disease conditions are diverse in nature and may elicit a wide range of responses. The final outcome of individual infections will depend upon the combination of physiological and immunological host factors and the virulence properties of the pathogen. It is also important to bear in mind that under certain circumstances, fish can act as asymptomatic carriers, passing the infective agent to susceptible animals. Currently, techniques involving molecular biology are being used to support diagnostic work, however, pathological assessment remains the 'gold standard' and pathologists will continue to play a unique role in diagnosing, understanding and interpretation of the pathological changes putting into context the results of other laboratory tests.

Common signs of disease include abnormal swimming, dark skin, inappetence and lethargy. Exophthalmia, distended abdomen, fin rot, skin ulcers and petechial haemorrhage, especially at the base of the pectoral and pelvic fins, may also be encountered. Gross pathology associated with fish diseases can be frustratingly similar, and difficult for inexperienced personal to distinguish between the different conditions, as very few are pathognomic. Notably where overlapping infections may be present, this challenges the pathologist to differentiate between 'dying of' from 'dying with', a given agent.

Non-infectious diseases also raise important ethical questions, particularly as the affected fish also become more susceptible to infectious diseases. There is also increasing evidence that intensively reared fish significantly alters some aspects of cardiac anatomy and physiology. Other factors contributing to the outcome of disease in both wild and farmed fish include stress related factors, which are recognized to increase the susceptibility of fish to infectious and non-infectious diseases. Stress in fish production can be summarised as 'an effect produced by any environmental or other alteration which requires an adaptation and response by the individual beyond their normal limits, such that the chance of survival is reduced'.

Continued research and development into the science of fish health management has been conducted in many countries alongside the growth of the fish farming industry. Stocking densities are generally lower, the water quality has been improved, sites within management areas are often fallowed and a policy of year-class segregation is applied. Statutory health surveillance and restrictions on the movement of fish with certain categories of infectious agents have also contributed to the containment of disease outbreaks. Considerable effort has also been made to reduce the impact of infectious conditions such as vibriosis, furunculosis, enteric redmouth, infectious pancreatic necrosis and pancreas disease through the introduction of effective vaccines. Overall an improved understanding of diseases to which salmonids are vulnerable have resulted in a decrease in the incidence of disease among farmed fish.

In 1996 we published 'A Colour Atlas of Salmonid diseases' and now completely revised, this new edition represents our current knowledge and the significant advances in the field of fish pathology that have occurred during this time. For instance, several conditions of unknown aetiology have been confirmed as viral diseases, for example cardiomyopathy syndrome (CMS) and heart and skeletal muscle inflammation (HSMI). Furthermore, new diseases or different manifestations of disease have been highlighted, while the significance of other diseases is fading.

We have set our main objective for this book as 'informative' so that the reader can familiarise themselves with the wide range of conditions that we consider as 'abnormal' among wild and farmed salmonids. This is supported through the inclusion of a chapter covering functional anatomy, namely the normal histological structure and function of all major organs. Furthermore, a necropsy guide and the recognition of tissue abnormalities as prerequisites to disease diagnosis, the use of appropriate terminology covering cell injury to tissue and organs, as well as disturbances in circulation, inflammation and healing process, have been included. We have chosen to classify diseases according to their aetiology and reference is made accordingly to disease conditions attributed to viruses, bacteria, fungal-like (i.e. oomycetes) and parasites (Protist and Metazoa). A range of non-infectious conditions which may also be encountered during the examination of farmed and wild salmonids are discussed, including those associated with farm production methods, so called 'production diseases'. Furthermore, diseases of obscure, complex or unknown aetiology are also discussed. Gross and light micrographs representing key characteristics of each condition are incorporated.

It is important to remember that wild salmonids are susceptible to the same infectious diseases recorded among their farmed relatives, and several conditions have been known or reported from wild fish prior to significant developments in aquaculture (e.g. furunculosis, bacterial kidney disease, sea lice, oomycete infections and *Gyrodactylus salaris*). Conversely, some agents initially identified in farmed fish are now being reported in wild non-salmonid fish (e.g. reovirus and salmonid alphavirus), highlighting the complex relationship involving carriers and virus reservoirs. Furthermore, there are conditions that appear to be exclusive to wild fish e.g. red vent caused by *Anisakis simplex* and the idiopathic ulcerative dermal necrosis (UDN).

It is impossible to include every disorder, but it is believed that the material selected for this book represents a comprehensive coverage of conditions found in wild and farmed salmonids and this will provide an invaluable guide to those involved in diagnosing fish diseases. It should be emphasized that the fish health status is dynamic and therefore, curiosity and an open mind combined with thorough knowledge and experience will remain an important attitude for the fish pathologist.

Further Reading

Amin AB, Mortensen L, Poppe T (1992) Histology atlas, normal structure of salmonids. Akvapatologisk Laboratorium AS, Bodø

Behnke RJ (2002) Trout and salmon of North America. The Free Press, New York

Bergh Ø (2007) The dual myths of the healthy wild fish and the unhealthy farmed fish. Dis Aquat Org 75:159–164

Branson EJ (2008) Fish welfare. Blackwell Publishing, Oxford

Farrell AP (2011) Encyclopedia of fish physiology, vol 1–3. Academic Press, London

Ferguson HW (2006) Systemic pathology of fish. A text and atlas of normal tissues in teleosts and their response in disease. Scotian Press, London

Johansen L-H, Jensen I, Mikkelsen H, Bjørn P-A, Jansen PA, Bergh Ø (2011) Disease interaction and pathogens exchange between wild and farmed fish populations with special reference to Norway. Aquaculture 315:167–186

Kent M, Poppe TT (1998) Diseases of sea water netpen-reared salmonid fishes. Fisheries and Oceans Canada, p 58. ISBN: 0920225101

Kristoffersen AB, Jensen BB, Jansen PA (2012) Risk mapping of heart and skeletal muscle inflammation in salmon farming. Prev Vet Med 109: 136–143. http://dx.doi.org/10.1016/j.prevetmed.2012.08.012

Leatherland JF, Woo PTK (2010) Fish diseases and disorders, noninfectious diseases, vol 2, 2nd edn. CABI Publishing, Oxfordshire

Marcos-López M, Gale P, Oidtmann BC, Peeler EJ (2010) Assessing the impact of climate change on disease emergence in freshwater fish in the United Kingdom. Transbound Emerg Dis 57:293–304

Mitchell SO, Rodger HD (2011) A review of infectious gill disease in marine salmonid fish. J Fish Dis 34:411–432

Noga EJ (2010) Fish disease diagnosis and treatment, 2nd edn. Wiley-Blackwell, Ames

Ostrander GK (2000) The laboratory fish. Academic Press, San Diego

Poppe TT (2000) Husbandry diseases in fish farming – an ethical challenge to the veterinary profession. Nor J Vet Med 112:15–20

Roberts RJ (2012) Fish pathology, 4th edn. Wiley-Blackwell, London

Rodger HD, Henry L, Mitchell SO (2011) Non-infectious gill disorders of marine salmonid fish. Rev Fish Biol Fish 21:423–440

Soares S, Green DM, Turnbull JF, Crumlish M, Murray AG (2011) A baseline method for benchmarking mortality losses in Atlantic salmon (*Salmo salar*) production. Aquaculture 314:7–12

Woo PTK (2006) Fish diseases and disorders, vol 1, 2nd edn, Protozoan and metazoan infections. CABI Publishing, Oxfordshire

Woo PTK, Bruno DW (2011) Fish diseases and disorders, vol 2, 2nd edn, Viral, bacterial and fungal infections. CABI Publishing, Oxfordshire

Woo PTK, Buchmann K (2012) Fish parasites pathobiology and protection. CABI Publishing, Oxfordshire

Woo PTK, Bruno DW, Lim LHS (2002) Diseases and disorders of finfish in cage culture. CABI Publishing, Oxfordshire

Yasutake WT, Wales JH (1983) Microscopic anatomy of salmonids: an Atlas. United States Department of the Interior, Fish and Wildlife Service, Resource Publication 50, Washington, DC

2 Functional Anatomy

Abstract

The microscopic examination of stained sections and the ability to interpret the relationship between fine structure and function is essential. The recognition and interpretation of physiological and pathological processes requires a thorough understanding of normal tissue structure and microanatomy, and importantly the variations within species are crucial for correct interpretation. This chapter covers the physiological changes, sexual maturation and aging processes that are 'normal', and inherently different from those resulting from injury, infection or disease.

Keywords

Histology • Anatomy • Normal structure • Salmon • Trout

Microanatomy or histology, the science of tissues, is the microscopic examination of thin, stained sections that allows the interpretation of the relationship between fine structure and function. Individual cells and tissues may undergo changes during physiological responses, sexual maturation and aging, that are 'normal' and largely different from those resulting from injury, infection or disease processes. Therefore, knowledge of the normal structure and variations within species is of crucial significance for correct interpretation and understanding of pathological changes.

Within the body, the coelomic cavities (pericardial and abdominal), various organs and components, are surrounded or held in position by layers of a serous membrane, the peritoneum, a mesothelium and connective tissue containing blood and secondary (lymph) circulation which also lines the septum transversum that separates both cavities. In this way the outermost layer covering the different regions of the alimentary canal known as the 'serosa', is effectively the 'visceral peritoneum', and covers other viscera lying within the 'peritoneal cavity or space', hence they are not in the body cavity but enveloped within a double layer of peritoneum, and analogous to pushing a finger within a balloon. This also determines that some organs are 'retroperitoneal', for example, the kidney.

Throughout this chapter we will refer to the functional unit of an organ as the parenchyma, while the space in between these parts, cells or functional units, is referred as the interstitium. The latter may contain various cellular and extracellular elements, supporting structures or secretions. For example, in the kidney the nephrons and the blood vascular system are embedded in the meshwork of the interstitium, with non-renal elements such as haematopoietic, secretory and supportive cells; whereas the liver consists of hepatocytes, with a meshwork of connective tissue, namely the framework of the organ.

In this chapter an overview of the anatomy of salmonids is presented by 'systems' (e.g. respiratory system, excretory system) and organised by a functional approach.

The staining of light micrographs are only identified when haematoxylin and eosin (H&E) is not used, and the magnification for the images are given either as a scale bar or if without a scalebar are identified as low power ×4–10; medium power ×20–40; high power ×60–100 (Fig. 2.1).

D.W. Bruno et al., *A Colour Atlas of Salmonid Diseases*,
DOI 10.1007/978-94-007-2010-7_2, © Springer Science+Business Media B.V. 2013

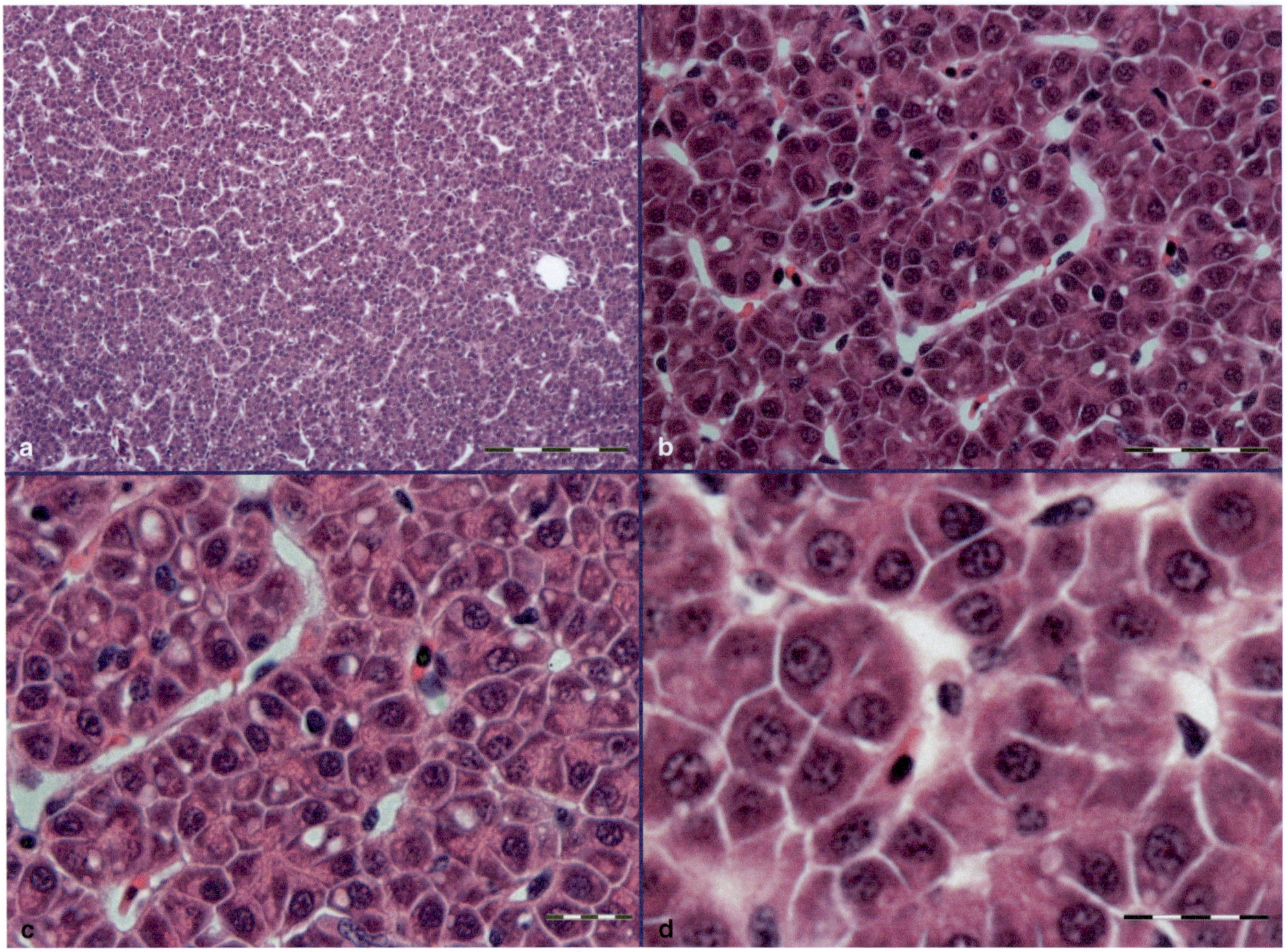

Fig. 2.1 Sections to illustrate different magnifications with representative bar scales. (**a**) Bar scale = 100 μm, ×20. (**b**) Bar scale = 50 μm, ×40. (**c**) Bar scale = 20 μm, ×60. (**d**) Bar scale = 20 μm, ×100

2.1 Respiratory System (including the operculum and pseudobranch)

Salmonids have four gill arches bilaterally placed on each side of the head, each supporting a holobranch with its two hemibranchs, the double vertical rows of gill filaments (Fig. 2.2). A series of cartilaginous or bony projections, the gill rakers, protrude forwards from the pharyngeal margin of the gill arch. These are relatively sparse in most salmonids but show a wide variation of morphologies in different species, and may form a fine grid that helps filter planktonic organisms from the water, at the same time preventing food particles from entering the gill chamber. Each hemibranch comprises a row of posterior-laterally oriented filaments with its respiratory epithelium covered lamella on each side (Fig. 2.3). Anatomically the filaments looks like a 'feather', supporting on each side a continuous symmetrically-spaced individual lamella. The filaments are supported along their proximal half by an interbranchial septum of connective and muscle tissues, but the septum is reduced to about a third of the filament length or even absent in more advanced fish. Each lamella consists of a supportive scaffold of pillar cells among which blood supply enters and leaves the lamella, which are covered by a thin double-layer epithelium separated by a space in which migrating inflammatory cells may be seen. The inner layer of the epithelium sits on a basement membrane that traverses the opposing face of the lamella in grooves located within the pillar cells, and in this way provides additional tensile support. However, the bulk of the respiratory epithelium obvious through light microscopy, is the outer squamous layer that provides a large and intimate interface with the water for exchange of gases, acid–base regulation, osmoregulation and excretion of nitrogenous waste products. Chloride and mucous cells, normally found near the base of the lamellae, may also be found distally under pathological conditions, especially the mucous cells (Fig. 2.4). Chloride cells are highly rich in

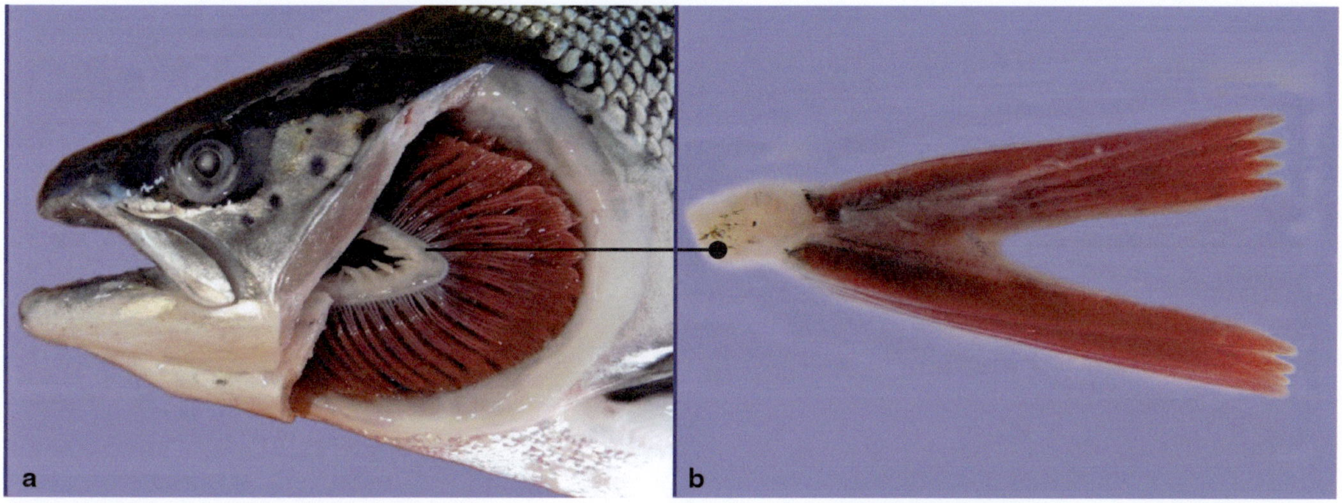

Fig. 2.2 (**a**) Gills of an adult wild Atlantic salmon, left operculum is removed. The complete gill arch with its gill rakers can be seen. (**b**) Transversal section of one holobranch showing the hemibranchs with the vertical rows of gill filaments

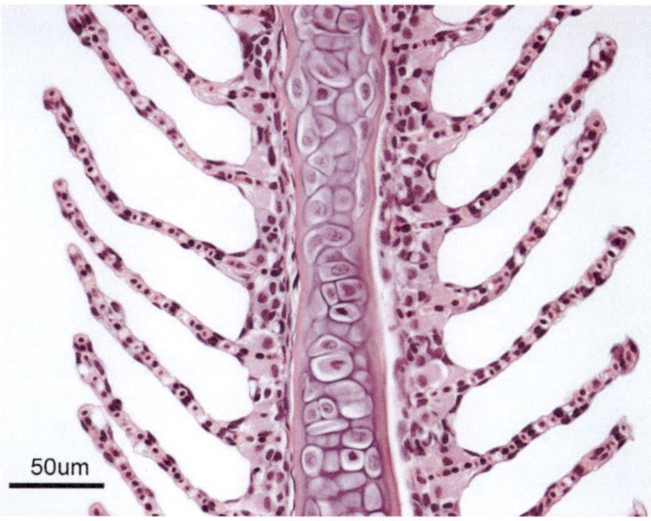

Fig. 2.3 Gill filament, central cartilage and lamellae from adult Atlantic salmon. Chloride cells are located near the base of the lamellae

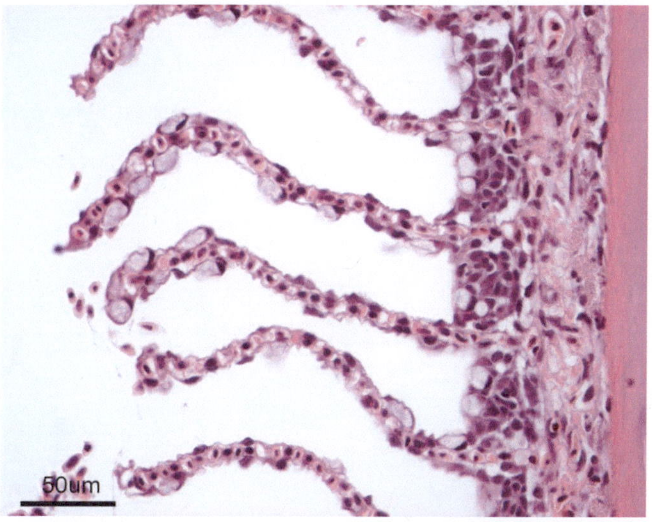

Fig. 2.4 Goblet (mucous) cells on lamellae of farmed adult Atlantic salmon

mitochondria and responsible for the secretion of sodium chloride from the blood, they are prominent in fish living in the marine environment and characteristic during smoltification, but their number may also increase in several pathological conditions. Other cells within the filament interstitium include lymphocytes, eosinophilic granular cells (EGCs) (see Fig. 4.15), macrophages, neuroepithelial cells and rodlet cells. Granular and neuroepithelial cells are obvious at the base of the lamella and more frequent in marine than in freshwater fish. Intraepithelial lymphocytic cell accumulations can be seen at the caudal edge of interbranchial septum in Atlantic salmon, and shown to be a lymphoid tissue and probably of significance for surveillance of gill infections (Fig. 2.5).

Blood flow in the lamella capillaries is opposite to the ventilatory water flow ('counter current') ensuring effective gas exchange between blood and water across the respiratory epithelium. Venous blood from the ventral aorta diverges to the afferent arterioles and capillaries of the lamellae where gas exchange takes place and becomes arterial blood. Arterial blood is drained through the efferent branchial artery and into the dorsal aorta. The water flow over the lamellae is continuous and achieved by means of the buccal-opercular pump.

Dorsally, on the inner surface of each operculum is the pseudobranch, a rudimentary first gill arch (Figs. 2.6 and 2.7). The pseudobranchial cells are in close proximity to a network of blood vessels and may play a role in the blood supply to the retina and in osmoregulation and sensing.

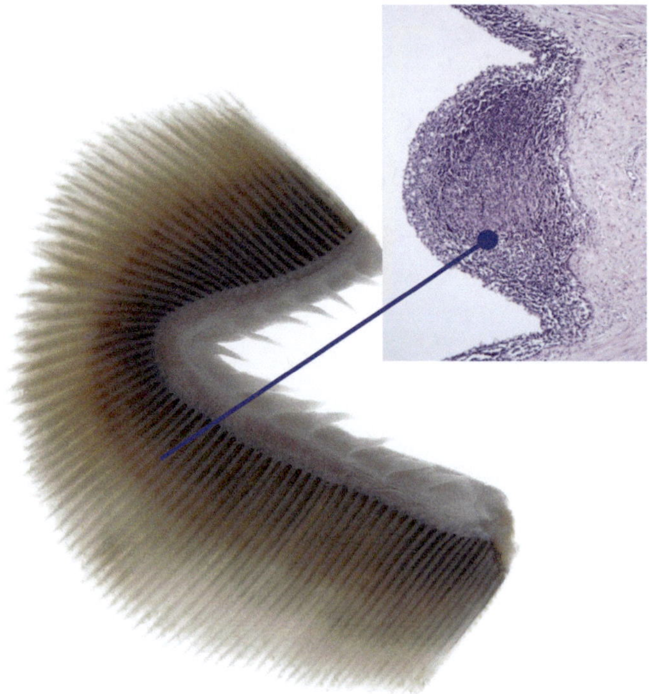

Fig. 2.5 Gills and location of interbranchial lymphoid tissue from Atlantic salmon. *Insert* medium power

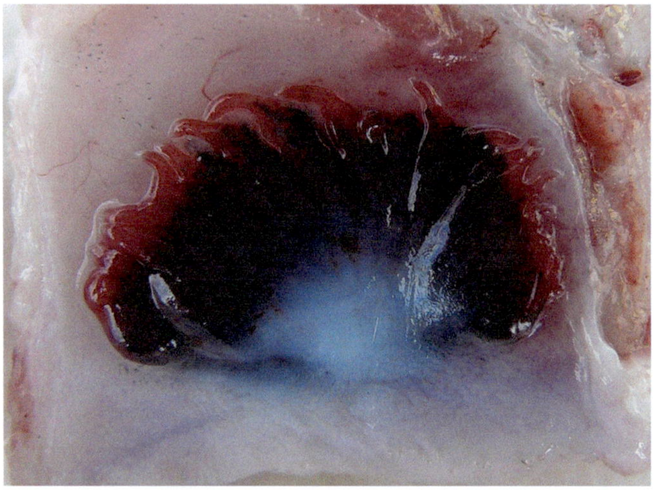

Fig. 2.6 Pseudobranch from an adult sea water farmed Atlantic salmon

The gills are frequently involved in several pathological conditions of diverse aetiology, but lesions should be differentiated from artefacts and post-mortem changes. Due to their delicate structure, exposed location, abundant blood supply and large surface, gills quickly undergo post-mortem changes that can make histopathological interpretation difficult (see Fig. 4.31).

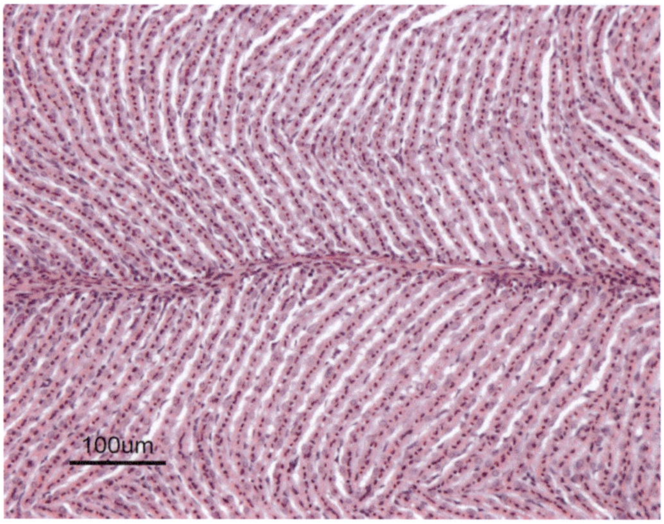

Fig. 2.7 Pseudobranch lamellae of adult Atlantic salmon

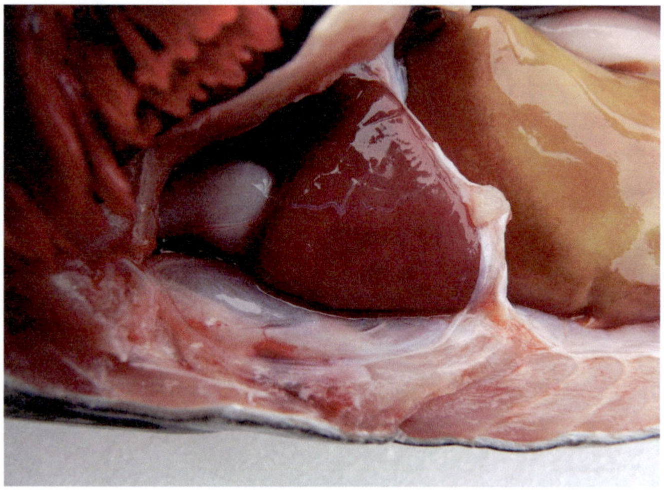

Fig. 2.8 Position of normal heart in the pericardial cavity of adult Atlantic salmon

2.2 Circulatory System (Cardiovascular and Secondary Circulation)

2.2.1 The Cardiovascular System

The cardiovascular system is a simple loop with the heart, gills and systemic circulation in series. The deoxygenated blood is pumped from the ventricle to the gills where it leaves as oxygenated blood to be delivered directly to body organs and tissues.

The heart is located in the pericardial cavity antro-ventrally to the peritoneal cavity, and is separated from the latter by the septum transversum (Fig. 2.8). It consists of four compartments: the sinus venosus, the atrium, the ventricle

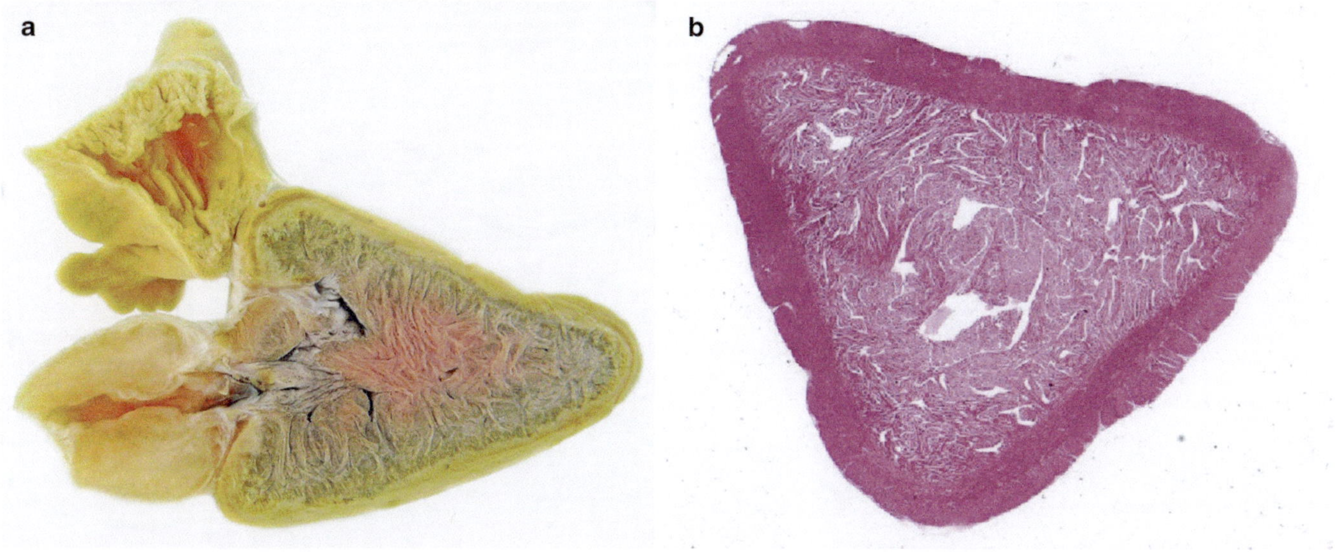

Fig. 2.9 (**a**) Sagittal section through the heart of wild Atlantic salmon showing atrium, ventricle with spongy and compact myocardium and bulbous arteriosus. Formalin fixed specimen. (**b**) Stained transverse section of ventricle from Atlantic salmon

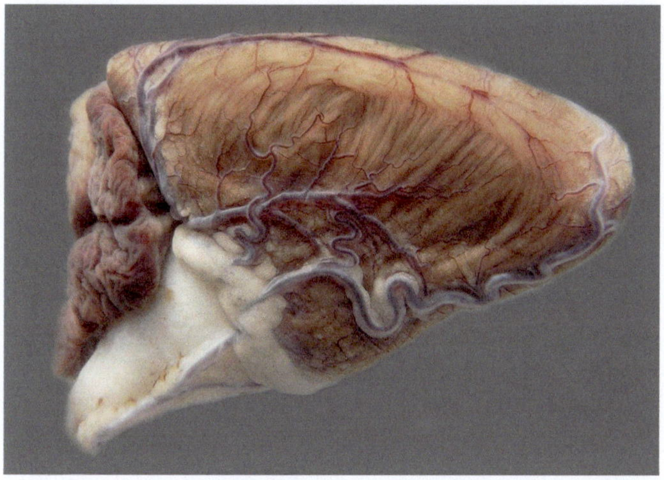

Fig. 2.10 Coronary vessels on the surface of bulbus arteriosus and the ventricle in adult farmed Atlantic salmon

and the bulbus arteriosus, interposed by the conus arteriosus. The former was considered to have been lost through evolution but recent work has clearly shown its presence and fundamental role in the heart outflow tract, where it supports the conus valves previously named bulbo-ventricular valves. Deoxygenated blood from the cardinal and hepatic veins flows into the thin-walled sinus venosus and passes through the sino-atrial valve and then into the thin-walled spongy atrium. Blood is drawn from the atrium via the atrio ventricular valves into the thick-walled and muscular ventricle. The ventricle has a pyramidal shape and consists of two muscular layers: the outer compact myocardium with its own supply of oxygenated blood, and the inner spongy myocardium (Fig. 2.9). The blood supply to the outer myocardium is via the coronary artery (Fig. 2.10), a branch of the hypobranchial artery from the second gill arch that runs caudally on the ventral side of the bulbus arteriosus, before it bifurcates and spreads over the ventricle surface. The inner spongy myocardium has no blood supply of its own, but is supplied with oxygen and nutrients from the venous blood being pumped through the organ. An example of the microanatomy of an artery and vein is shown in Fig. 2.11. The last chamber is the highly compliant bulbus arteriosus with thick walls composed of fibro-elastic and connective tissue. This chamber functions as a depulsator and delivers a steady blood flow to the ventral aorta. The adventitia of the organ consists of blood vessels and large nerve bundles in a collagen matrix. All chambers of the heart are covered by a flat epithelium called the epicardium which fuses with the pericardium that covers the inner surface of the pericardial cavity. All inner surfaces are covered by the endocardium. The thickness of the outer, compact myocardium may vary with age, sex and the habitat of the fish (Fig. 2.12). The absolute and relative thickness of the compact myocardium increases with age and is thicker in males than in females, and thicker in fish living in running water than those living in lakes. Generally, wild fish have a thicker compact myocardium than farmed fish of the same size. The cardiac striated muscle (myocardium) is differentiated from skeletal striated muscle by the branching structure of the fibres and centrally located nuclei (Fig. 2.13).

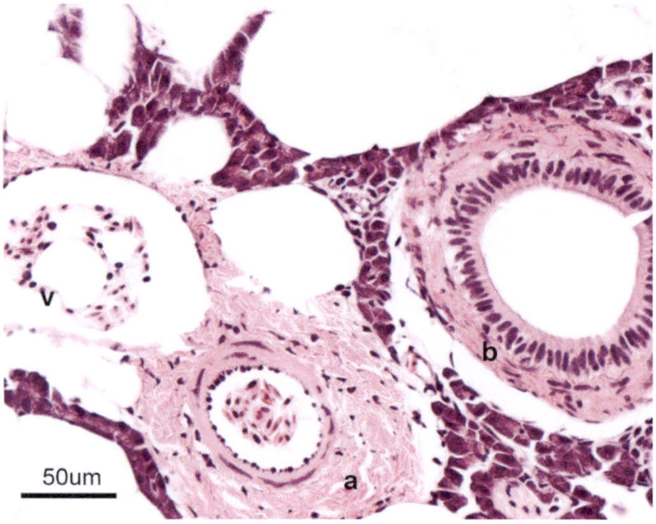

Fig. 2.11 Transverse section through thick walled artery (*a*) and thin walled vein (*v*) from Atlantic salmon, note presence of bile duct (*b*)

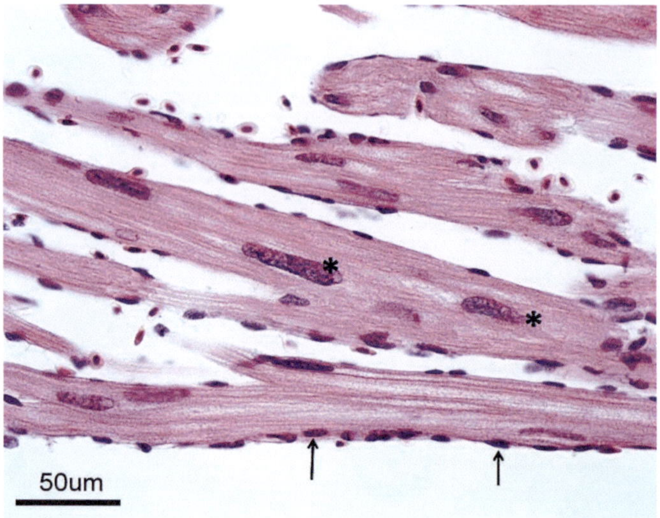

Fig. 2.13 Longitudinal section of spongy myocardium of Atlantic salmon; note central location of nuclei (*) and endocardial cell nuclei (*arrows*)

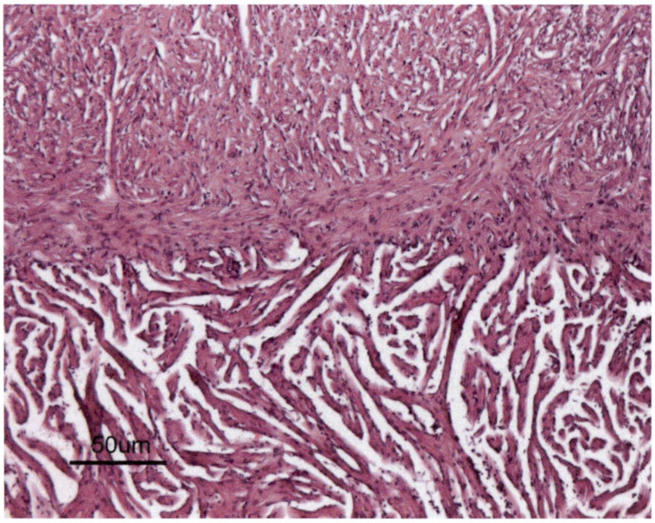

Fig. 2.12 Interphase between compact and spongy myocardium of ventricle of Atlantic salmon parr

Blood Cells

Blood is composed of humoral (plasma) and cellular (blood cell) components. In contrast to mammals, fish red blood cells (erythrocytes) are nucleated and ovoid in shape, 13–16 μm long and 7–10 μm broad. Erythrocyte numbers may vary, but are usually in the range of 1.05×10^6–3.0×10^6 mm^3. Giemsa staining shows that mature cells have a dense chromatin, purple-red centrally-located nucleus, and a clear homogenous, light red cytoplasm. The later reflects the absence of organelles and the quantity of haemoglobin present which in mature cells is very abundant. The peripheral blood is mainly composed of mature erythrocytes, although immature and developmental stages can be distinguished. Immature erythrocytes are known as reticulocytes, they are rounder with a relatively larger nucleus. Five categories can be recognized based on structure, distribution and quantity of basophilic substances within their cytoplasm. Normally, these represent about 1 % of the total count in healthy fish.

Lymphocytes in fish constitute 70–90 % of the total number of leucocytes. Lymphocytes are arbitrarily separated into categories of large (10–15 μm diameter) and small (7–10 μm diameter), which may represent different functional stages. The round or oval nucleus virtually occupies the whole cell leaving a narrow margin of basophilic cytoplasm. The cytoplasm may show pseudopodia-like projections on the surface. Both T and B lymphocytes forms are recognised in fish, playing a most significant role in the innate and the acquired immune responses.

Thrombocytes are responsible for blood clotting and are important in homeostasis and defence. Typically they are elongated but can be also be spindle-shaped and ovoid, with an indentation. They are of variable size (5–8 μm long) with a light basophilic rim of cytoplasm and a densely staining nucleus that occupies most of the cell. Between 1 and 6 % of the total white cells in rainbow trout are thrombocytes.

Neutrophils are morphologically similar to those in mammals and are commonly found at sites of inflammation. The eccentric nucleus is often kidney-shaped, although in mature cells two or five lobed nuclei may be recognized which are connected to each other by threads of nuclear material. These cells vary from 4 to 13 μm in diameter. There is evidence of phagocytic capacity in Atlantic salmon however much lower than the 'professional' macrophages, therefore unlikely that phagocytosis is their primary function. Their role seems to be more important in extracellular killing through enzymatic and other antimicrobial secretions.

Monocytes form about 0.1 % of the circulating leucocytes and are partially differentiated end cells which under appropriate conditions will develop into mature cells of the mononuclear phagocyte system. Monocytes are 9–25 μm in diameter with a lighter staining cytoplasm than small lymphocytes and contain small granules with a large nucleus.

Differentiation of phagocytes into macrophages usually takes place when they become extravascular, migrating from the vessels into the tissues and therefore usually not seen in circulation. Their phagocytic capability is well documented, but resting macrophages in tissues are difficult to distinguish from fibrocytes in H&E stained sections.

2.2.2 Secondary Circulation

Studies indicate that fish lack a true lymphatic system but a second vascular system derived from and connected to the primary circulation is present. Analogies between the lymphatic and the secondary system have been noted and observed in skin, fins, gills, oral mucosa and lining of the peritoneum, however it remains to be determined whether the secondary system is the antecedent of a lymphatic system or a coincidentally similar structure.

The spleen is usually dark red or almost black in colour and a discrete organ with sharply defined edges located near the greater curvature of the stomach. A thin serous capsule covers the surface. The spleen functions as a haematopoietic organ, a temporary blood bank and as a remover of circulating antigens and effete blood cells. Occasionally, two or more spleens may be recorded and the organ may also be located elsewhere in the abdominal cavity (Fig. 2.14). The spleen structure is provided by a capillary and a connective tissue meshwork among which the cells fill up the interspaces i.e. erythroblasts, mature and immature erythrocytes, lymphocytes, monocytes and macrophages. The parenchyma of the spleen is composed of white pulp, namely lymphoid tissue, surrounding small arteries which diffusely intermeshes with the haematopoietic red pulp, composed of a reticular cell network and supporting blood-filled sinusoids. There is no sharp demarcation between red and white pulp as the parts rich in erythrocytes and those rich in lymphocytes are intermingled (Fig. 2.15). The ellipsoids form the main elements of the spleen and are a thick-walled filter capillary network gradually forming from the artery which enters the organ. Each ellipsoid comprises a thick basement membrane-bound tube within which the vessels run and is separated from the membrane by a layer of sheathed components. Degradation products of senescent erythrocytes are stained yellow by H&E and known as haemosiderin, a common feature in the spleen parenchyma. Perl's staining is used to differentiate haemosiderin deposits (see Fig. 4.22). Variable numbers of melanomacrophages may be found scattered in the spleen tissue. Phagocytic cells capable of trapping large quantities of particulate matter from the circulating blood, once replete migrate to the ellipsoids to the melanomacrophages.

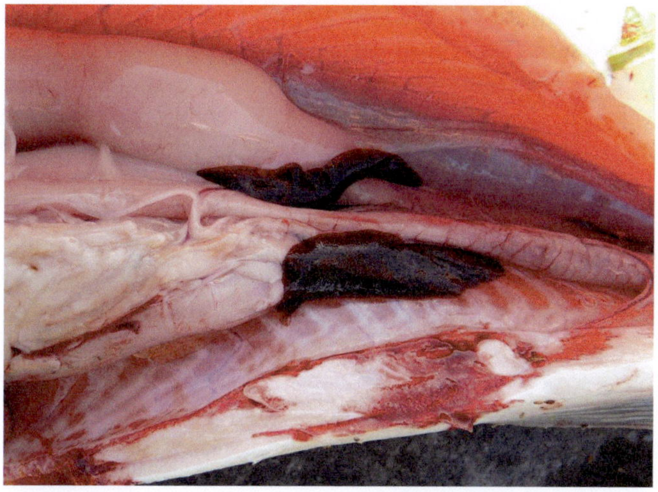

Fig. 2.14 Spleen in abdominal cavity of Atlantic salmon, note duplicate organ in this case

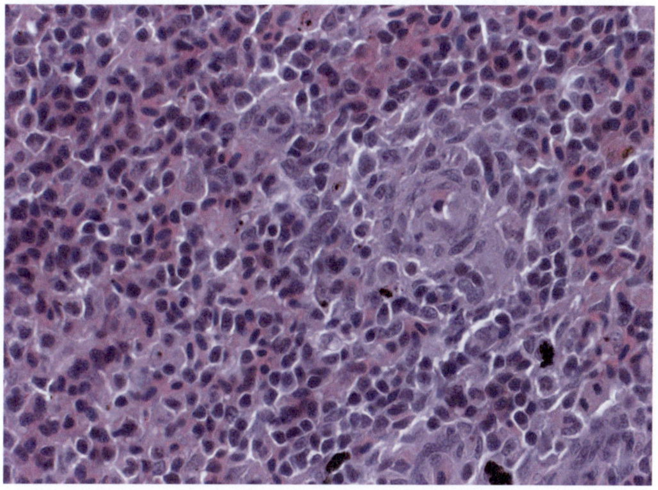

Fig. 2.15 Spleen of adult Atlantic salmon. Medium power

2.3 Integument System

The structure of the skin varies to some extent among species but is basically composed of two layers: an outer epidermis and an underlying dermis. The epidermis constitutes the barrier between the body and the aquatic environment and can be divided into two additional layers. The outer epithelium is composed of stratified squamous cells and a basal layer of undifferentiated cuboid germinal cells. The depth is greatest on the head and over the fins where scales are absent. The thickness of epidermis and number of mucous-secreting goblet cells in the epidermis varies between species and location on the body. Their number may increase during sexual maturation and spawning migration, and lymphocytes can also be present in the epidermis (Fig. 2.16). Mucous cells have characteristic basal, compact nuclei. The epidermis sits on a thin acellular layer that can be observed by light microscopy and is the fusion of the basal and the reticular lamina known as the basement membrane, although the components, i.e. the basal lamina, can only be observed by electron microscopy.

The dermis is mainly composed of collagenous connective tissue. Two layers can be distinguished histologically, the upper stratum spongiosum comprised of a loose network of collagen and reticulin fibres and also containing pigment cells (chromatophores), and the lower stratum compactum composed of a collagenous dense matrix providing structural strength to the skin.

Scales are translucent acellular plates of dermal origin that project into the epidermis. They are composed of a mineralized matrix anchored in dermal pockets between layers of collagen in the stratum spongiosum, and the epidermal basement membrane. Scales have variable size, for example, they are small in Arctic char but large in whitefish. In addition, grayling scales differ from those of other salmonid species by their large size and shape with characteristic indentations on the caudal edge (ctenoid type scale in contrast to cycloid type in other salmonids).

A thin, cellular layer covers the entire scale and is distinct from other epidermal tissues, the scleroblasts which are a rich source of calcium. During periods of starvation or sexual maturation calcium from the scales may be reabsorbed by osteoclasts leaving a scar in the outer margin of the scale ('spawning scar').

The hypodermis is the layer between deeper layer of the dermis and the underlying muscle and is composed of loose connective tissue and some fat cells. In the head region, the hypodermis is indistinguishable from the stratum compactum of the dermis.

2.4 Musculoskeletal System

2.4.1 The Skeletal System

The skeletal system includes the bones of the skeleton and the cartilage, ligaments, and other connective tissue that stabilize or connect the bones. The bulk of the body muscle is organized in four quadrants, with the striated muscle further organized in blocks or myotomes or myomere (Figs. 2.17, 2.18 and 2.19). The myosepta separates but also holds the myotomes together. The bulk of the skeletal muscle consists of anaerobic white fibres with a relatively poor vascularisation and few mitochondria and is used for bursts and strong swimming activity. The red aerobic muscle

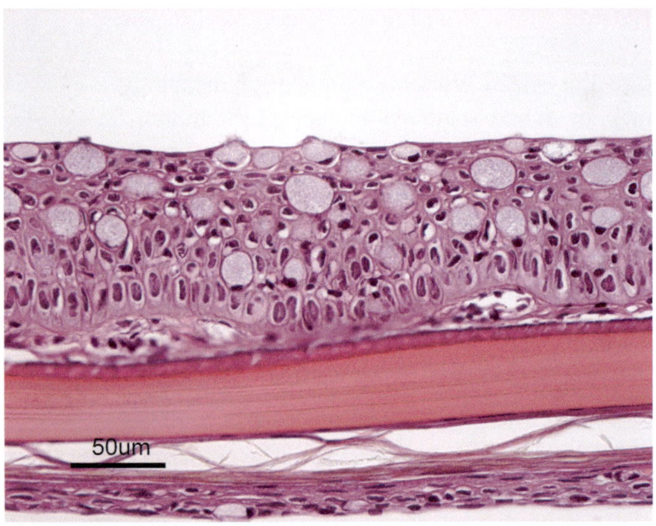

Fig. 2.16 Epidermis with goblet cells on top of a scale. Adult Atlantic salmon

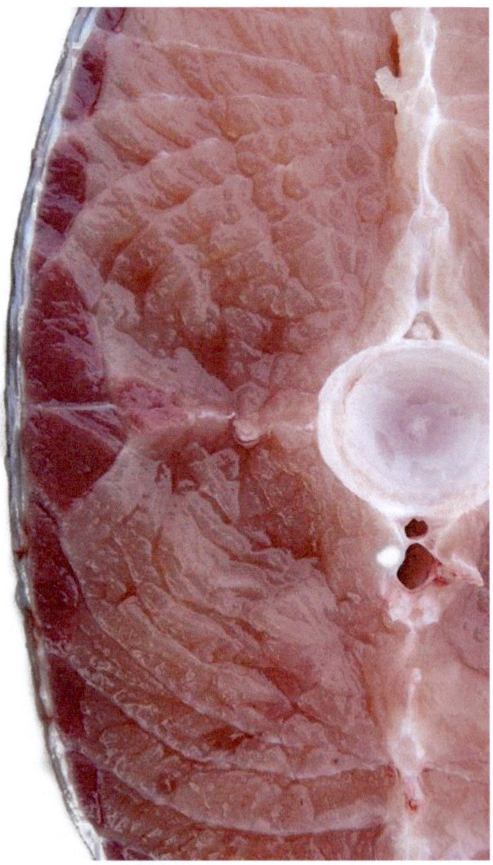

Fig. 2.17 Transverse section through body muscle of adult Atlantic salmon showing red and white muscle

is organized as a triangular band along the flank just beneath the lateral line and is highly vascularised, rich in myoglobin, glycogen and lipids with numerous mitochondria (Fig. 2.20). Red muscle is used for long-term sustained swimming and moderate speed swimming activity. Skeletal striated muscle differs from the heart striated muscle with a peripheral instead of a central nucleus.

Fig. 2.18 Myotomes in the belly flap of an adult Atlantic salmon

2.4.2 Fins

Salmonids have a complete set of fins. Each fin is covered by stratified squamous epithelium continuous with the epidermis of the body. The dermis has a reduced stratum compactum and a thicker hypodermis compared to that of the main body. Median fins of include the dorsal, adipose, caudal and anal, and the paired sets are the pectoral and pelvic fins supported by bony girdles, which are embedded in the ventral body musculature as floating structures. Unpaired fins are generally supported on small bones within the musculature septa. The caudal fin is supported by a greatly modified, posterior most caudal vertebrae, flattened into an almost symmetrical plate against which the flexible fin rays of the caudal fin articulate. The adipose fin located between the dorsal and caudal fins, despite its name, has no adipose tissue or bony support. A clearer idea of its role has emerged with evidence of sensory function reported, suggesting it may act as a precaudal flow sensor, therefore its removal can be detrimental to swimming efficiency.

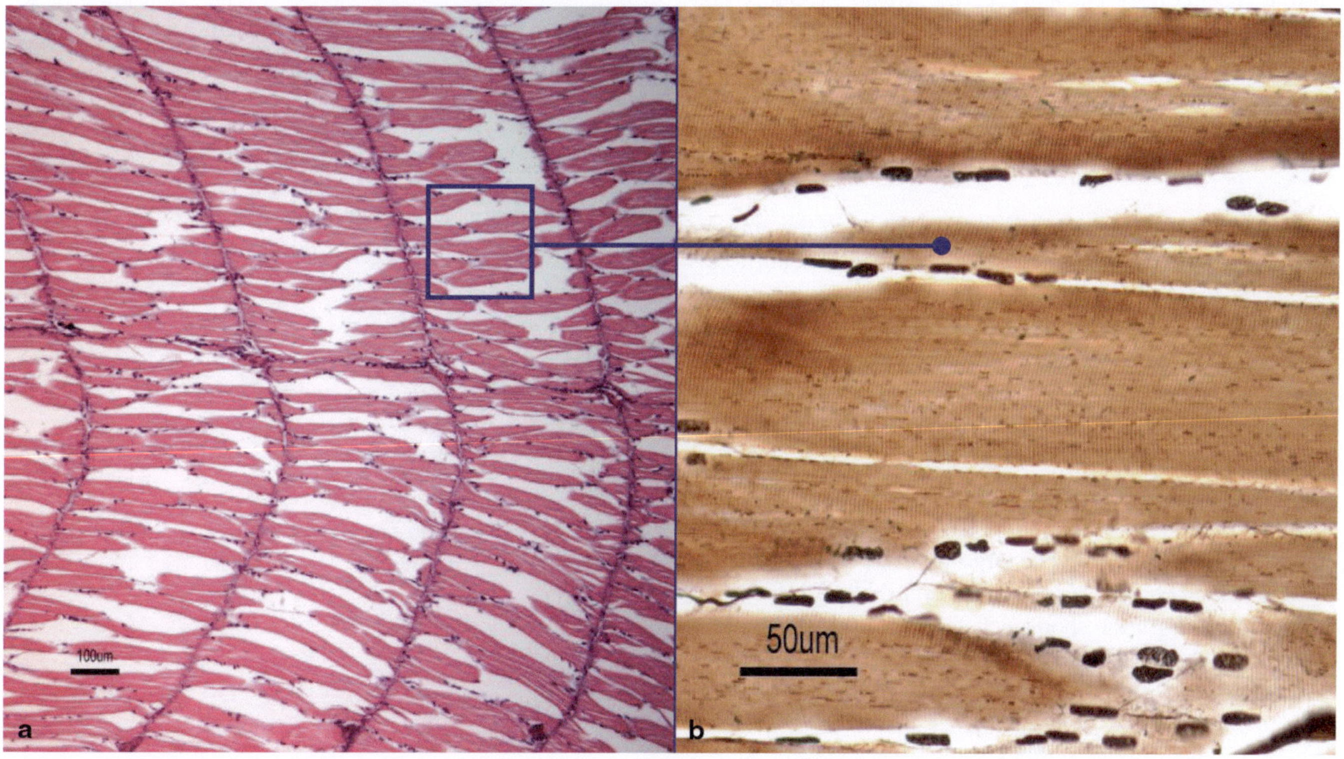

Fig. 2.19 (**a**) Longitudinal section through body muscle of Atlantic salmon fry showing arrangement of myotomes. (**b**) *Insert* is stained with Wilder silver stain

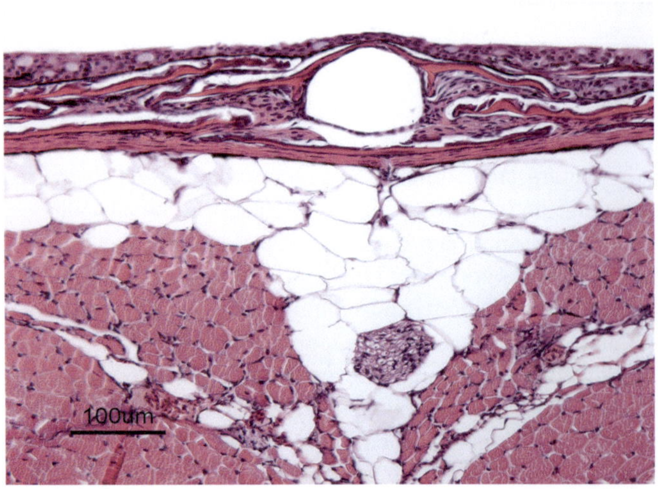

Fig. 2.20 Transverse section through Atlantic salmon parr showing epidermis, scales, lateral line, red and white muscle. Low power

Fig. 2.21 Horizontal section of rainbow trout showing head kidney. Note Y-shaped portion at cranial end

2.5 Excretory System

Salmon, like most fish, release their nitrogenous wastes as ammonia and the gills play an important role by excreting this compound through diffusion into the surrounding water. However the core of the excretory system remains in the kidney, with primary functions performed by filtering wastes from the blood to maintain the body fluid levels, collect and excrete the waste products and maintain pH.

2.5.1 Kidney

The kidney comprises tightly fused units giving the appearance of a single organ. The kidney is located retroperitoneally along most of the length of the body cavity, ventral to the vertebral column and dorsal to the swim bladder. The anterior section, also known as the cranial or head kidney is composed entirely of haematopoietic and lymphoid tissue (Fig. 2.21), while the posterior section has the excretory role and the functional unit are the nephrons (glomeruli and tubules) embedded in haematopoietic tissue (Fig. 2.22).

The typical nephron of salmonids living in fresh water is characterized by a relatively large glomerulus that fills up the Bowman's capsule (Fig. 2.23). From the latter, the renal tubule begins with the short neck segment characterized by low cuboidal epithelium with long cilia. This section is divided into a first segment with eosinophilic, cuboidal to columnar epithelium with a distinct brush border, and a second segment with a taller columnar epithelia and a centrally located oval nucleus. The latter has a prominent brush border but lack the extensive tubular system in the

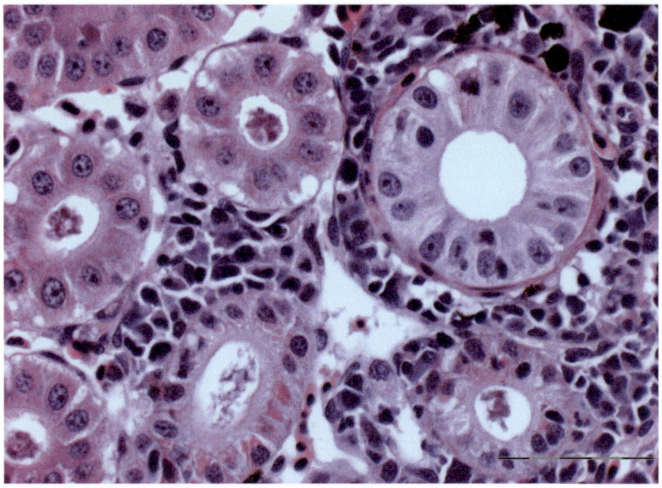

Fig. 2.22 Kidney tubules and interstitial tissue

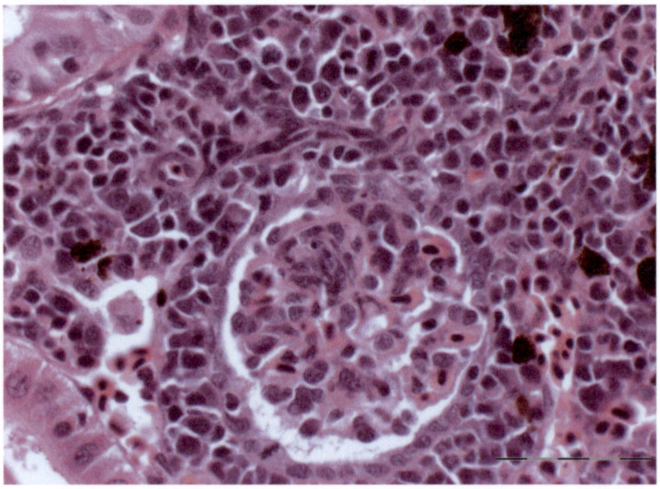

Fig. 2.23 Glomerulus in the interstitial renal tissue of adult Atlantic salmon

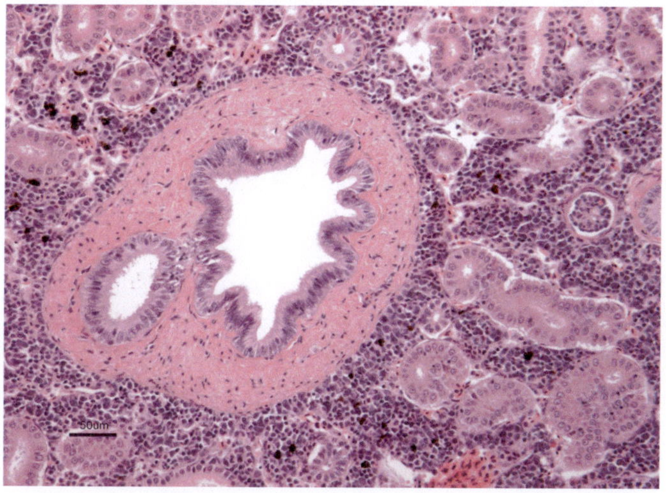

Fig. 2.24 Collecting duct and urether with wall of smooth muscle in farmed rainbow trout

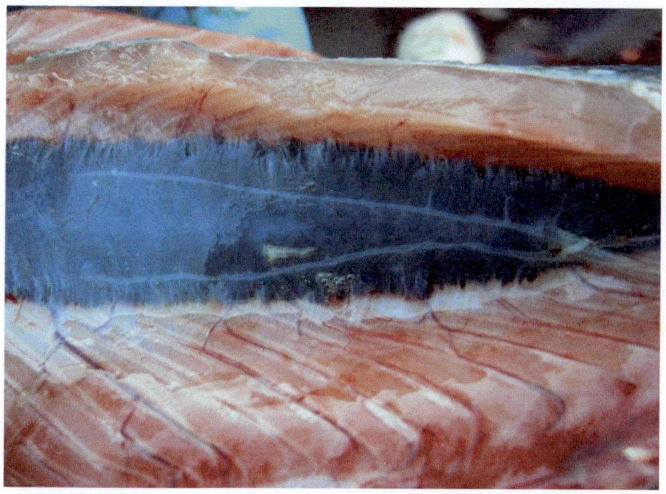

Fig. 2.25 Ureters of an adult Atlantic salmon fusing to form a small urinary bladder

epithelium of the first segment. A variable intermediate segment may be distinguished, with a lower and more cuboidal epithelium. The brush border becomes intermittent and as it reaches the distal segment, they are absent. Each collecting duct system terminates in a mesonephric duct (Fig. 2.24). Histologically the proximal tubules have a wider lumen compared to that of the neck region and the distal tubules. Within the glomerulus, erythrocytes can be distinguished within the capillary lumen as well as the nuclei of mesangial cells, capillary endothelial cells and the podocytes of the visceral epithelium of the Bowman's capsule. Salmonids from the marine environment have fewer and smaller glomeruli and the distal part of the tubule is lacking. Collecting ducts pass urine into two ureters which fuse to form the urinary bladder (Fig. 2.25).

Functionally, the principle role of the posterior kidney is maintenance of a stable internal environment with respect to

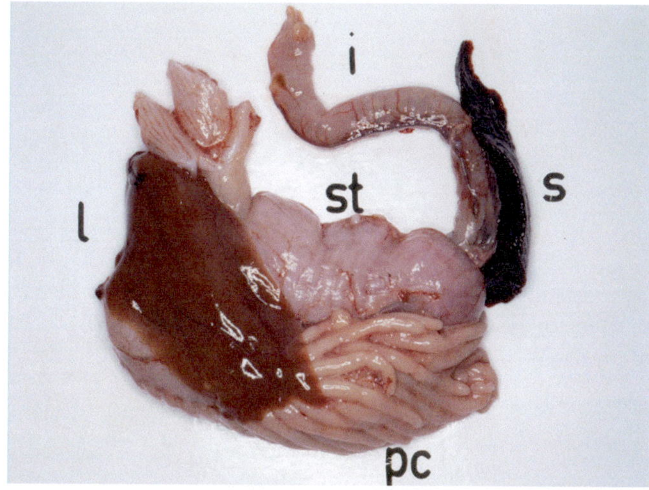

Fig. 2.26 Visceral organs of the digestive system from Atlantic salmon; liver (*l*), pyloric caeca (*pc*), spleen (*s*), stomach (*st*), intestine (*i*)

water and salts, therefore it needs to adapt to the external water conditions. Accordingly, in the freshwater the fish is hypertonic and the nephron must conserve salts and eliminate excess water which enters the body through the gills. Conversely, in the marine environment, the fish is hypotonic, the urine produced is scant and contains various di- and trivalent electrolytes as well as nitrogenous end-products. The nephron must conserve water through a reduction in urinary volume in order to prevent dehydration. This function is accomplished by a high glomerular filtration rate, reabsorption of salts in the proximal tubules and further concentration of the urine in the distal segment. Ammonia, urea and monovalent electrolytes are mainly excreted through the gills.

2.6 Digestive System

The digestive system is composed of the alimentary canal and digestive glands (gastric glands, pyloric caeca, liver, pancreas and intestinal glands, Fig. 2.26). The following regions are generally distinguished: oral cavity, pharynx, oesophagus, stomach and intestine. Functionally, the role of the digestive tract is the hydrolysis of food items.

The oral cavity contains the tongue and teeth. The tongue is relatively poorly developed in and is typically a rather rigid structure of connective tissue covered with epithelium and many unicellular glands (Fig. 2.27). The mucosal epithelium of the tongue consists of stratified epithelium and contains many taste buds and mucous cells. A lamina propria and a thin submucosa are present in the oral cavity wall, but the muscularis mucosae and submucosa are not recognized. The teeth are joined by connective tissue to the bone. The pulp of teeth is composed mainly of connective tissue and

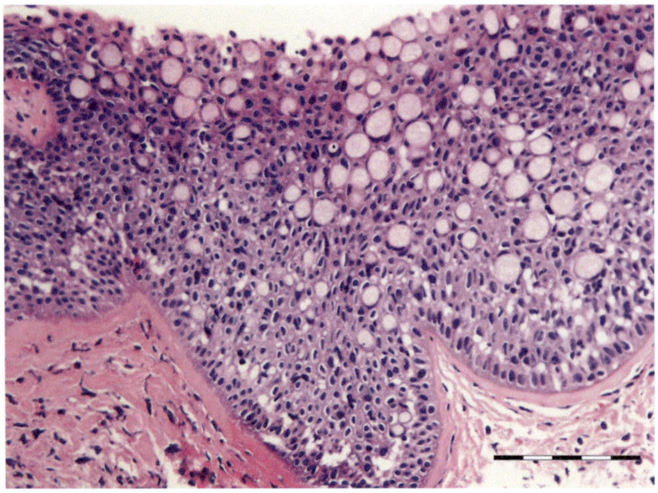

Fig. 2.27 Transverse sections of tongue from rainbow trout

occupies the centre of the tooth. Odontoblasts are arranged at the outermost region of the pulp and secrete dentin. The presence of teeth does not imply chewing activity but often have a role in grabbing and tearing food.

Although the histological structure of the alimentary canal varies along its length, it is basically composed of the following layers from inside out: mucosa, submucosa, muscularis and serosa (Fig. 2.28). The mucosa comprises a simple columnar epithelium with mucous cells; loose connective tissue makes up its lamina propria, richly supplied with blood capillaries. The submucosa supports the mucosa and joints it to the underpaying muscle layer and is composed of dense irregular connective tissue. The muscularis mucosa is a thin layer of longitudinal smooth muscle and the serosa is effectively a sheet of the visceral peritoneum.

The mucosa of the pharynx consists of shallow folds of stratified epithelium. Epithelial cells of the outermost layer are flat, but those at the base are columnar. Mucous cells are present and especially numerous at the bottom part of the

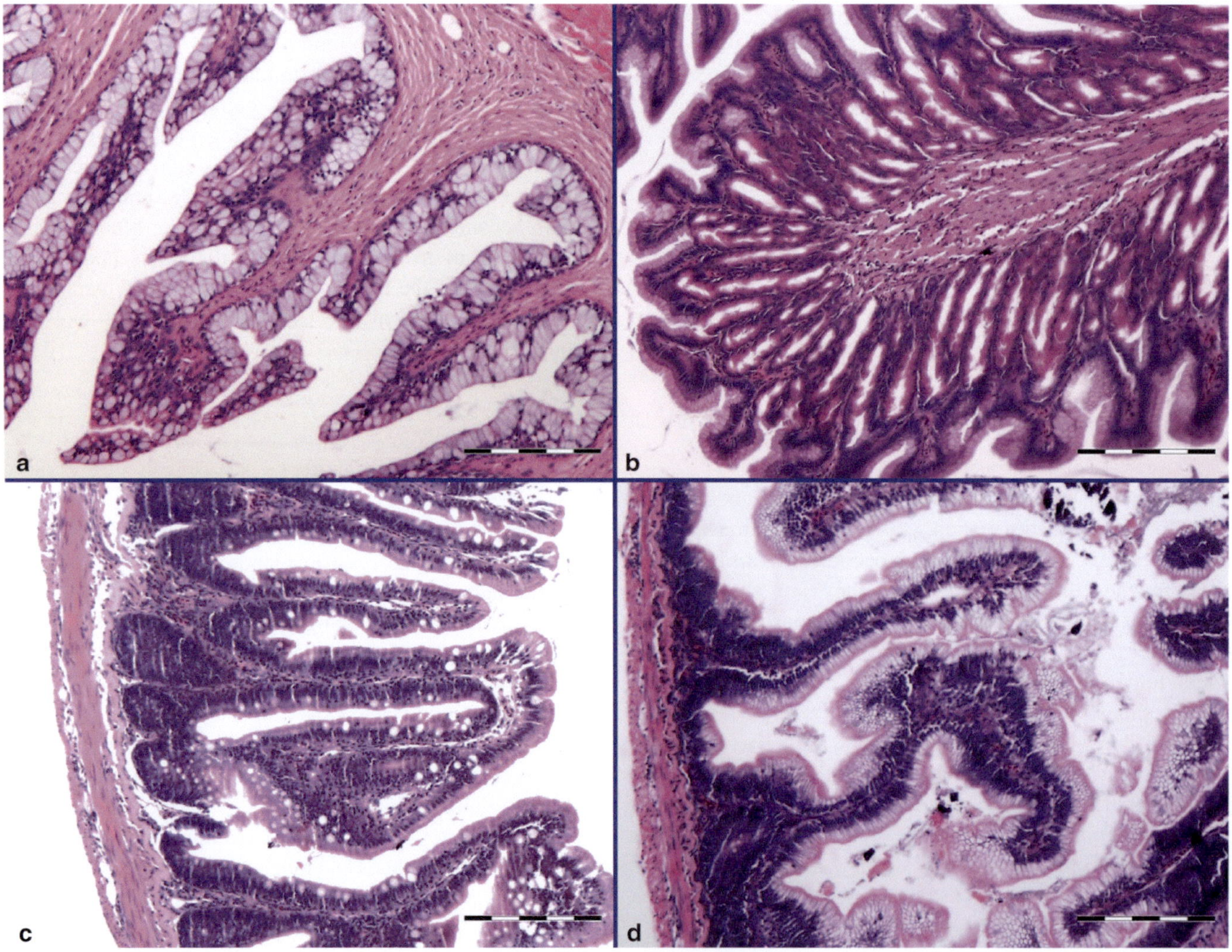

Fig. 2.28 Transverse sections of gut from rainbow trout. (**a**) Oesophagus. (**b**) Stomach. (**c**) anterior gut. (**d**) Anterior gut. Bar = 200 μm

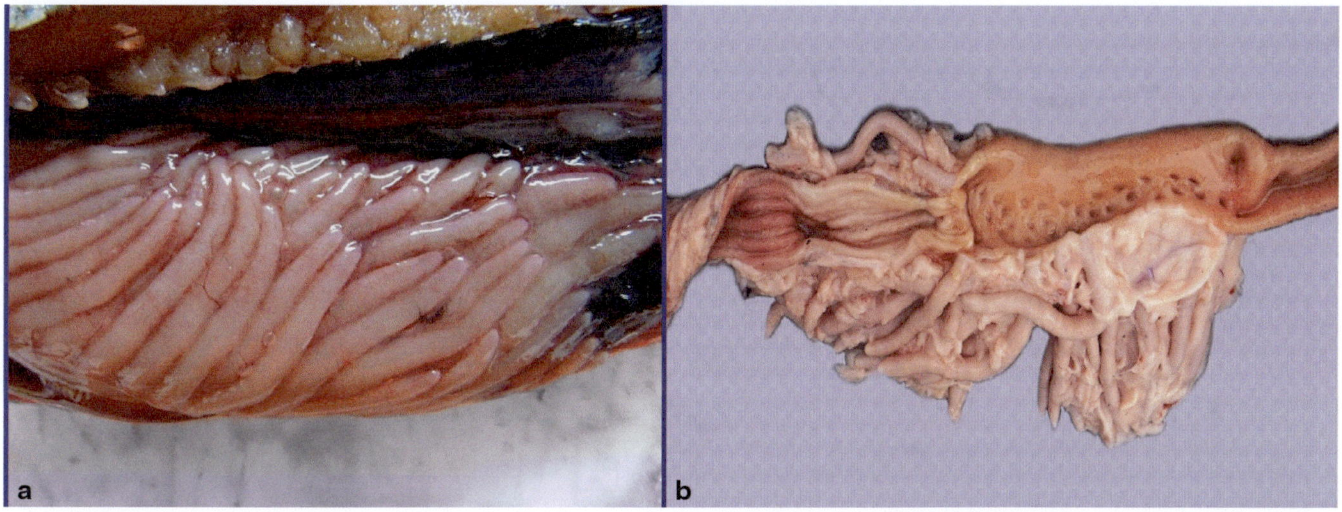

Fig. 2.29 (a) Pyloric caeca of adult Atlantic salmon. (b) Opened stomach and pyloric region showing gastric folds and the openings into the pyloric caeca

folds (crypts). The pharynx mucosa contains a lamina propria, but is devoid of a muscularis mucosa. The muscularis is composed of a thick outer layer of circular muscle and a thin layer of longitudinal muscle. Both layers of the muscularis are striated.

The oesophagus is a short, muscular thick-walled tube with longitudinally arranged folds of mucosa to facilitate swallowing and propulsion of food particles. At the entrance to the oesophagus the mucosa contains many mucous cells and taste buds. This region usually lacks a muscularis mucosa, distinguishing it from the rest of the digestive tract.

In the stomach and intestine the stratum compactum is located between the lamina propria and the muscularis mucosa which is composed of dense collagen fibres. Eosinophilic granular cells (EGC's) may be present and form a layer at the inner and outer sides of this stratum and are considered to be closely related to fish mast cells. This layer of EGC's is sometimes referred to as the stratum granulosum.

The stomach is U-shaped and composed of the cardiac portion (anterior), fundus and pyloric portion. Each region has a simple folded mucosal epithelium. At the cardiac portion the folds are shallow and become deeper at the fundus and pyloric portion. The epithelium varies between cuboidal and highly columnar, and the nuclei are generally located in the basal region of the cell. Gastric glands are located in the lamina propria and often into the crypts of the mucosal folds.

The fundic region of the stomach is a blind sac, pouching off from the main tube of the organ and characterized by numerous gastric glands although, in contrast, they may be absent in the pyloric region.

Pyloric caeca are blind-ended, finger-like projections that extend outwards from the pyloric valve region of the stomach and the anterior intestine (Fig. 2.29). Their structure and function resemble that of the intestine with a multi-folded intestinal type epithelium, and regions where fats are broken down into fatty acids and glycerine. They expand the nutrient absorption surface but also contribute to the salt and water balance, therefore playing a functional role in osmoregulation.

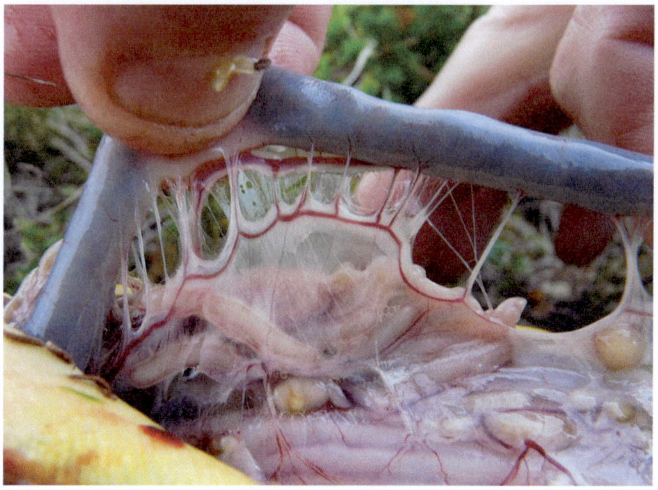

Fig. 2.30 Mesentery with blood vessels of the posterior intestine in adult Arctic char

The intestine extends from the end of the pyloric portion of the stomach to the vent and includes the duodenum, anterior intestine, posterior intestine and rectum. The main function of the intestine is uptake of lipids, proteins and ions. The mesentery, a double layer of peritoneum, represents the peritoneal fold that attaches the small intestine to the posterior body wall. Blood vessels and nerves for the intestine are located in the mesentery (Fig. 2.30). The bile and pancreatic ducts and pyloric caeca open into the duodenum. The anterior and posterior intestine can be distinguished from each other by the shape of their respective mucosal

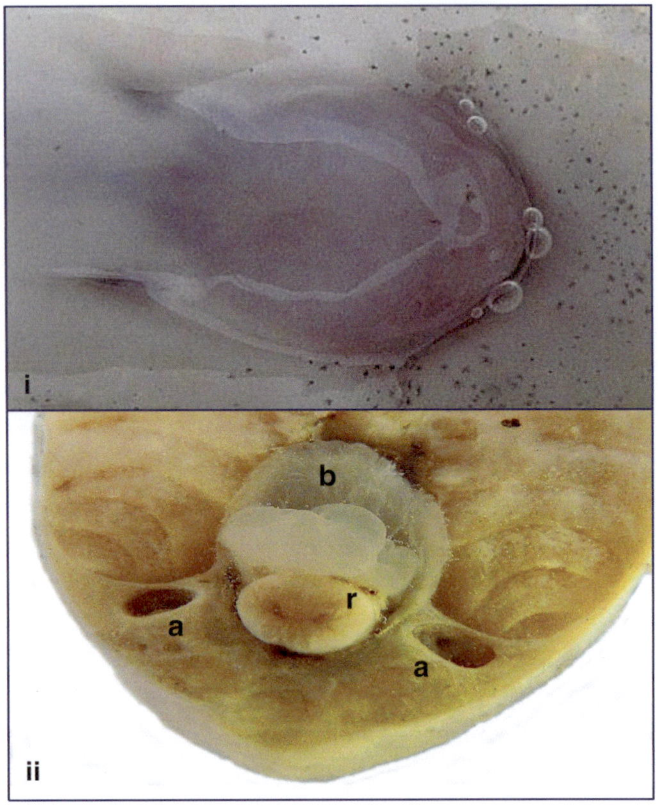

Fig. 2.31 (**i**) External view of vent area of Atlantic salmon. (**ii**) Transverse section through formalin fixed Atlantic salmon, vent area. End of bladder (*b*), rectum (*r*), abdominal pores (*a*)

folds. In rainbow trout for example, the anterior region has shallow folds and the posterior region deep folds with a thicker muscle layer with many mucous cells.

The vent is equipped with a muscular sphincter. The rectum connects with the vent located cranial to the anal fin and has a thicker muscular wall than the intestine and is capable of considerable distension. The term 'vent' specifically defines the external opening of the alimentary canal or the anus, although in a wider sense the term is used to refer to the region that includes the rear portion of the alimentary canal, the urogenital papilla, genital cavity and pore, and last portion of the urinary canal and bladder, surrounded by tissues of the posterior abdominal wall, abdominal pores and underlying adipose and muscle in the immediate area (Fig. 2.31). Abdominal pores are a paired communication between the abdominal cavity and the exterior at the rear of the abdominal cavity, and they lead to the exterior through the body wall one at each side within or behind the vent and urogenital region.

2.6.1 Liver and Gall Bladder

The liver is a large reddish-brown organ normally located in the left anterior part of the abdominal cavity, with its cranial part close to the septum transversum. The parenchyma is composed of chords of cuboidal hepatocytes supported by lattice fibres and some connective tissue (Fig. 2.32). Each cell has a roundish polygonal cell body containing a clear spherical nucleus, usually with one nucleolus and contains variable amounts of lipid and glycogen, depending on normal variation or the nutritional status of the fish. The normal appearance of a liver from Arctic char is included for comparative purposes (Fig. 2.33). Blood is filtered through a network of sinusoids running between poorly defined, cord-like structures of hepatocytes. Phagocytosis and particulate antigens presentation represent an important immune function. Both the hepatic artery and portal vein enter the liver.

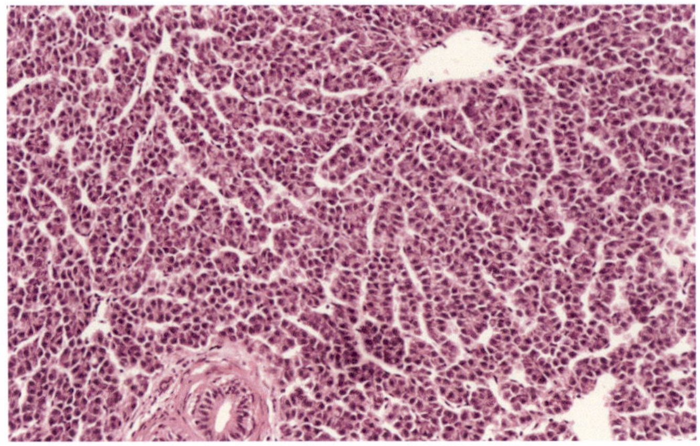

Fig. 2.32 Normal liver parenchyma from Atlantic salmon. Note a vein (*top*) and bile duct (*bottom*). Low power

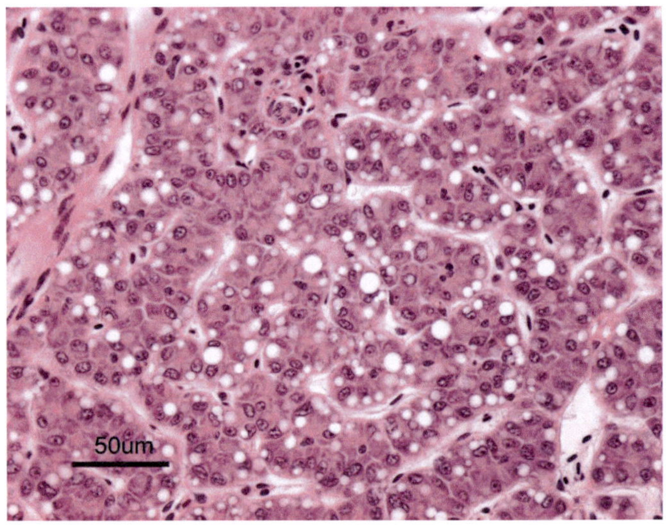

Fig. 2.33 Normal parenchyma of Arctic char liver

The hepatocytes secrete bile into the bile canaliculi where it is carried into the extracellular bile canaliculi to form the bile duct, which subsequently joins with the hepatic duct and opens into the duodenum. A branch of the hepatic duct called ductus cysticus, leads into the gall bladder where the

Fig. 2.34 Full gall bladder in adult farmed Atlantic salmon

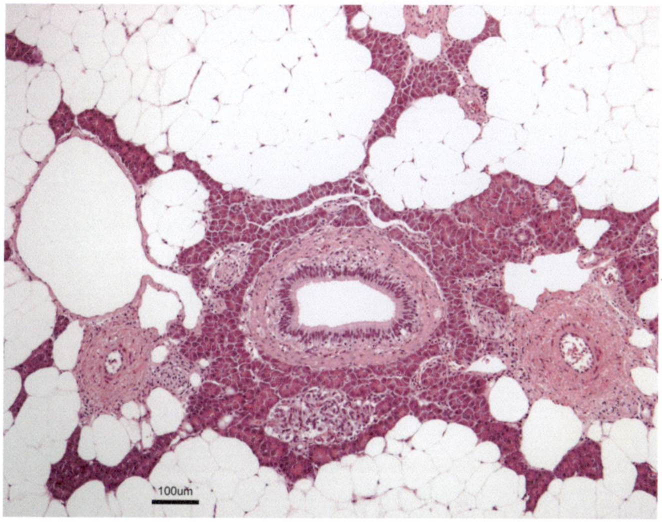

Fig. 2.35 Normal pancreas with endocrine, exocrine tissue and ducts

bile is stored (Fig. 2.34). The wall of the gallbladder is thin, contractile, and will contract when food, especially fatty food, passes through the duodenum.

2.6.2 Pancreas

The pancreas is a diffuse organ which is interspersed throughout the adipose tissue and surrounds the mesenteric fat mainly among the pyloric caeca. Functionally, the pancreas is both a digestive organ and an endocrine gland. The exocrine tissue is organized in distinct clusters ('nests') of 'acinar cells' of strongly basophilic cytoplasm and therefore stains purple with H&E (Fig. 2.35). The triangular or polygonic cells have basally located, well-defined nuclei and nucleoli. In actively feeding fish, bright eosinophilic secretory zymogen granules are present in the glandular cytoplasm. Dark staining is apparent when the quantity of zymogen granules is high and pale when the quantity of zymogen granules is low. A cell which has atrophied as a result of starvation or disease, contains few zymogen granules and consequently becomes basophilic and small in size. Digestive enzymes secreted from these cells are carried into the ascending intestine through a series of ducts and participate in the breakdown of proteins, fats and carbohydrates.

The endocrine portion of the pancreas is lightly capsulated and occur in clusters of glandular cells, the Islets of Langerhans', occurring among fat cells or surrounded by exocrine tissue. It is composed of cords of cells and generally recognized as having three functionally independent cell types: alpha (A), beta (B) and delta (D) cells that secrete hormones, including glucagon and insulin.

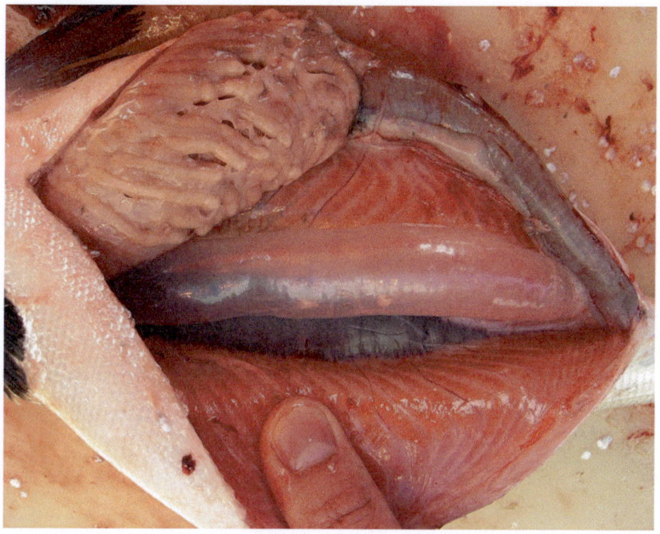

Fig. 2.36 Swim bladder from Arctic char

2.6.3 The Swim or Air Bladder

The swim or air bladder develops as an out pushing from the anterior part of the gastrointestinal tract and is a hydrostatic organ that can be filled or emptied to regulate buoyancy. In primitive fish it remains connected to the oesophagus by a tube, the ductus pneumaticus (physostomous fish), while in higher teleosts the connection is lost during development (physoclistous) and the swim bladder filling depends on a 'gas gland' (Figs. 2.36 and 2.37). Anatomically, the gas-filled swim bladder is a conspicuous organ located along the dorsal wall of the peritoneal cavity. Histologically three layers can be distinguished, the inner mucosa, the muscle layer and the outermost fibrous connective tissue.

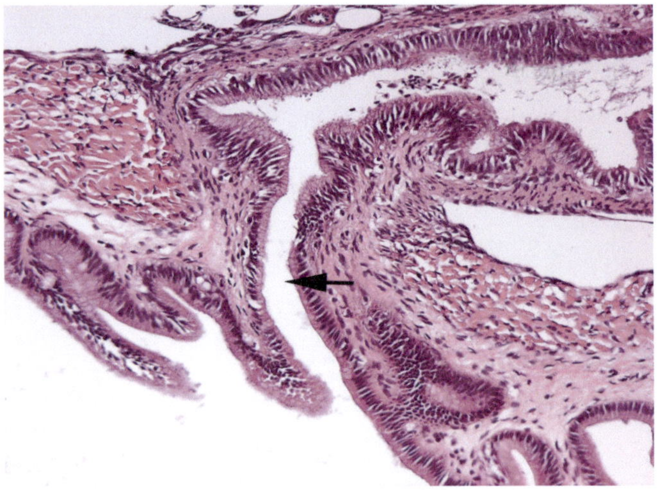

Fig. 2.37 Pneumatic duct (*arrow*) between oesophagus (*bottom*) and swimbladder (*top*). Low power

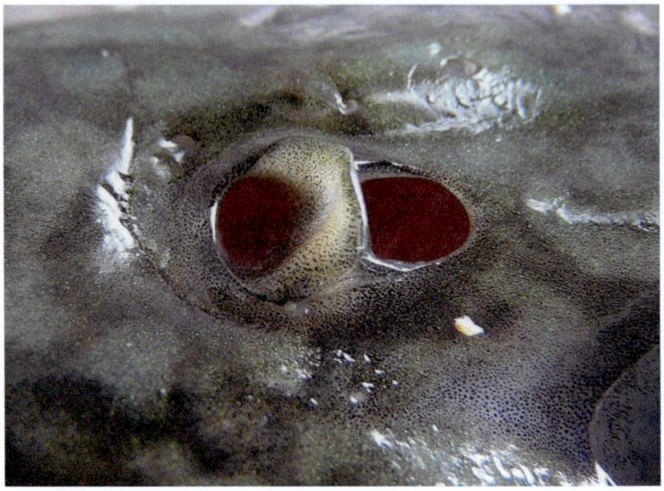

Fig. 2.38 Nostrils of an adult lake trout

2.7 Sensory System

Fish sensory systems include mechano, olfactory and taste receptors, equilibrium, hearing as well as vision organs. The olfactory sensory neurons give a sense of smell mediated by specialized sensory cells of the nasal cavity (Fig. 2.38). For the purpose of this chapter we will only refer to the mechano reception, exemplified by the lateral line and vision systems, as the most important regions for tissue pathology and infection.

2.7.1 The Mechanosensory Lateral Line

The mechanosensory lateral line system comprising a series of receptor organs, composed of neuromasts, located on the epithelium or within canals on the head and trunk, and innervated by several lateral line nerves which project to the hindbrain. This sensory system can detect movement and vibrations in the surrounding water, allowing the fish to respond to unidirectional or oscillatory movements, at relatively short distance. The trunk lateral canal is easy to observe as a fairly straight and clearly defined line, running along the middle top section of the flanks (Fig. 2.39). It comprises short segments of overlapping tubed scales (the lateral line scales) with a pore present at each end of the canal segments that links to adjacent overlapping scales to form a continuous viscous fluid-filled canal. Additional pores piercing the canal walls might be present and provide additional access to the external environment. A neuromast in located within each lateral line scale and additional superficial ones or 'accessory neuromasts', may also be located in the epithelium in proximity to the trunk canal.

Fig. 2.39 Normal skin with scales of grayling, lateral line is visible

2.7.2 Visual

The fish eye is a delicate and highly specialised structure that is particularly vulnerable because of its exposure to the environment and absence of protective eyelids. The eye is similar to that of other vertebrates (Fig. 2.40) and its function is to collect and focus the light and convert it into a nervous impulse. The components of the eye are the cornea, iris, lens, sclera, choroid and retina (Fig. 2.41). The cornea is non-pigmented and consists of a stratified squamous epithelium on a thick basement membrane and has a refractive index similar to water and therefore, irrelevant as an optical surface. The lens is a spherical ball consisting of three layers: a first encapsulating sheath of non-transparent material which is secreted by the second layer, an underlying tissue of physiologically active cells which are nucleated and capable of division and secretion, and a third tissue immediately

Fig. 2.40 Eye of a grayling

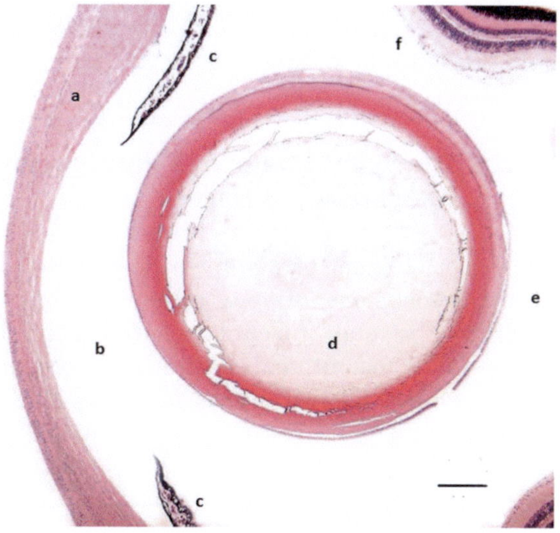

Fig. 2.41 Lateral view of the salmonid eye. (*a*) Cornea. (*b*) Anterior chamber. (*c*) Iris. (*d*) Lens (*e*) Vitrous body. (*f*) Retina

beneath consisting of non-nucleated long, slender, transparent cells in parallel rows that occupy the greatest volume. The lens protrudes through the iris providing a wide angle of view. In order to accommodate changing sight requirements, the retractor lentis muscle must be drawn inwards towards the retina. The iris is fixed as the sphincter and the dilator muscles are poorly developed. The innermost element of the eyeball is the photo-sensitive retina. Choroid vessels form a subscleral network of capillaries which provide nourishment for the retina. Eight layers are recognized: pigment epithelium elements, rods and cones, outer nuclear layer, outer plexiform layer, inner nuclear layer, inner plexiform layer, ganglion cell layer and nerve fibre layer (Fig. 2.42).

A counter current system is present in the choroid layer of the retina. This vascular structure (the choroid rete mirabile) is supplied by a branch of the ophthalmic artery and blood

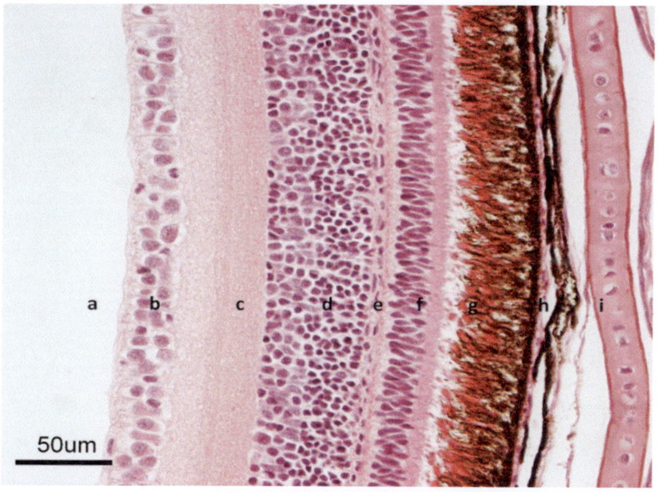

Fig. 2.42 Retina of coho salmon showing the various layers. (*a*) Vitreous body. (*b*) Ganglion cell axons. (*c*) Inner plexiform and ganglion cell layer. (*d*) Inner nuclear layer. (*e*) Outer plexiform layer. (*f*) Nuclei of rods and cones cells. (*g*). Rods and cones. (*h*) Pigmented epithelium. (*i*) Scleral cartilage

from of the choroidal vessels initially passes through the pseudobranch. The correlation between the oxygen pressure and development of the rete suggest that the choroid rete mirabile plays a role in establishing the high oxygen pressure of the retina.

2.8 Nervous System

The nervous system can be divided into the cerebrospinal system, i.e., the brain, spinal cord, ganglia, cranial and spinal nerves, and the autonomous or vegetative system, comprising ganglia and sympathetic and parasympathetic nerves that works in close integration and interdependency with the endocrine system. Overall the main function is integration and control of organs and the communication with the outside environment. Histologically, the nervous system cellular elements are the neurons and the neuroglia and consist of two basic types of cell: neurons and glial cells.

2.8.1 Brain

The brain has the same basic regions as that of other vertebrates, but the proportions between the anatomical units are different, particularly the mesencephalon with the optic lobes being very conspicuous (Fig. 2.43). The brain is traditionally divided into five different units: telencephalon, diencephalon, mesencephalon, metencephalon and myelencephalon. The protective layers around the brain are known as the meninges, and represent an important barrier against

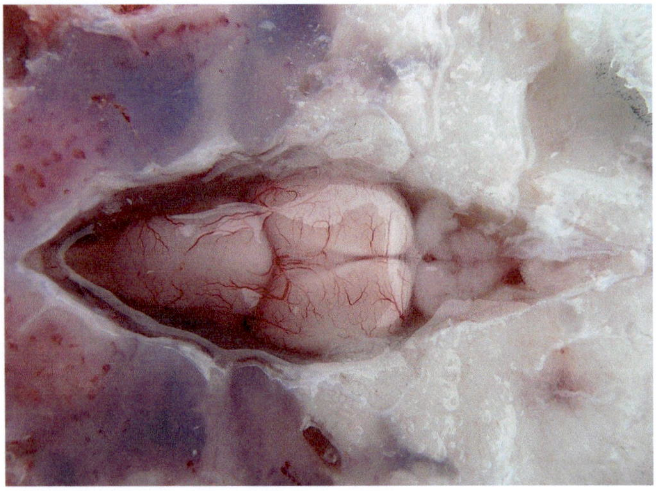

Fig. 2.43 Exposed brain of adult Atlantic salmon

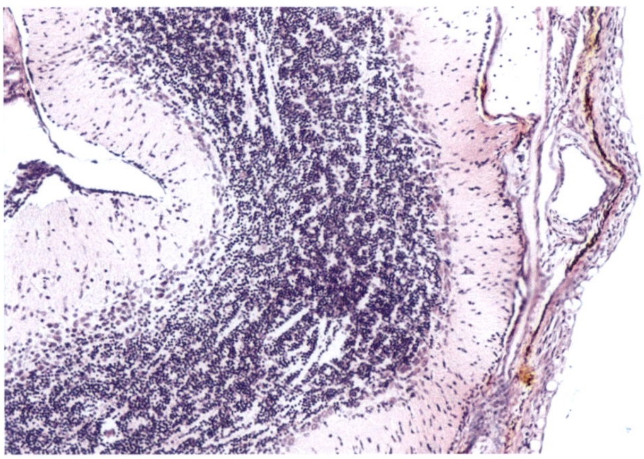

Fig. 2.44 Granular layer of cerebellum. Low power

pathogens between the blood and the cerebrospinal fluid (Fig. 2.44). They are small compared to its mammalian counterparts and are dominated by the olfactory lobes. In contrast to mammals, the histological conspicuous six-layered neocortex is absent.

The ventral part of the diencephalon, including the hypothalamus, is termed the infundibulum and is well developed in salmonids. Three important proteins are produced here and important components of the cerebrospinal fluid.

The telencephalon structures are located in the ventral midline of the diencephalon: chiasmaopticus, the pituitary gland (hypophysis) and a ventral choroid plexus called 'saccus vasculosus'. In the optic chiasma, the nerve fibres from the retina are crossing before entering the brain. The complex pituitary gland is a neuro-epithelial structure orchestrating several other endocrine organs and is involved in osmoregulation, gonadal development, growth and melanization. The saccus vasculosus is a sac-like structure with blood sinusoids quite similar to the choroid plexus of the eye, and communicates with the lumen of the third ventricle. On the dorsal side of the diencephalon is the pineal gland (epiphysis). This is a well vascularised non-image forming photoreceptor structure, located under a thin non-pigmented area of the skull that permits light to reach the structure ('pineal window'). This structure can detect alterations in ambient light (decreasing light in fall, increasing light in the spring) and is of importance for regulation of seasonal physiology such as smoltification. The mesencephalon is highly developed and conspicuous with its two optic lobes (tectumopticum) reflecting the importance of vision for these fish. Centrally, there is a large lumen (ventriculus mesencephali) filled with cerebrospinal fluid. The metencephalon has a well-developed, partly folded dorsal part called the cerebellum, responsible for movement coordination. The myelencephalon is the origin of the cranial nerves and the beginning of the spinal cord (medulla oblongata).

2.8.2 The Sympathetic and Parasympathetic Ganglions and Nerves

The sympathetic and parasympathetic ganglions and nerves of the autonomous nervous system are responsible of regulating several functions and mostly antagonist responses to that of the cerebrospinal system, as both innervate most internal organs including digestive tract, heart and gills.

2.9 Endocrine System

Glandular derivatives such as thyroid, thymus and ultimobranchial bodies are formed from the pharynx during embryological development. Thyroid hormones play a supportive role in sea water acclimation and follicles are distributed throughout the connective tissue of the pharyngeal area. They may also be observed around the eye, ventral aorta, hepatic veins and anterior kidney, and are similar histologically to mammalian thyroid tissue (Fig. 2.45). The thymus is located on the dorsolateral wall of the pharynx with its ventral surface covered with mucosal epithelium (Fig. 2.46).

At the junction of the head and posterior kidney are the corpuscles of Stannius, a sac-like body which has an endocrine role in calcium metabolism. The corpuscles of Stannius are not always visible macroscopically and may be embedded deep in the renal tissue (Figs. 2.47 and 2.48). Variable amounts of individual or clustered pigment-

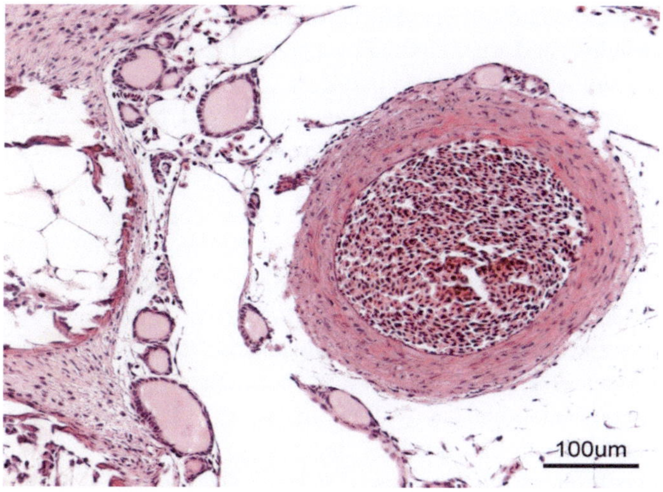

Fig. 2.45 Thyroid follicles located near the ventral aorta of an Atlantic salmon parr

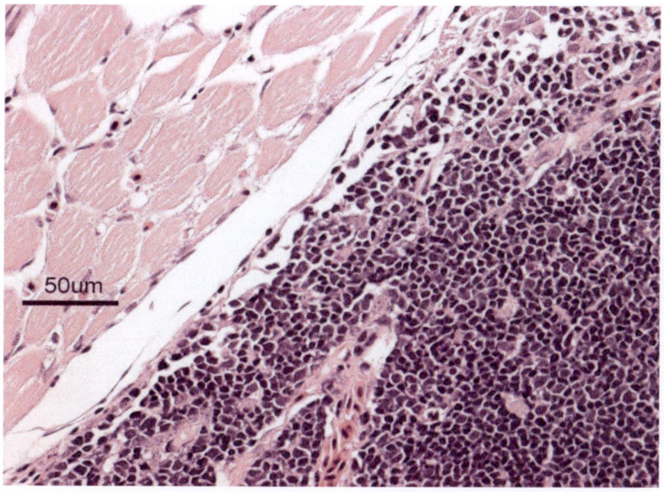

Fig. 2.46 Thymus of Atlantic salmon parr

containing melanomacrophages are normally present in the interstitial renal tissue. Their number typically increases with age and with disease conditions. The kidney also contains endocrine elements such as chromaffine cells located in the wall of the posterior cardinal vein that release adrenaline and noradrenaline into the circulation, and interrenal tissue which in most teleosts is located around major veins (Fig. 2.49).

Ventral to the oesophagus in the septum transversum separating the heart from the abdominal cavity is the small endocrine ultimobranchial gland, derived from the pharynx. This secretes the hormone calcitonin which lowers serum calcium levels and acts with hypocalcin (secreted by the corpuscles of Stannius) to regulate calcium metabolism. The epiphysis projecting from the epithalamus produces melatonin.

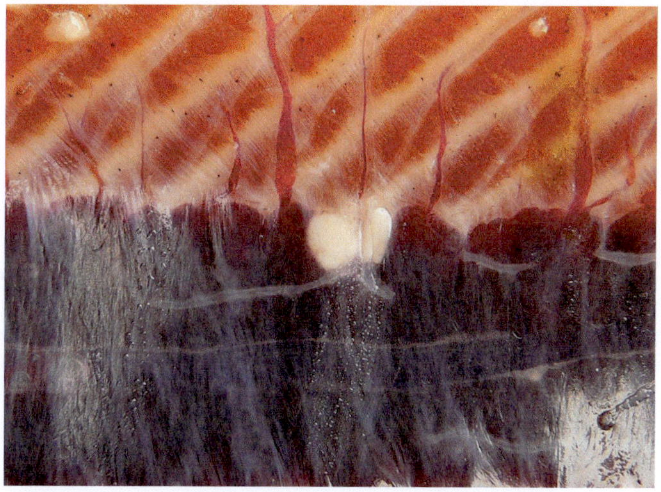

Fig. 2.47 Corpuscle of Stannius near the margin of the kidney in brook trout

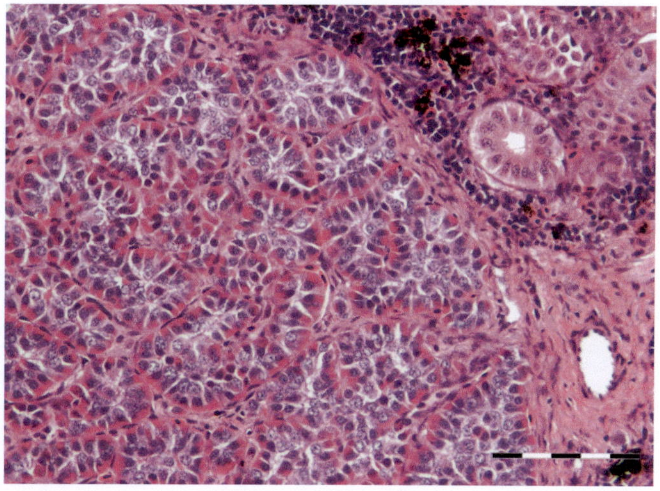

Fig. 2.48 Corpuscle of Stannius (*left*) within the kidney of Atlantic salmon. Bar = 100 µm

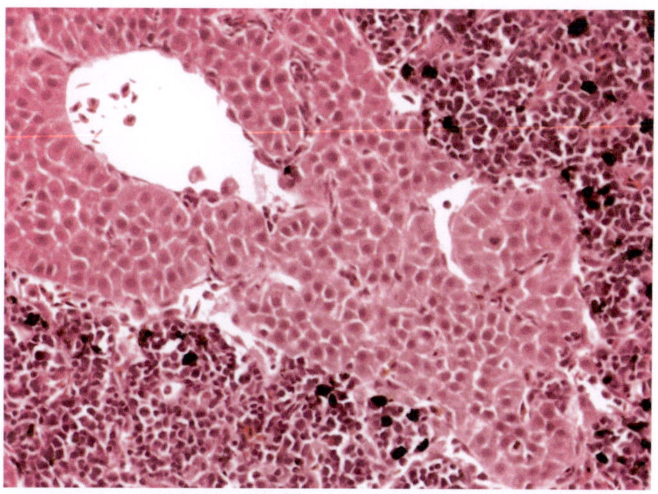

Fig. 2.49 Adrenal cortical tissue in head kidney from rainbow trout. Note melanomacrophages in renal interstitium. Medium power

2.10 Reproductive System

2.10.1 Ovary

The ovaries are paired sac-like organs in the dorsolateral part of the abdominal cavity (Fig. 2.50). They are composed of germinative, stromal, vascular and nervous tissues and suspended in a mesenterium called the mesovary, from the roof of the abdominal cavity. The oviduct is not complete and the ripe eggs are shed into the posterior part of the abdominal cavity before they are funnelled through the urogenital papillae.

In immature fish the ovaries appear as small yellowish or orange spheres. The ovarian follicles line a hollow cavity and ova are passed into this cavity as they mature. The maturing oogonia are surrounded by a single layer epithelial cells and this aggregate of ova and epithelial cells is known as the ovarian follicle. The epithelial cells grow as the ovum develops and are separated by a gradually thickening hyaline capsule called the 'zona pellucida'.

During maturation the oogonia are referred to primary and secondary oocytes. The oocytes enlarge as yolk granules are included into the cytoplasm (vitellogenesis) and the follicular epithelium thickens. With continued oocyte growth the ooplasm becomes impregnated with yolk granules. Several other morphological changes also take place towards the end of the growth phase of the eggs. For example, the eggs become translucent due to the coalescence of yolk globules. At sexual maturation, the eggs may almost fill the abdominal cavity. Examination of ovaries undergoing active oogenesis indicates that oocyte development is not synchronous, i.e. oocytes of varying sizes and in different phases of vitellogenesis are present. Histologically, the ripe egg is characterized by a translucent cell membrane and a distinct animal pole with the nucleus at the micropyle.

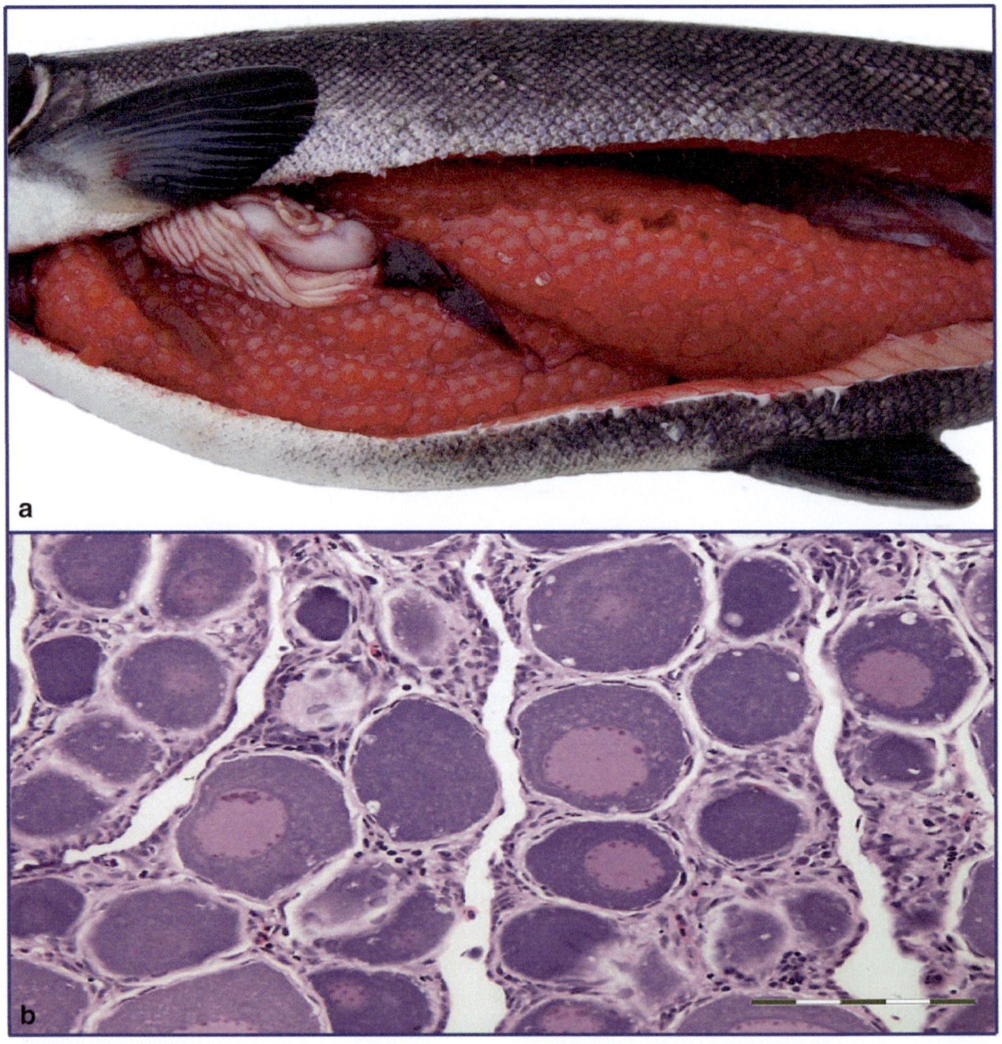

Fig. 2.50 (**a**) Ovaries in mature Atlantic salmon. (**b**). Developing oocytes in a rainbow trout fry. Bar = 100 μm

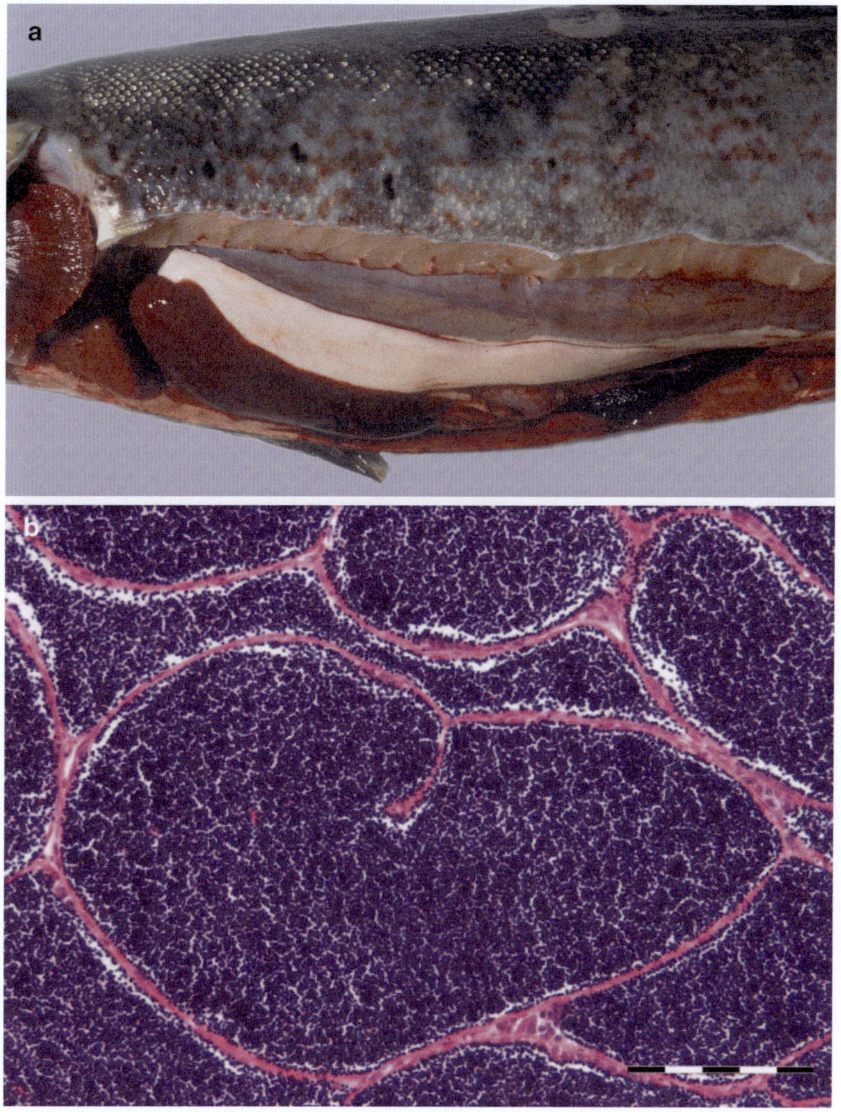

Fig. 2.51 (**a**) Mature testes from Atlantic salmon. (**b**) Lobules packed with spherical spermatogonia of varying diameter

2.10.2 Testes

The testes are a pair of sac-shaped organs surrounded by a capsule of connective tissue suspended from the abdominal roof by a mesenterium called the mesorchium. In juveniles, they are thin strands while they may be large white flabby organs in mature males (Fig. 2.51). The process of maturation of the male gamete involves the multiplication of spermatogonia or sperm mother cells, which develop from the spermatogenic epithelium to form spermatocytes. Many of these cells eventually undergo a meiotic division to become haploid spermatozoon with head, middle section and a long tail. The earliest stages of spermatogenesis are the primordial germ cells. Some of these divide to form primary spermatogonia and then divide to form cysts of spermatocytes, while others remain quiescent. During maturation lobules packed with spherical spermatogonia of varying diameter become evident as the lobules divided into cysts contain spermatogonia in different stages of development. Cells with mitotic figures are apparent. Most cysts contain primary and secondary spermatocytes and a few contain spermatids ready to be discharged into the lumen of the lobule. A main collecting duct termed the 'vas deferens' collects the mature spermatozoa to an excretory meatus at the urogenital papilla.

Further Reading

Buckland-Nicks JA, Gillis M, Reimchen TE (2012) Neural network detected in a presumed vestigial trait: ultrastructure of the salmonid adipose fin. Proc Biol Sci 279:553–563

Icardo JM (2006) Conus arteriosus of the teleost heart: dismissed, but not missed. Anat Rec 288A:900–908

Nematollahi A, Shadkhast M, Shafeie S, Majidian F (2011) Circulatory system of rainbow trout *Oncorhynchus mykiss* (Walbaum): a corrosion cast study. Appl Ichthyol 27:916–919

Pieperhoff S, Bennett W, Farrell AP (2009) The intercellular organization of the two muscular systems in the adult salmonid heart, the compact and the spongy myocardium. J Anat 215:536–547

Pulcini D, Russo T, Reale P, Massa-Gallucci A, Brennan G, Cataudella S (2012) Rainbow trout (*Oncorhynchus mykiss*, Walbaum) develop a more robust body shape under organic rearing. Aquac Res. doi:10.1111/j.1365-2109.2012.03236.x

Savić N, Rašković B, Marković Z, Poleksić V (2012) Intestinal histology and enterocytes height variation in rainbow trout (*Oncorhynchus mykiss*) grown in cages: effects of environmental conditions. Biotechnol Anim Husb 28:323–332

Witten PE, Huysseune A (2009) A comparative view on mechanisms and functions of skeletal remodelling in teleost fish, with special emphasis on osteoclasts and their function. Biol Rev 84:315–346

Wittenber G, Jonathan B, Wittenberg BA (1974) The choroid rete mirabile of the fish eye. I. Oxygen secretion and structure: comparison with the swimbladder rete mirabile. Biol Bull 146:116–136

3 Post-mortem Examination and the Recognition of Tissue Abnormalities

Abstract

Necropsy is an essential part of an investigation of fish health of wild and farmed fish. Information on management practices and diet, detailed history of mortality, changes in fish behaviour, stock weight and length, management practices (recent or former transport, grading or treatments) as well as feeding response, are all important factors normally available with farmed fish. For wild fish, as much information as possible should also be collected and in both scenarios, water temperature, chemical and physical characteristics should be recorded with notes of any concurrent or recent event affecting other species in the area. The chapter describes the procedures of necropsy with particular reference to obtaining samples of the most common tissues collected for histological examination.

Keywords

Fish necropsy

3.1 Introduction

Post-mortem examination or necropsy (from the Greek 'nekros': 'corpse, dead' and 'opsis': 'eye', 'to see'), corresponds to the term autopsy when performed on the human body, namely the medical procedure of examining a body with the objective of assessing the cause of death and the lesions present. This is achieved through a systematic approach and observation of external and internal structures, organs or tissues, assisted by the collection of samples for further analysis. Necropsy plays a major role in the investigation of fish health of wild and farmed fish, both at the individual or the population level. Fish health assessment begins when the fish is alive in their habitat, when important observations can be made covering aspects related to clinical signs, the water and the environment. Under farming conditions, information on management practices and diet also become of particular relevance. Records of the number or the best estimate of affected individuals within the population should be ascertained to establish the morbidity rate and pattern of the spread of the disease or abnormality observed. If disease is associated with mortality, a detailed history of the daily and total mortality is required, taking into account age, class, and stock origin. Changes in fish behaviour including swimming pattern, position in the water column and respiratory patterns should be noted. Additional information on the stock average weight and length, management practices (recent or former transport, grading or treatments) as well as feeding response, are all important factors normally available with farmed fish. For wild fish, as much information as possible should also be collected and in both scenarios, water temperature and chemical and physical characteristics should be recorded with notes of any concurrent or recent event affecting other aquatic or terrestrial species in the area.

Compared to terrestrial animals there are limited laboratory tests applicable for live fish and therefore, generally clinical diagnosis is not enough for a conclusive diagnosis. This emphasises the need of the post-mortem examination as an essential step towards diagnosing disease in fish. The description provided in this chapter describes the procedures of necropsy with particular reference to obtaining adequate samples of the most common tissues collected for histological examination. This is on the understanding that during necropsy, other samples will also be taken e.g. for microbiological

D.W. Bruno et al., *A Colour Atlas of Salmonid Diseases*,
DOI 10.1007/978-94-007-2010-7_3,

analysis, as well as blood or tissue samples for immunological, serological or molecular studies. However, these procedures will not be covered or discussed in detail in this book.

3.2 Sample Size and Euthanasia

The number of fish sampled for a health assessment will vary according to the objectives of the study. For example, certification of freedom of a notifiable disease generally follows the guidelines from the Office International des Epizooties (OIE). Here the sample size is based upon an assumed prevalence of the specific pathogen to an agreed level of confidence. To obtain a 95–98 % probability of detecting at least one infected fish in a clinically healthy population this translates to a minimum of 30 individuals. Conversely, for disease investigations 5–10 fish showing abnormal behaviour or the characteristic signs of the condition will be adequate for necropsy. Fish removed for examination should, where practical, be placed into a smaller container where further observations can be made before any procedure or the removal of tissues or body fluids. The fish should be euthanized by a humane method, ideally through an overdose of anaesthetic and ideally maintained at cool temperatures throughout the necropsy.

3.3 Necropsy Procedure

3.3.1 External Examination

Fish should be placed on a surface that prevents further contamination to help the prosector to perform the work. A steel tray is ideal due to containment, easiness for disinfection and durability. Normally fish are placed on the right flank with head to the left, a convention based on fusiform fish such as salmonids, where internal organs become readily observable and easily accessed from the left flank, therefore requiring minimal displacement of organs to observe or access other structures. Different species and body forms (e.g. Pleuronectiformes) will require a different approach to achieve the same objective. The fish should be examined in a cool environment and case notes taken throughout the process of the post-mortem examination, and noting any deviation from normality for the species. The provision of a reference to the relative position of the abnormality or the sample taken is an essential part of the report; a few useful anatomical terms of location applicable to whole animals, tissues or histological sections are outlined in Fig. 3.1.

For fish less than 2.0 cm in length they can be examined under a dissecting microscope and when necessary, samples of skin and mucous or gill scrapes can be taken for immediate analysis. However, for the purpose of histological examination fish of this small size can be preserved whole, provided an abdominal flap has been cut open or removed, a practice that allows the fixative to penetrate within the body cavity and guarantee proper fixation. For larger fish, fresh samples of tissues or body fluids can also be taken for initial *in situ* analysis, but for histological examination tissues need to be dissected using scalpel, scissors and forceps.

Gills are a delicate tissue that requires prompt examination as once exposed to air, changes occur rapidly and they quickly dry. The gills are protected by the opercula which should be lifted, assisted by forceps and occasionally may need to be cut to facilitate access to the gill arches. An inspection for evidence of anaemia, increased mucous, blood clots or parasitic infection of the gill filaments or the arch is carried out. A sample of the whole first or second arch (from relative small fish), or a portion (usually at the curved level) of gills (see Fig. 2.2) from larger fish should be removed. For this, the arch can be held using forceps by gripping the gill arch at a location that will not become part of the sample, and carefully cutting the portion to be fixed with scalpel or scissors. Special attention should be paid to avoid any compression of the filaments. Tissue should be promptly placed in the fixative. The pseudobranch (see Fig. 2.6), located, on the inner surface of the operculum, is a target tissue in salmon for the Myxozoa parasite *Parvicapsula* spp. and should be checked and sampled if required.

A thorough check of the entire body external surface should follow noting the integrity of the skin and fins, changes of normal pigmentation for the species, excessive mucus, raised or lost scales, erosion, ulcers, haemorrhage (e.g. petechial), exophthalmia, grossly visible parasites, evidence of skeletal deformity or muscle atrophy. As fish skin is in contact with the external environment, it is vulnerable to damage from a variety of sources, including primary and opportunistic pathogens. Moreover, factors such as handling and net damage will also contribute to the health of the fish. During the post-mortem examination some abnormalities will become apparent, but others will only be recognised through examination of tissues by light microscopy. A skin sample should represent the affected area or lesion including an edge of normal tissue. Both tegument and some of the underlying musculature is usually included in the skin sample. A standard 'normal' skin sample that also includes red and white muscle is also advisable and a default area is usually at the level of the lateral line just below the dorsal fin. Different parts of the fish surface have a particular structure, e.g. head skin lack scales, and therefore samples representing discrete areas of interest should be included. Carefully remove a small piece approximately 1 cm^3 and place immediately in fixative such as 10 % buffered formal saline. The eye will normally be sampled next to prevent excessive drying before analysis and if of interest for the study. A gross examination of the eyes should include reference to corneal opacity, cataract or exophthalmia which, although not necessarily pathognomonic, can indicate a minor infection or a sign of a more serious condition. To remove the eye carefully dissect the skin around the orbit

Fig. 3.1 Orientation planes

using small curved scissors or a scalpel until sufficient tissue is available to grip with forceps. Pull the eye forwards in order to expose the associated muscles and optic nerve and then cut free the entire eye ball. Briefly examine to note presence of haemorrhage or exudate in the anterior chamber or visible parasites. Some laboratories use Carnoy's instead of formalin for eye fixation. Where the main interest is parasitological examination this will usually require fixation in 70 % alcohol. It is advisable to make an incision on the eye ball to allow proper fixation.

The mouth and the oral cavity should be examined recording the presence of any possible petechial haemorrhage, vesicles, parasites or abnormality associated with these structures.

Finally, the cranium needs to be opened in order to expose and examine the brain, an organ relatively 'protected' from contamination within the cranium. The brain is not routinely sampled for histological assessment but when required, it can be performed either at this point of the external examination, or at the end of the necropsy, after the internal examination is finished and provided dissection has not been extensively delayed. In salmonid fish up to ~500 g the cranial structures are sufficiently soft that they can be cut with a sharp scalpel. Holding the head firmly with forceps introduced into the oral cavity, cut open the top of the head with a single decisive movement in a horizontal plane across the head just above the eye level. Continue the cut along the top edge of the opercula and upwards into the dorsal musculature. The severed top of the cranium will contain the brain, already separated from the ten pairs of nerves below and the connection to the medulla oblongata. Carefully remove and place in the fixative. Alternatively, the 'box' of cartilage surrounding the brain can be fixed *in situ* for 24 h before doing the delicate work of removing it from the cranium. Occasionally, when the cut is performed too high, half the brain remains in the lower portion, nevertheless the tissue is sufficiently exposed as to become readily accessible. When larger fish are examined the opening of the cranium requires a sharp strong knife to perform the same cut. A record of blood staining or other discolouration of the cerebrospinal fluid or the organ itself should be recorded.

3.3.2 Internal Examination

To access the internal organs the body cavity is opened. There are several ways to approach dissection, however, the choice must prevent or reduce the likelihood of the process introducing artefacts, damaging tissues, compressing, cutting, moving or displacing organs as well as avoiding the risk of contamination.

The instruction provided in this section is a guide for easy internal examination with the naked eye of fish from ~>15 cm and above. Smaller fish can also be dissected, however, the tools required should be adapted and working under the dissecting microscope should be considered.

One of the most common approaches is to tilt the fish so the belly is facing upwards and, using a scalpel, carefully cut a small incision through the skin and underlying tissues either ~1–2 cm in front of the vent, or in the isthmus, the fleshy part between the opercula beneath the head. The incision is to allow the blunt end of the scissors to be introduced to cut open along the mid-ventral line of the belly sectioning the harder structures encountered at the pectoral and pelvic girdles. The cut should not start with the scissors within the vent as this will damage the hind gut and contaminate the rest of the organs; for the same reason if the cut starts at the isthmus approaching from the opposite direction, it should not go any further than ~1–2 cm in front of the vent. As experience increases all the procedures can be performed using a scalpel.

With this single cut it is possible to start the internal examination and the sampling by lifting the flap with forceps but without exposing the cavity, a practice deemed to contribute to the protection of the internal organs from contamination. Generally exposure of the entire body cavity is required for assessment and allows access to organs for sampling. A second cut to dissect and remove the body wall starts at the caudal end of the first, moving upwards and slightly backwards to reach a level just below the lateral line. From there, the cut turns towards the head almost horizontally just below the lateral line reaching the opercula. Lifting the sectioned flap with forceps will help to guide the last cut behind the opercula, downwards to the isthmus region to completely dissect free the body flank (fillet). The cranial boundary of the body cavity is defined by the septum transversum, separating the peritoneal and the pericardial cavities. If the heart needs to be accessed at the same time, continue anteriorly with the ventral cut slightly further to expose the pericardial cavity and heart, just cranial to the septum transversum. Notes on the general appearance of the body cavity may include references to the extent of the body fat, tissue growth or colour changes, swelling, ascites, adhesions and absence of encysted parasites. For tissue sampling and depending of the fish size, whole organs may be fixed from small individuals (e.g. heart or the entire gastrointestinal tract); conversely from larger fish portions of ~1 cm^3 of each organ should be removed and fixed.

The heart is dissected by lifting the cranial part of the bulbous arteriosus and cutting the connection to the ventral aorta and by holding this end, pulling gently to enable the heart to be moved sufficiently forward to expose and cut free the sinus venosus, connecting with the cardinal veins and hepatic sinus. The heart can show lesions involving the myocardium and associated blood vessels, including blood clots filling the pericardial cavity (haemopericardium) and occasionally, parasites can also be found on or around the heart. For improved fixation of large hearts it may be necessary to divide the organ longitudinally before placing into the fixative.

Most of the abdominal organs can be removed by cutting the oesophagus and lifting with forceps from that end, pulling the gut and associated organs and making a further cut near the vent. In this manner the intestinal tract, liver, spleen, pancreas, and swim bladder are removed out of the fish, leaving the gonads, attached anteriorly, and the kidney remaining within the carcass. Complete removal of the viscera is practical for easier assessment under a dissecting microscope e.g. for parasitological analysis, however inappropriate, if aseptic microbiological samples are required. All the organs can be sampled for histology without removing them from the body cavity.

The salmon liver may vary in colour depending of the type of diet (farm and wild fish), as well as the health status of the fish. A sample should include a portion of the capsule (e.g. the tip of a lobule) and should be collected using a sharp scalpel rather than scissors. Avoid tearing liver tissue or accidentally puncturing the gall bladder, as rough handling can release bile which results in degenerative changes and artefacts (see Fig. 4.31). Record any abnormal colour, suspicious or new growths, fatty changes, the aspect of the cut surface, as well as the appearance of the bile. Inspect the spleen, usually at the posterior curve of the stomach or slightly further back and similarly, take a sample for histology. An examination of the gastrointestinal tract, pyloric caeca and associated pancreatic tissues can be carried out once relevant microbiological sampling has been performed. The gastrointestinal tract can be opened to expose the lumen and allow examination of the mucosa and irregularities should be noted. Several pyloric caeca with associated fat should be removed, thus allowing examination of that portion of the gastrointestinal tract. Additionally, a piece of the stomach and other portions of the intestine may also be sampled. The swim bladder is normally a transparent to opaque, whitish coloured organ. Any change in colour, thickening of its wall, haemorrhage or presence of fluid should be recorded. Evaluation of the gonads provides information on the sex of the fish and degree of maturation and a sample for histological assessment can be performed slicing a section not thicker than 1 cm. Bouin's fixative has been recommended for this tissue as gonads in an advanced stage of development can be hard to cut after routine formalin fixation.

The kidney can be examined after the swim bladder has been moved aside. Many infectious and non-infectious conditions impact on the kidney and possible changes include colour abnormalities, swelling, haemorrhage or a granular appearance. Both the cranial and caudal areas should be sampled for light microscopy avoiding compressing the tissue by cutting first the peritoneal sheet before attempting to lift the sample.

Maintaining an organised and systematic approach to the necropsy is an important aspect of the procedure, and careful observations made during this examination will provide valuable information not only immediately, but consequently during the interpretation of the histological sections. All tissues samples must be clearly identified with a reference code when sent for processing to ensure that there is no risk of incorrect reporting.

Words of wisdom: **'A poorly performed necropsy cannot be improved at a later stage'**

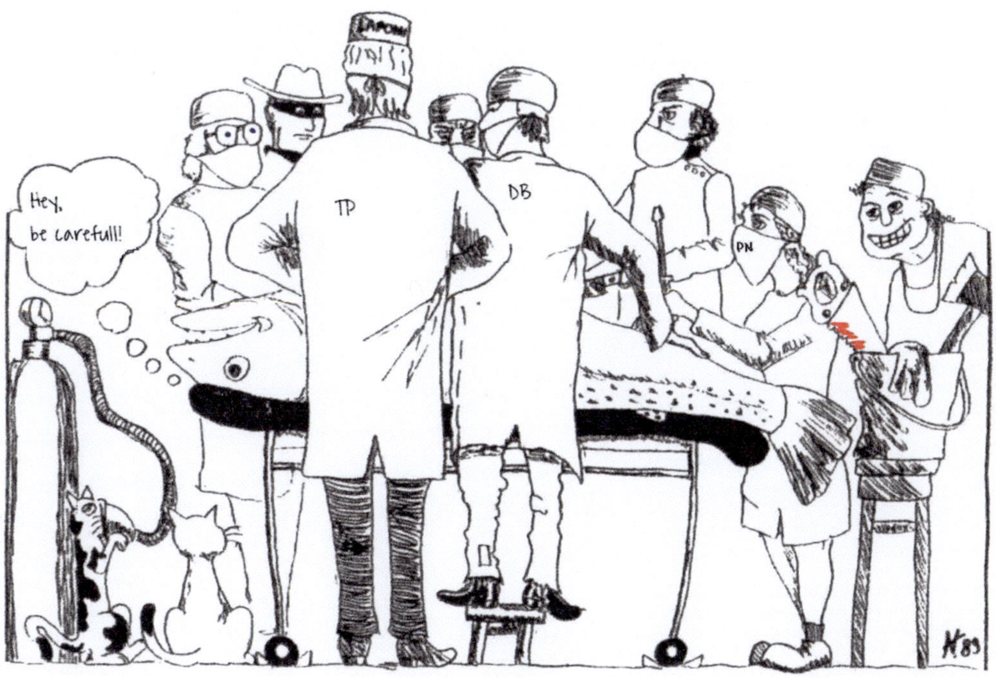

Pathology and Disease Diagnosis

4

Abstract

A methodical approach is a prerequisite for an accurate diagnosis and requires a description of the tissue changes that occur following infectious and non-infectious conditions. Cells have a limited repertoire of morphological response to injury which is linked to biochemical mechanisms that determine the outcome of cell damage, thus accounting for the appearance of cells within lesions inducing general pathological changes, rather than those that are pathognomonic. This chapter covers the different types of cell and tissue responses to acute or chronic injury.

Keywords

Fish disease • Inflammation • Proliferation • Circulatory disturbances • Necrosis • Pigments • Neoplasia

Pathology is the study and diagnosis of disease, and the recognition and interpretation of the physiological and pathological processes. This requires a thorough understanding of normal tissue structure and microanatomy. Normal structure varies widely among species, age and physiological and developmental stages, and even within a population variations may occur, therefore understanding these changes and how they relate to the status of the species under investigation is essential. Furthermore, many diseases look similar at the gross level and to illustrate this aspect, images from skin, kidney and liver showing a range of lesions of different causes are presented in Figs. 4.1, 4.2 and 4.3.

Cell types in fish are, in principle, the same as those found in mammals and similarly, many direct and indirect pathological stimuli induce general pathological changes rather than being pathognomonic. Cells have a limited repertoire of morphological response to injury and are linked to biochemical mechanisms that determine the outcome of cell injury, accounting for the appearance of cells within lesions.

A methodical approach is a prerequisite for an accurate diagnosis and this requires a description of the tissue changes that occur in relation to infectious and non-infectious agents, response to acute or chronic injuries, nutritional imbalance and other causes of disease or abnormality, followed by histopathology. The following areas will be covered in this chapter: inflammation, proliferation, circulatory disturbances, cell injury and necrosis, pigments and mineralization and neoplasia. Finally, a brief description of artefacts is provided to help distinguish these from tangible pathological changes.

Common prefixes and suffixes used in compounded words are provided in Table 4.1, and a glossary appropriate to veterinary terminology is provided in Chap. 13. Reference should also be made to Chap. 2 which discusses normal tissues.

4.1 Inflammation and Proliferation

Lesions can be classified according to their onset, namely acute and chronic. However the pathogenesis of an inflammatory lesion and the histological appearance can be similar. A major component of acute inflammation involves changes in plasma proteins. Serous and fibrinous exudates are a feature of inflamed tissues and histologically consist of eosinophilic staining in the intercellular space and presence of eosinophilic strands, respectively.

Inflammatory foci are characterized by exudation comprising a vascular response, vasodilatation, interstitial fluid changes and cell migration, all of which are common in

D.W. Bruno et al., *A Colour Atlas of Salmonid Diseases*,
DOI 10.1007/978-94-007-2010-7_4, © Springer Science+Business Media B.V. 2013

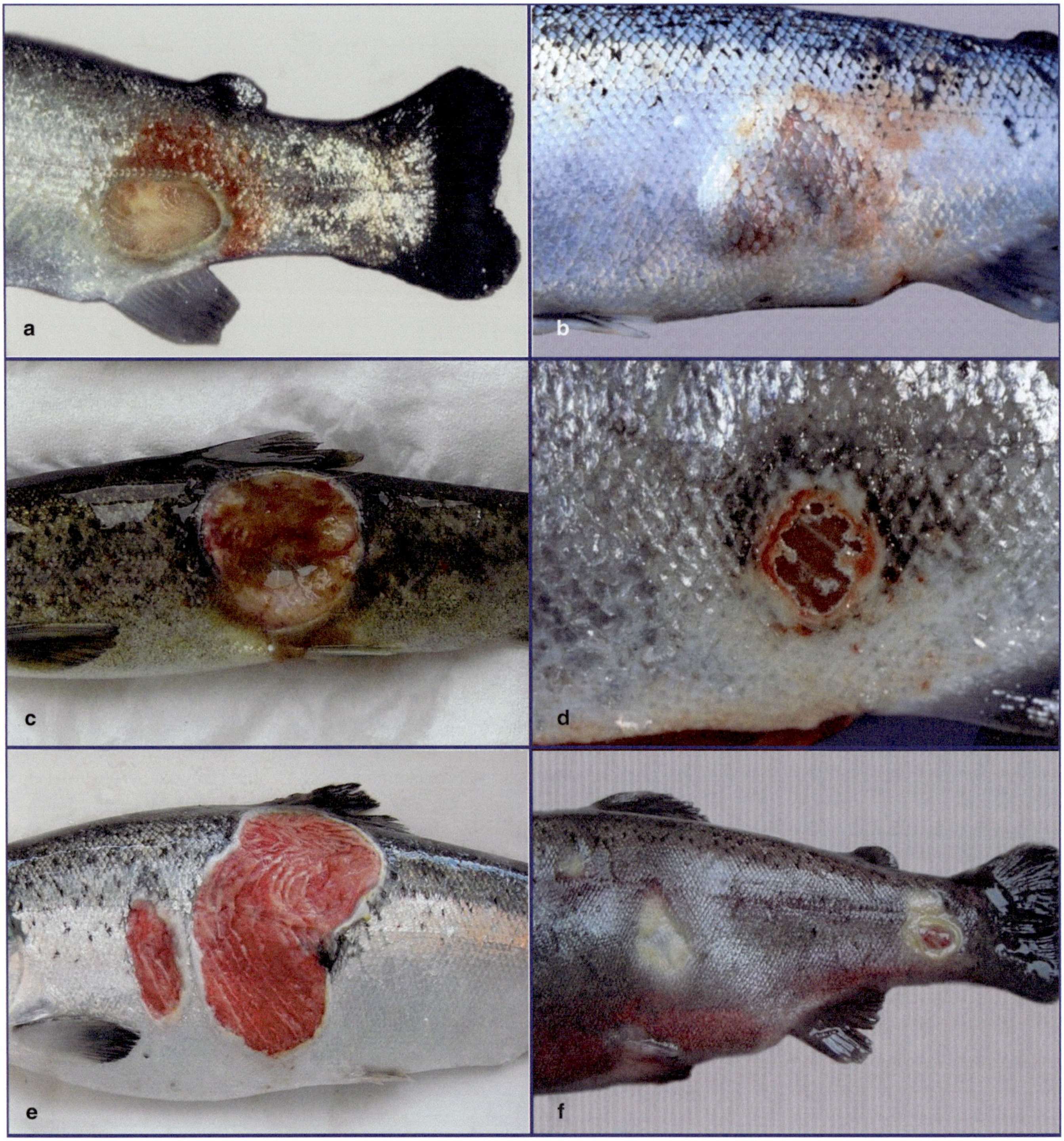

Fig. 4.1 (**a**) Skin ulcer caused by *Aeromonas* sp. in farmed rainbow trout. (**b**) Boil lesion caused by *Aeromonas salmonicida* subsp. *salmonicida* in farmed Atlantic salmon. (**c**) Deep skin ulcer caused by *Pseudomonas fluorescens* in Atlantic salmon smolt. (**d**) Healing *Pseudomonas* ulcer in Atlantic salmon. (**e**) Winter ulcer associated with *Moritella viscosa* in Atlantic salmon. (**f**) Skin ulcers caused by *Flavobacterium psychrophilum* in rainbow trout. (**g**) Classical *Vibrio* infection in Atlantic salmon. (**h**) *Tenacibaculum maritimum* in Atlantic salmon. (**i**) Early red mark syndrome in rainbow trout. (**j**) Advanced red mark syndrome in farmed rainbow trout. (**k**) Lesion associated with cormorant strike in Arctic char. (**l**) Peduncle disease caused by *Flavobacterium psychrophilum* in rainbow trout. (**m**) Ventral skin haemorrhage in Atlantic salmon with CMS. (**n**) Systemic *Moritella viscosa* infection in Atlantic salmon. (**o**) Healed skin ulcer in Atlantic salmon. (**p**) Papillomatosis in wild Atlantic salmon. (**q**) Puffy skin with haemorrhage in rainbow trout. (**r**) *Saprolegnia* infection in Atlantic salmon

Fig. 4.1 (continued)

haemorrhagic septicaemias e.g. furunculosis and vibriosis which may also be accompanied by enteritis (Fig. 4.4). Peritonitis, results from inflammation of the thin tissue that covers most of the abdominal organs (Fig. 4.5), and an inflammatory cardiomyopathy is also present in several fish diseases (Fig. 4.6).

An early hyperaemia in the hypodermis and dermis can be preceded by infiltration of macrophages and other inflammatory cells with a liquefactive necrosis within the centre of the lesion. Lesions may also be designated as degenerative or proliferative. In the former, intracytoplasmic vacuolation, cysts, hyalinization and spongiosis may be seen, whereas

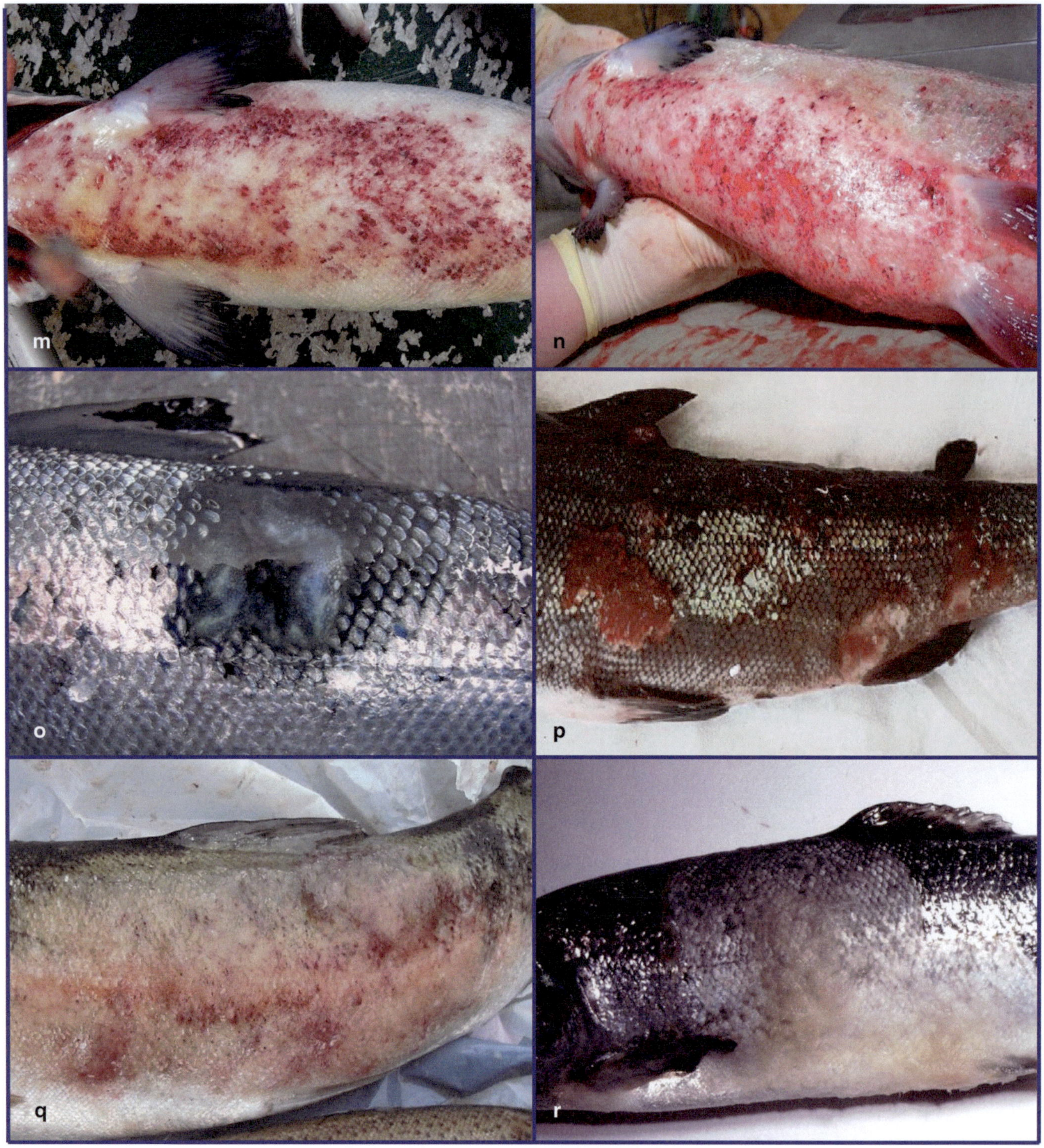

Fig. 4.1 (continued)

proliferative lesions include epithelial cell hyperplasia, fibrosis, focal lymphocytic infiltration and macrophage aggregates. An indication of inflammation or infection can be recorded as perivascular cuffing (Fig. 4.7).

The increase in size of individual cells is termed hypertrophy and can be attributed to increased work load, while physical, chemical irritation or infections, can lead to an increase in number of cells in a tissue, termed hyperplasia.

Fig. 4.2 (**a**) BKD in Atlantic salmon. Kidney capsule has been opened. (**b**) *Mycobacterium* infection in Atlantic salmon. Kidney capsule has been opened (**c**) PKD in rainbow trout. (**d**) *Phyllodistomum umblae* in wild Arctic char. (**e**) Multiple necrosis caused by *Spironucleus salmonicida* in Atlantic salmon. (**f**) Non-specified kidney neoplasia in Atlantic salmon

For example, damage to parts of the myocardium will result in compromised cardiac function and reduced output. Remaining and intact cardiomyocytes will therefore have to compensate for this insufficiency by hypertrophy and hyperplasia, often accompanied by grossly enlarged cardiomyocyte nuclei (compensatory hypertrophy) (Fig. 4.8).

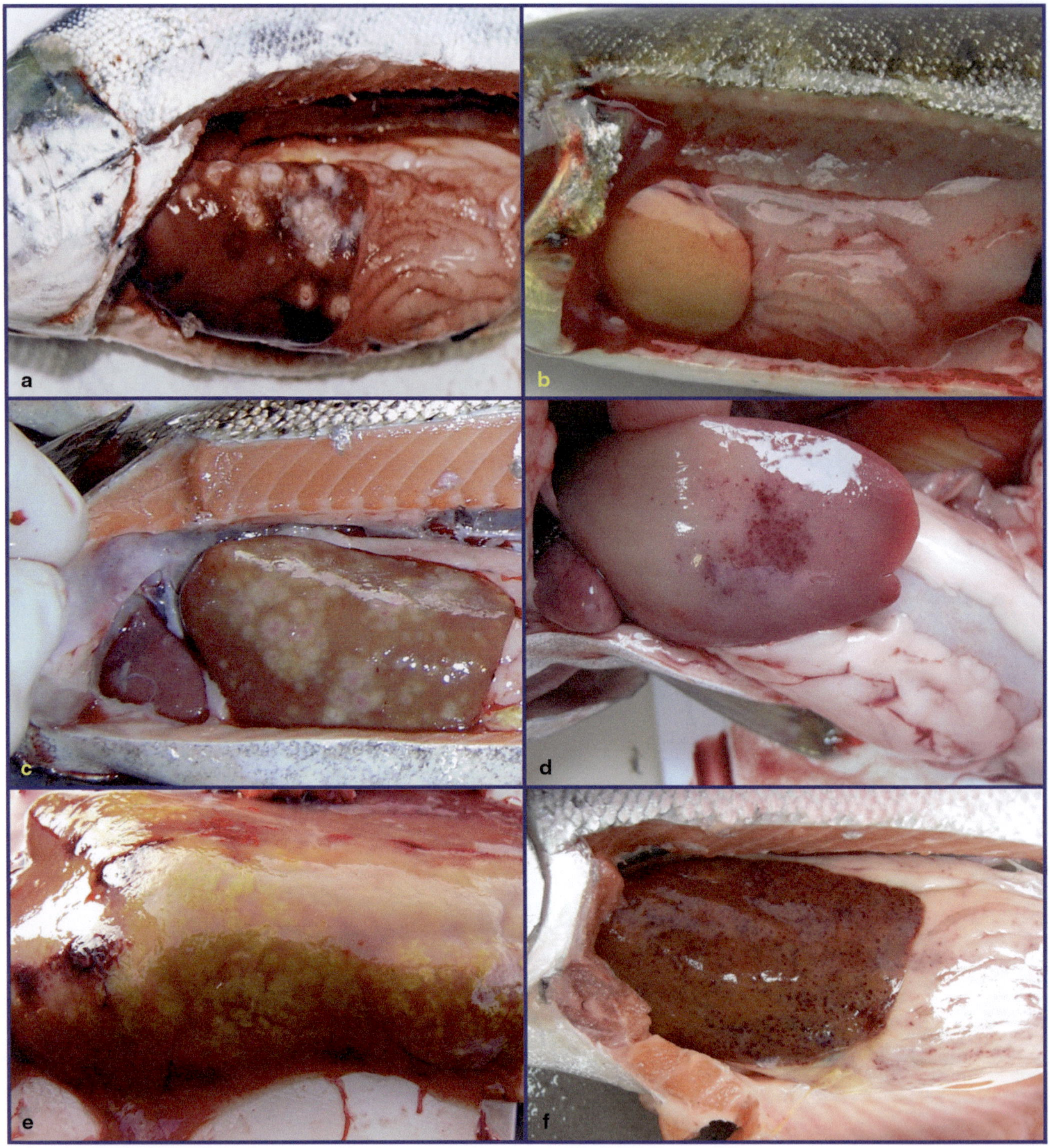

Fig. 4.3 (**a**) Infection with *Piscirickettsia salmonis* in Atlantic salmon. (**b**) Pale, yellowish liver in Atlantic salmon smolt with Infectious pancreatic necrosis. (**c**) Multiple necrosis caused by *Spironucleus salmonicida* in salmon. (**d**) Enteric red mouth in rainbow trout. (**e**) *Myxidium truttae* plasmodia in bile ducts of wild Atlantic salmon. (**f**) Petechiae in septicaemic Atlantic salmon. (**g**) Haemorrhage caused by *Listonella anguillarum* in Atlantic salmon. (**h**) *Anisakis simplex* larvae in wild Atlantic salmon. (**i**) Fibrinous coat on liver in salmon with cardiomyopathy syndrome. (**j**) Post mortem artifact. (**k**) *Philonema salvelini* in wild brook trout. (**l**) Polycystic liver in Atlantic salmon

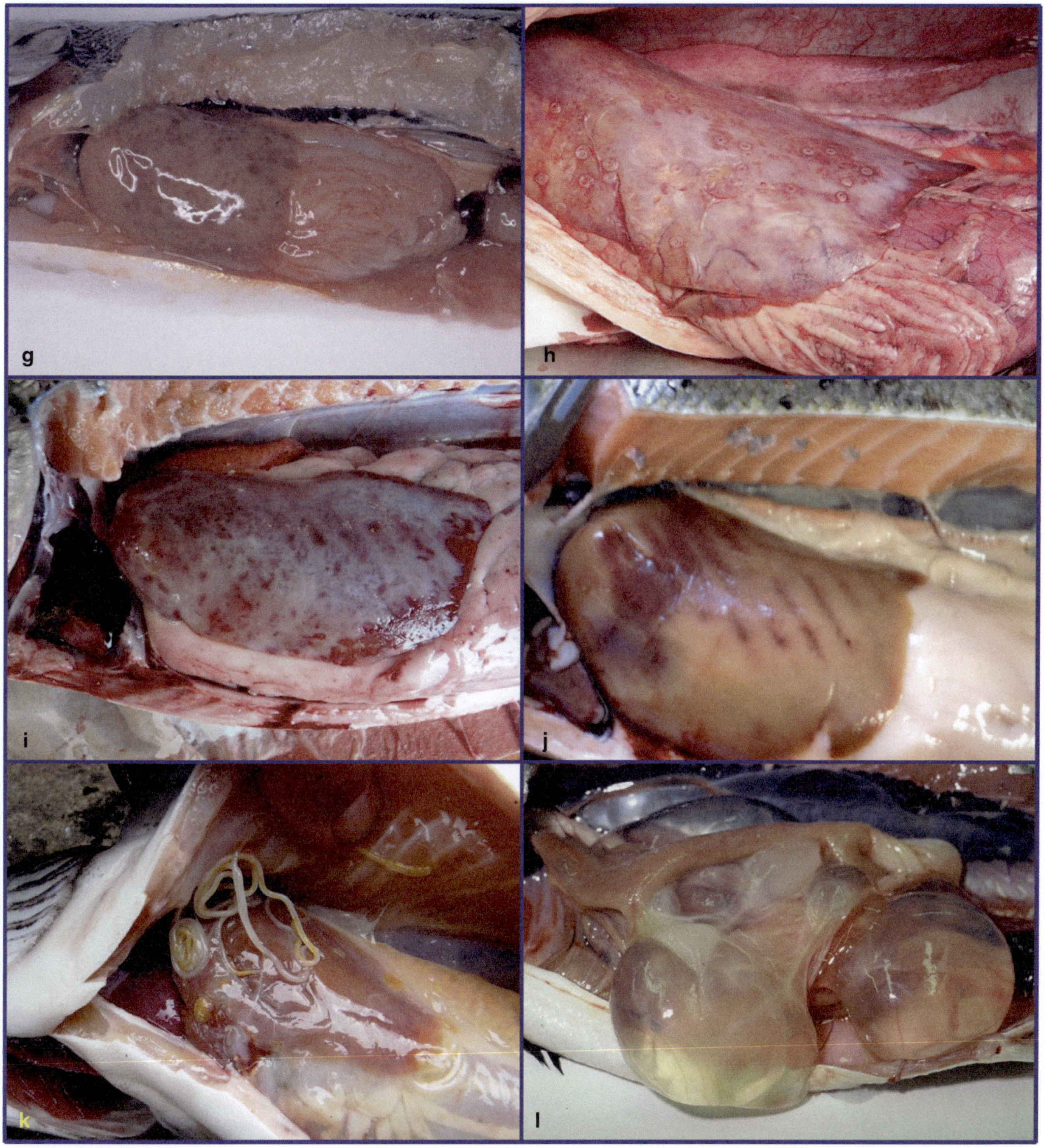

Fig. 4.3 (continued)

Table 4.1 Examples of prefixes and suffixes and their use in compounded words

Prefix	Meaning	Example
Adeno-	Glandular	**Aden**oma
An-	No, not	**An**aemia
Angio-	Blood or lymph vessels	**Angio**pathy
Anti-	Counteracting	**Anti**body
Apo-	Separated from	**Apo**ptosis
Auto-	Self	**Auto**immunity
Cardio-	Heart	**Cardio**myopathy
Chol-	Bile	**Chol**angitis
Con-	Together	**Confluent**
Cyto-	Cell	**Cyto**pathic
De-	Remove or loss	**Degeneration**
Derma-	Skin	**Dermatomycosis**
Dys	Abnormal	**Dys**plasia
Ect-	Outer or external	**Ecto**parasite
Endo-	Within or inner	**Endo**parasite
Enter-	The intestine	**Enter**itis
Epi-	Above, upon	**Epidermis**
Fibro-	Fibres or fibrous tissue	**Fibroplasia**
Gastro-	Stomach	**Gastro**intestinal
Haemo-	Blood	**Haemo**lysis
Hepato-	Liver	**Hepato**megaly
Hetero-	Difference	**Hetero**pagus
Histo-	Tissue	**Histology**
Homo-	Similar, like	**Homo**genous
Hyper-	Indicating an excess	**Hyper**pigmentation
Hypo-	Indicating a deficiency	**Hypoplastic**
Idio-	Self	**Idio**pathic
Inter-	Between	**Inter**stitial
Intra-	Within	**Intra**cellular
Karyo-	Cell nucleus	**Karyo**megaly
Leuco-	Lack of colour, white	**Leuco**penia
Lipo-	Fatty	**Lipo**idosis
Macro-	Large	**Macro**phage
Mal-	Disorder or abnormality	**Mal**ignant
Melan-	Black colour	**Melan**in
Micro-	Small	**Micro**cytic
Morpho-	Structure	**Morpho**logical
Multi-	Many	**Multi**cellular
Myco-	Fungus	**Mycosis**
Myo-	Muscle	**Myo**cardium
Necro-	Death or dissolution	**Necrosis**
Nephro-	Kidney	**Nephro**calcinosis
Osteo-	Bony	**Osteo**clast
Patho-	Disease	**Patho**gen
Peri-	Around or enclosing	**Peri**orbital
Phago-	Eat; devour	**Phago**cyte
Post-	After	**Post**erior
Poly	Many	**Poly**cystic
Pseudo-	False	**Pseudo**membrane
Retro-	Behind or turned backward	**Retro**bulbar
Sidero-	Iron	**Sidero**sis
Scolio-	Twisted	**Scolio**sis
Spleno-	Spleen	**Spleno**megaly
Steato-	Fatty tissue	**Steatosis**
Steno-	Narrow; constricted	**Steno**sis
Syn-	Union or fusion	**Syn**echiae
Vaso-	Vessel	**Vaso**dilation
Suffix	Meaning	Example
-iasis	Condition of, state	Helminth**iasis**
-iosis	Disorder	Scol**iosis**
-itis	Inflammation of an organ, tissue	Myocard**itis**
-logy	Science of, study of	Patho**logy**
-lysis	Breaking down	Karyo**lysis**
-megaly	Enlargement	Cardio**megaly**
-oid	Likeness, "of a kind"	Cer**oid**
-oma	Tumour or swelling	Sarc**oma**
-ous	Like, having the nature of	Granulomat**ous**
-pathy	Disease	Neuro**pathy**
-penia	Lack of, or deficiency	Leuko**penia**
-phage	Ingesting	Macro**phage**
-philia	Affinity for	Eosino**philia**
-phylaxis	Protection	Ana**phylaxis**
-stasis	Stagnation	Haemo**stasis**
-somatic	Of the body	Hepato**somatic**
-trophy	Nourishment	Dys**trophy**

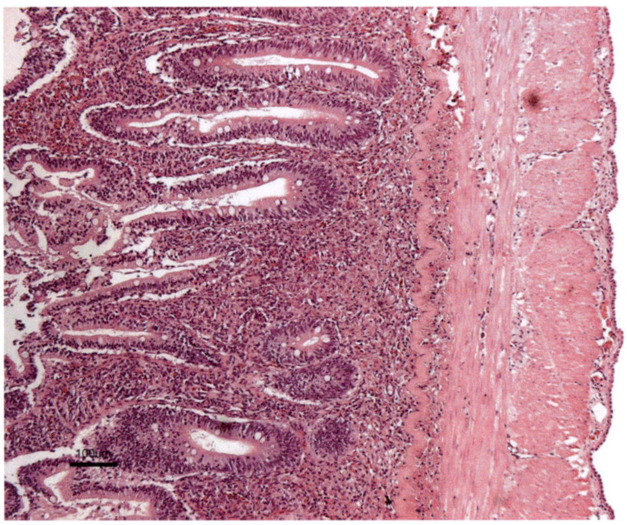

Fig. 4.4 Chronic enteritis in distal intestine of farmed Atlantic salmon. Note cellular infiltrates in stratum proprium. Low power

Gill synechiae refers to fusion of adjacent gill lamellae, and hyperplasia results from a generalised proliferation of cells (Fig. 4.9). The gills are often accompanied by

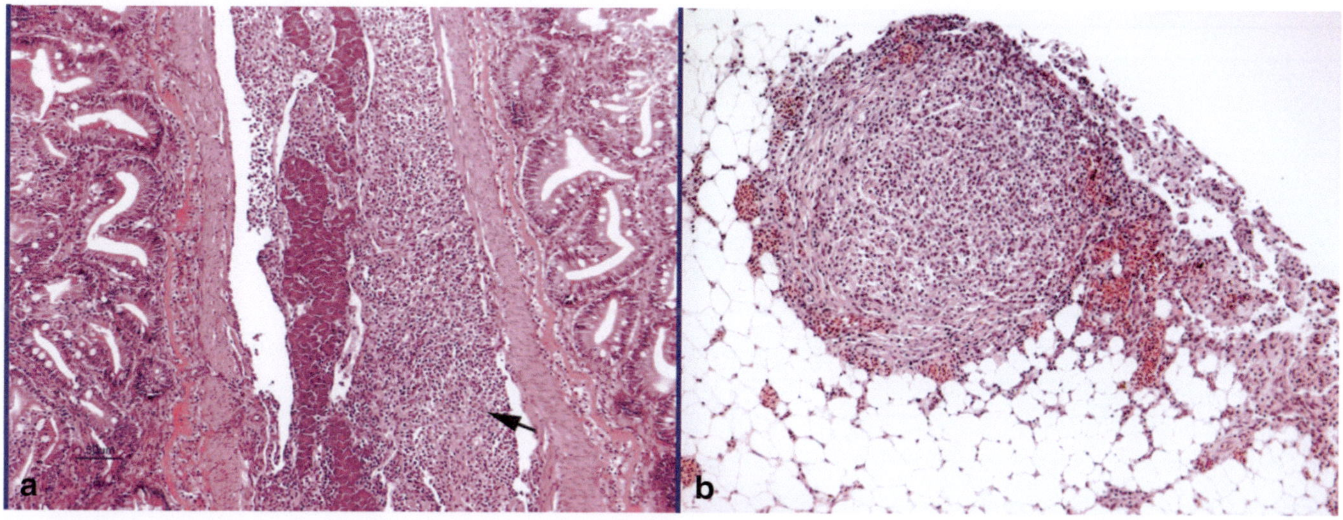

Fig. 4.5 (**a**) Vaccine induced granulomatous peritonitis (*arrow*) in farmed Atlantic salmon. (**b**) Granuloma resulting from intraperitoneal injection of oil adjuvanated vaccine

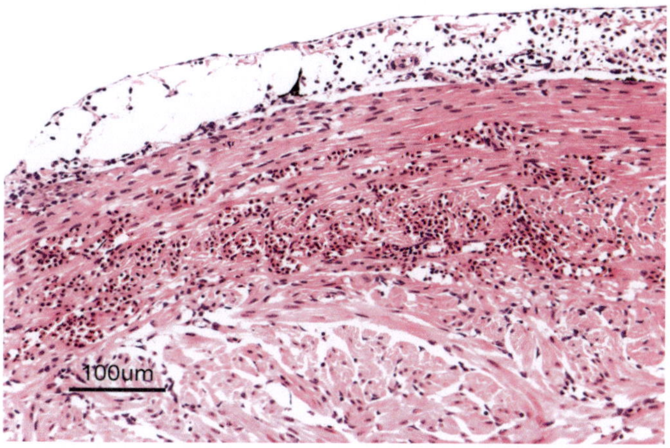

Fig. 4.6 Myocardial haemorrhage in farmed Atlantic salmon smolt

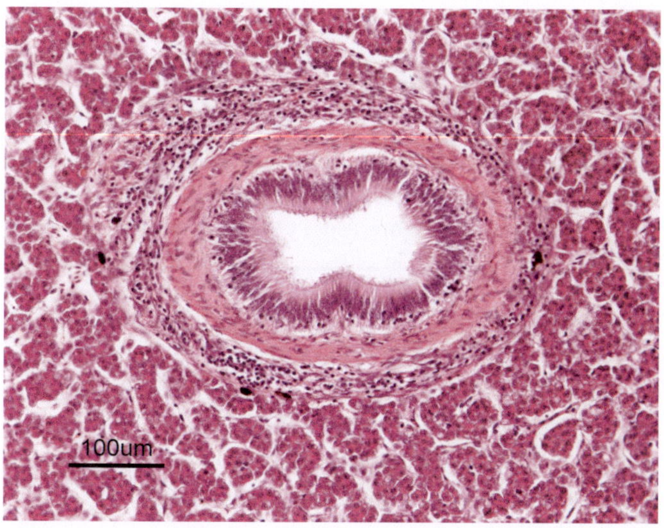

Fig. 4.7 Peribiliary lymphoctic infiltration in farmed Atlantic salmon liver

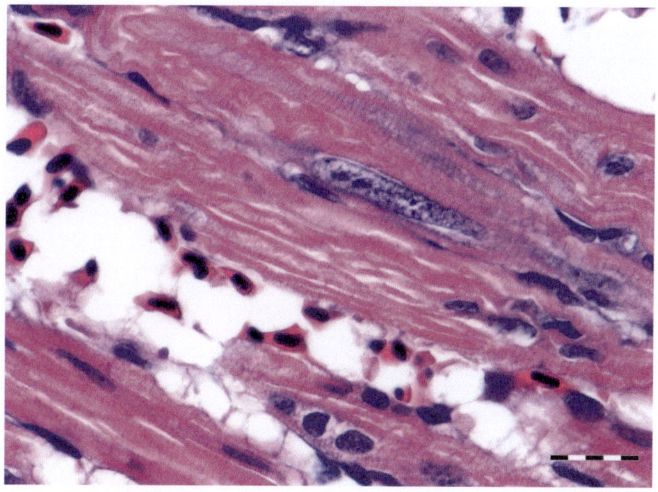

Fig. 4.8 Longitudinal section of spongy myocardium showing nuclear hypertrophy in farmed Atlantic salmon. Bar scale = 20 μm

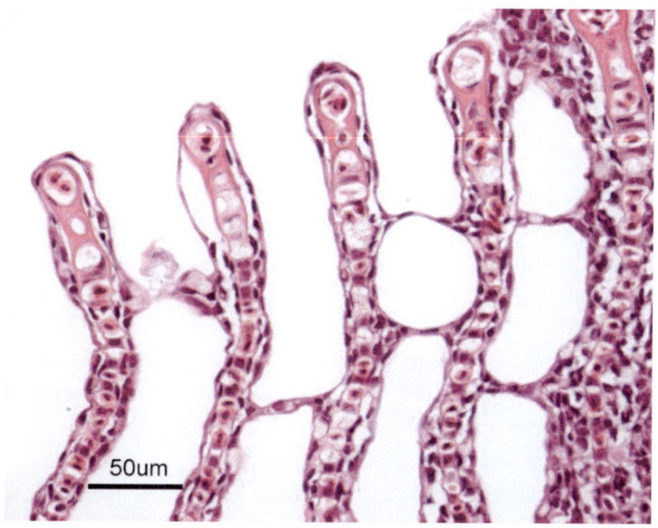

Fig. 4.9 Synechia between lamellae in adult farmed Atlantic salmon

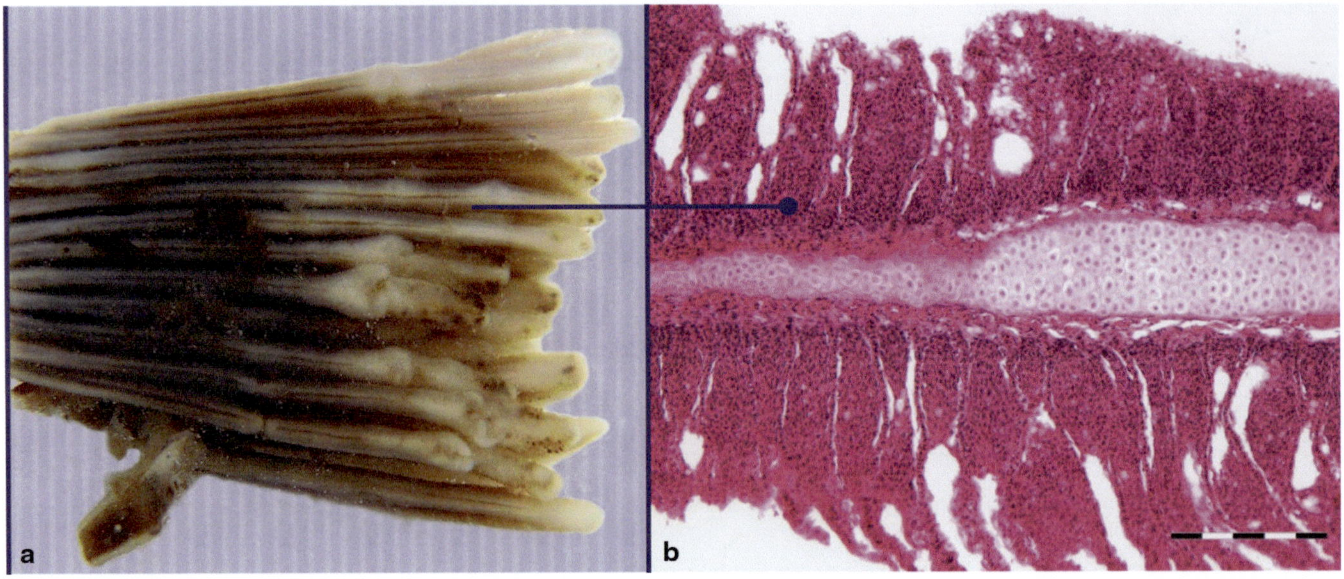

Fig. 4.10 (**a**) Thickened lamellae, fresh material. (**b**) Gill hyperplasia with extensive fusion of lamellae in farmed Atlantic salmon. Bar scale = 200 μm

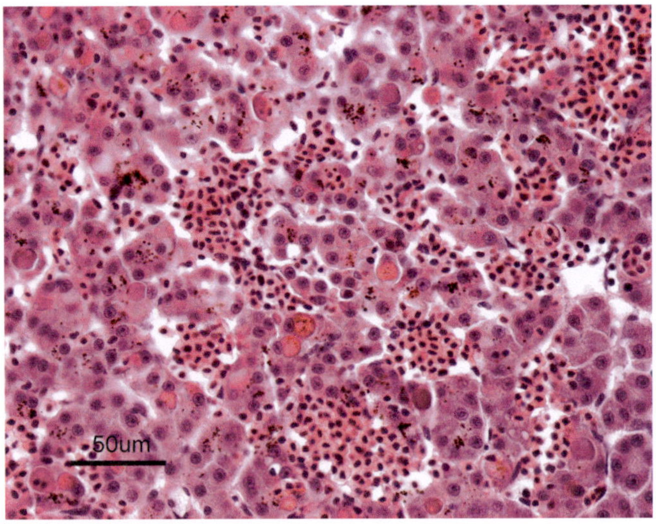

Fig. 4.11 Liver of farmed Atlantic salmon smolt with haemorrhagic smolt syndrome, note extensive erythrophagocytosis

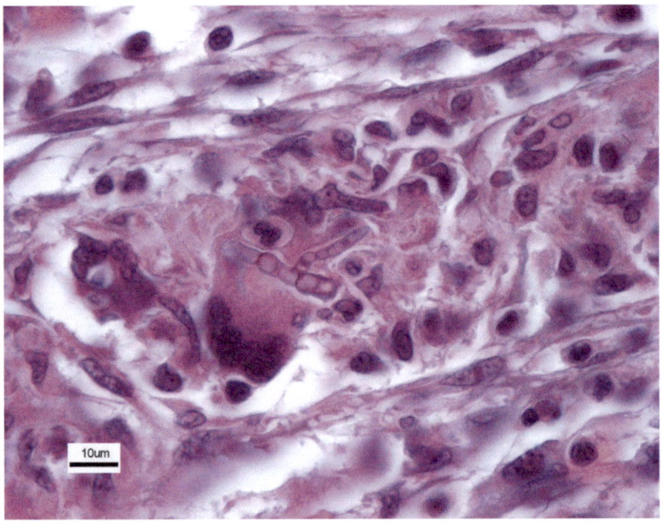

Fig. 4.12 Multinucleated giant cell in kidney of farmed Atlantic salmon with systemic fungal infection

spongiosis induced by an increased functional demand (e.g. gill lamellar or renal interstitial hyperplasia), physical or chemical irritant of the gill epithelium, excessive hormonal stimulation or infectious agents including viruses. Localised hyperplasia can also be seen entrapping parasites. Regenerative hyperplasia occurs when normal cells that survive toxic exposure, proliferate and regenerate necrotic tissues (Fig. 4.10). The biliary epithelium is responsive to various types of insults and toxicity which may induce a chronic bile duct hyperplasia.

Phagocytosis may occur in cases of chronic toxicity or infections, for example hepatocyte erythrophagocytosis (Fig. 4.11). A granuloma is formed when the immune system attempts to isolate a foreign substance or agent, and defined as a nodular cluster of macrophages and debris, encircled by a layer of lymphocytes, which are usually surrounded by a layer of fibroblasts, particularly in older lesions. In H&E sections the epithelioid cells have a pink granular cytoplasm with indistinct cell boundaries, often appearing to merge together. The coalescence of epithelioid cells or macrophages may fuse to form multinucleated giant cells within the granuloma. Granulomas are reported from bacterial infections, parasites and reaction to foreign material (e.g. oil component in adjuvant vaccines). For example, an extensive response involving macrophage, epithelioid and giant cell infiltration (Fig. 4.12), is observed following infection by *Exophiala* and *Pasteurella skyensis*.

A prominent feature of chronic inflammation is the presence of melanomacrophages or macrophage aggregates that are normally located in the stroma of the haematopoietic tissue of the kidney, liver and spleen. Chronic inflammation is characterized by cellular proliferation rather than exudation, and comprises resorption, formation of granulation tissue and fibrosis. Granulomatous inflammation, a distinctive pattern of chronic inflammation, and fibrinous epicarditis are recurrent findings in conditions such as bacterial kidney disease and mycobacteriosis (Fig. 4.13). Regardless of the origin of these inflammatory responses, granulomatous inflammation tissue may have resemblance to neoplastic changes.

Ulceration involves necrosis and an eroded epithelial surface with underlying acute and chronic inflammation following many types of tissue injury. Necrotic tissue takes on a different colour and consistency and dependent upon the age of the lesion and the quantity of blood. Focal necrosis develops to an ulcer when tissue is sloughed from the area. Once the epidermis is breached, loss of fluid, swelling, haemorrhage necrosis and potential involvement of secondary pathogens occur. An ulcer that starts from the epithelium often have a base-narrow shape, conversely an ulcer that begins as a lesion below the surface tends to be 'base-wide'. The inflammatory response is the start of the healing process and the skin shows a capacity for repair with hypertrophy, fibrosis and regeneration of the damaged tissue or scar formation. Damaged scales seldom recover their original pattern and newly formed scales are usually smaller and undulated and therefore easy to identify. In some cases of extensive tissue loss, a cavity may simply remain at the site of injury.

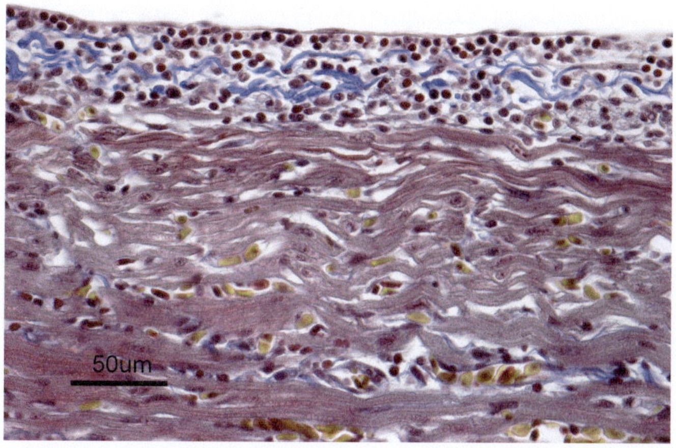

Fig. 4.13 Idiopathic highly cellular epicarditis in adult farmed Atlantic salmon. Martius, scarlet blue (*MSB*) stain

Exophthalmia refers to an excessive protrusion of the posterior aspects of the ocular globe and occur in several diseases. It is usually the result of increased intraocular pressure, inflammation in the well-vascularised rete behind the eye with resultant swelling, oedema and loss of connective tissue or granulomatous inflammation. These are common findings in moribund fish, occasionally with associated haemorrhage but overall non-specific clinical signs (Fig. 4.14). Histologically, oedema, infiltration of macrophages in the retrobulbar area, swelling and necrosis can be observed. A progressive panophthalmitis is primarily associated with bacterial infections and refers to inflammation of all coats of the eye including intraocular structures.

Eosinophilic granular cells (EGCs), also known as mast cells are present in most species of teleosts in a variety of tissues, including the gut, gills (Fig. 4.15), skin, brain, and surrounding major blood vessels. Acute tissue

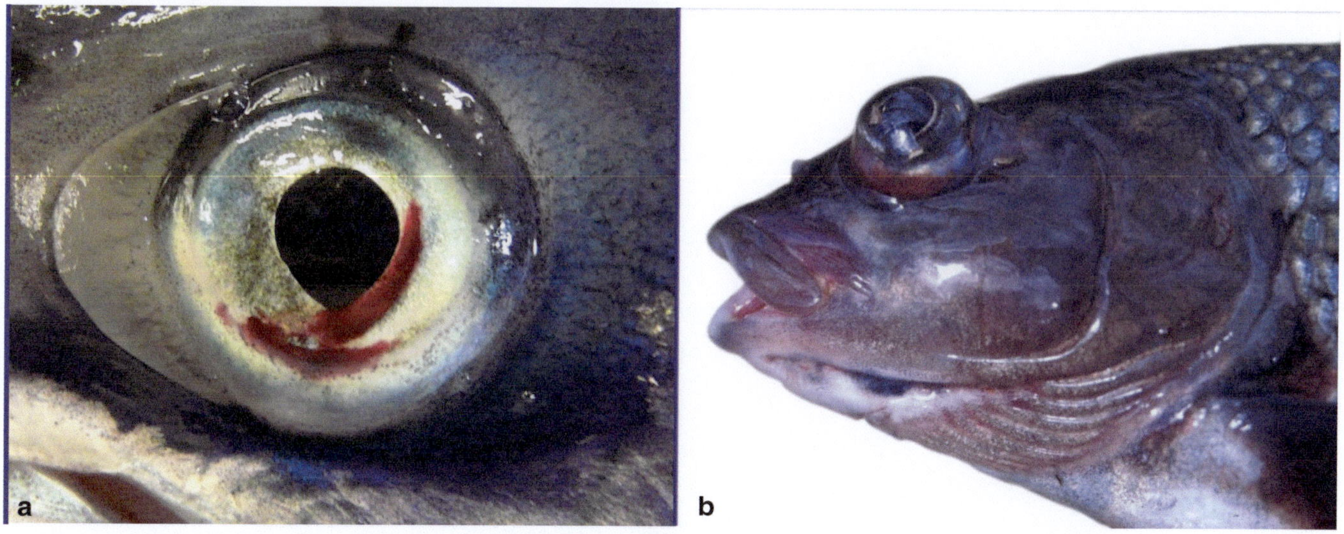

Fig. 4.14 (**a**) Eye haemorrhage in adult farmed Atlantic salmon. (**b**) Exophthalmia in wild grayling

damage can result in granule cell degranulation and the release of mediators of inflammation, whereas an increase in the number of these cells is reported in chronically inflamed tissues. There is diversity in their staining properties, with both basophilic and acidophilic components in their granules.

An immune complex-mediated glomerulonephritis with inflammation and subsequent dysfunction of the glomeruli is reported in farmed fish following bacterial infections such as *Renibacterium salmoninarum*, but also noted in returning Atlantic salmon where glomerular damage is severe and causes morbidity from osmoregulatory failure (Fig. 4.16).

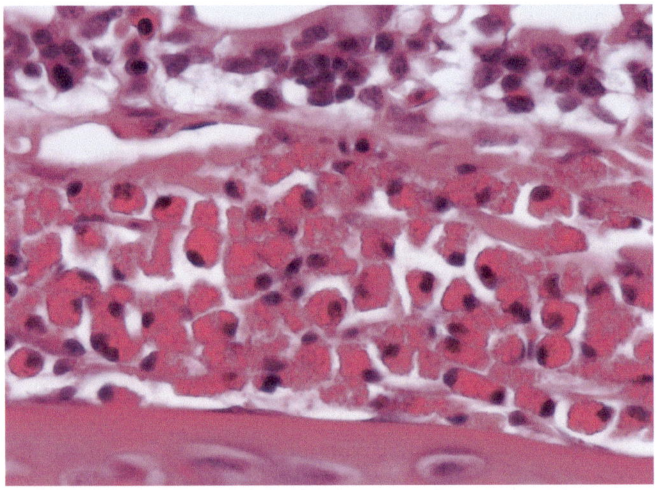

Fig. 4.15 Eosinophic granular cells between the top and the base of lamellae in gills from farmed Atlantic salmon. High power

4.2 Circulation Disorders

Common circulatory disorders include congestion, haemorrhage, hyperaemia, ischemia, stasis, aneurysms and thrombosis (Figs. 4.17, 4.18, 4.19 and 4.20). The cessation of blood flow to an organ may result in a coagulative necrosis, whereby the cell membrane and morphology of the organ is maintained, but accompanied by nuclear changes and a reduction in cytoplasmic basophilia.

Branchial pallor is often an indication of anaemia and is attributed to a deficiency of haemoglobin in the blood and/or a decrease in the number of erythrocytes. Anaemia may result from a decrease in red blood cell production, an abnormal blood loss or the excessive breakdown of red blood cells (Fig. 4.21). The most useful classification of such deficiencies is based upon pathophysiological mechanisms. A haemolytic anaemia is characterised by the accumulation of haemosiderin (Fig. 4.22), with an increase in the rate of destruction of erythrocytes and a corresponding release of immature blood cells to the circulation, e.g. infection by *Listonella anguillarum* and infectious salmon anaemia virus. Hypoplastic anaemia results from the failure of the haematopoietic tissue to produce adequate numbers of cells or a deficiency in haemoglobin synthesis, e.g. malnutrition. A haemorrhagic anaemia occurs as a result of bleeding and is evident in conditions including viral haemorrhagic septicaemia.

Haemorrhage is caused by injury to the vascular endothelium and can attributed to infection, inflammation, necrosis, neoplasia or trauma. Petechiae are associated with locally increased intravascular pressure or damage to the vascular endothelium and may indicate septicaemia when occurring

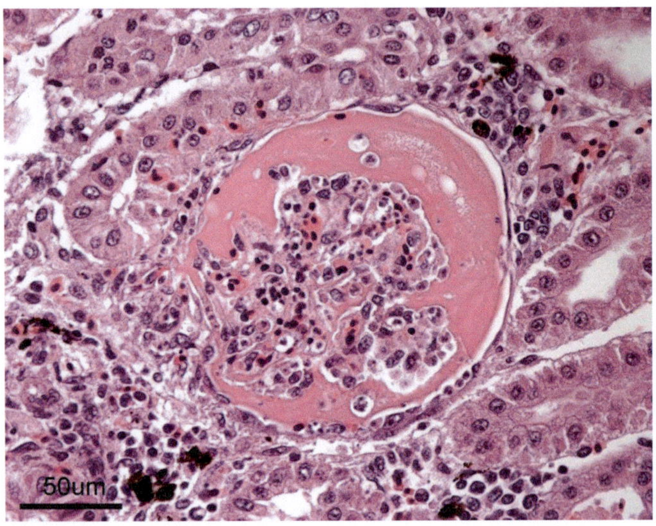

Fig. 4.16 Glomerulonephritis in Atlantic salmon. Medium power

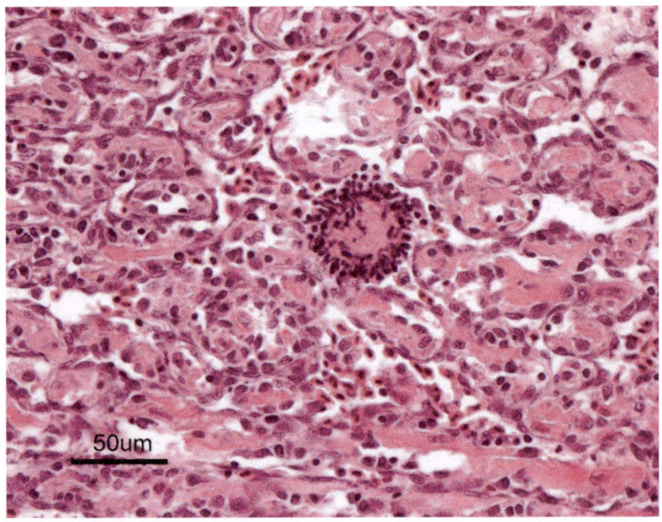

Fig. 4.17 Thrombus in spongy myocardium of farmed Atlantic salmon with cardiomyopathy syndrome. Medium power

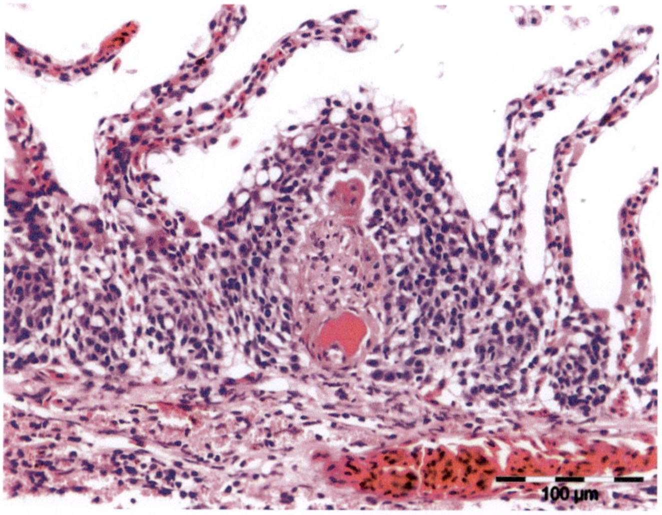

Fig. 4.18 Thrombus, fused lamellae and necrosis in gills of farmed Atlantic salmon

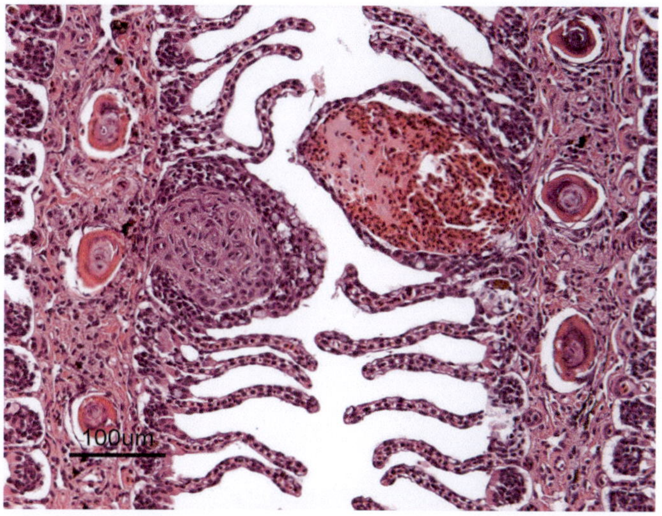

Fig. 4.20 Old (*left*) and recent (*right*) lamellar aneurysm of farmed Atlantic salmon

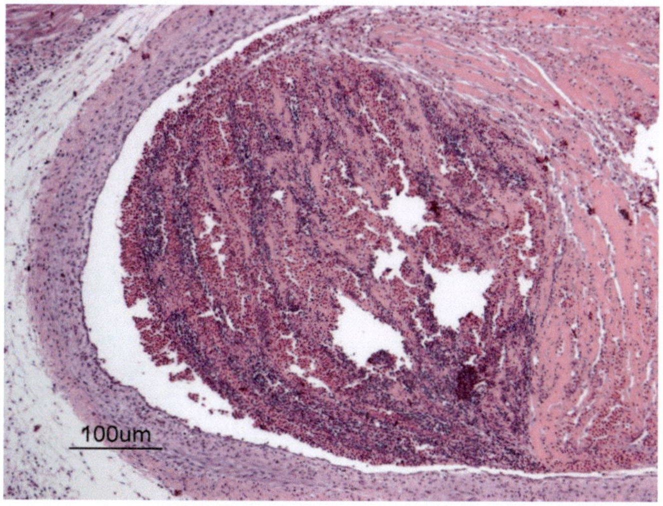

Fig. 4.19 Thrombosis of coronary artery of farmed Atlantic salmon

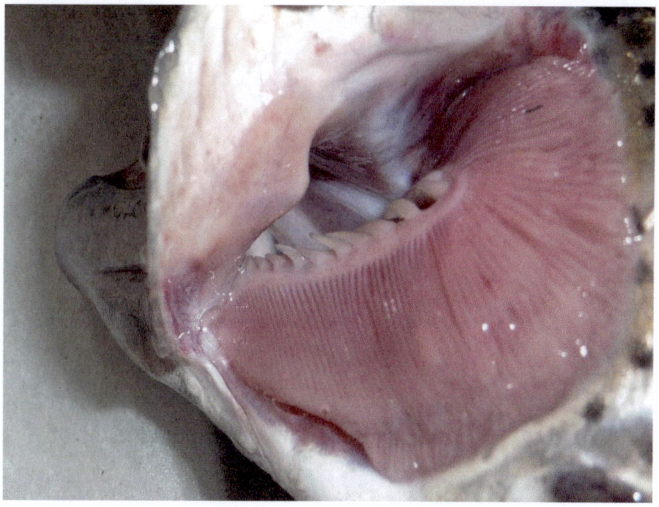

Fig. 4.21 Anaemic gills of farmed rainbow trout

on the ventral parts of the fish (Fig. 4.23). A haemorrhagic diathesis in the liver is reported for Atlantic salmon smolts suffering from haemorrhagic smolt syndrome (see Figs. 11.16 and 11.17).

Hyperaemia and congestion originate from locally increased blood volumes. Hyperaemia is an active process and implies arterial side engorgement of the vascular bed, while congestion indicates a passive process resulting from reduced outflow of blood from a tissue. Hyperaemia is usually accompanied by evidence of inflammation, and is associated with vascular dilation due to localized release of inflammatory mediators. Passive congestion is linked with reduction in venous outflow due to non-inflammatory events such as cardiac failure, constriction or obstruction of vascular outflow due to tissue torsion, neoplasia, or other compressive events. It is often difficult to distinguish hyperaemia from congestion. Fish can show a number of cardiovascular changes when exposed to hypoxia as result of depletion of oxygen in the water, and ultimately, this leads to vascular congestion and necrotic or apoptotic lesions.

Myointimal hyperplasia (arteriosclerosis) is usually restricted to the main coronary artery and occurs progressively in the majority of salmonids during sexual maturation and at spawning (Fig. 4.24). The initiating mechanism for coronary lesion formation appears to be vascular injury to the coronary artery, as a result of the bulbus arteriosus being markedly distended.

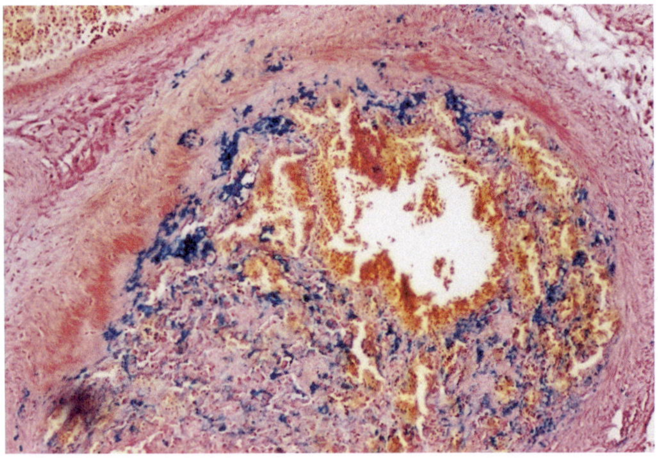

Fig. 4.22 Haemosiderin deposits in heart of rainbow trout. Low power

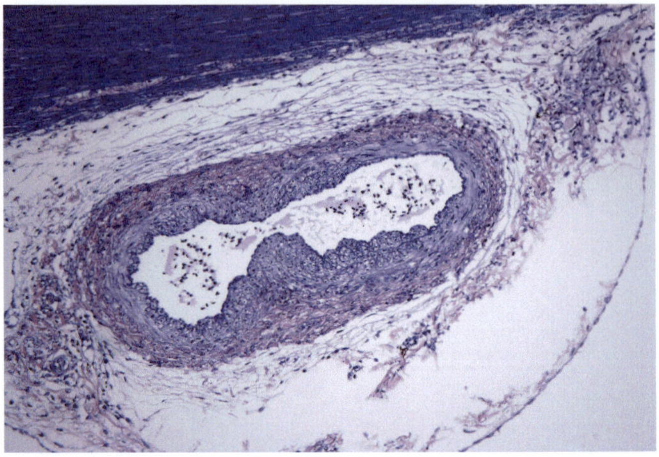

Fig. 4.24 Myointimal hyperplasia (arteriosclerosis) of the coronary artery in farmed Atlantic salmon. Medium power

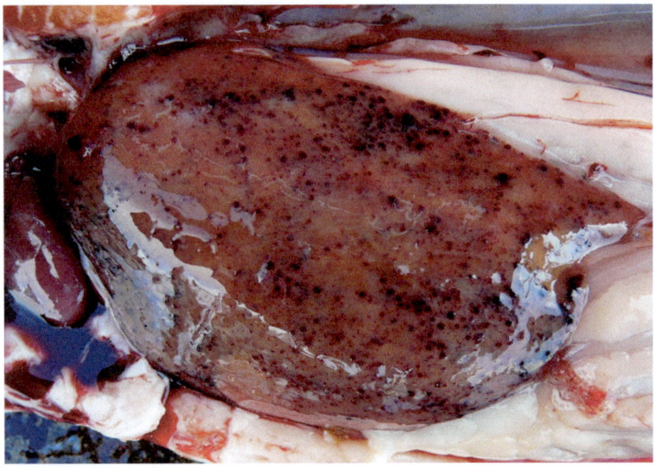

Fig. 4.23 Septicaemic petechiae in liver of adult farmed Atlantic salmon

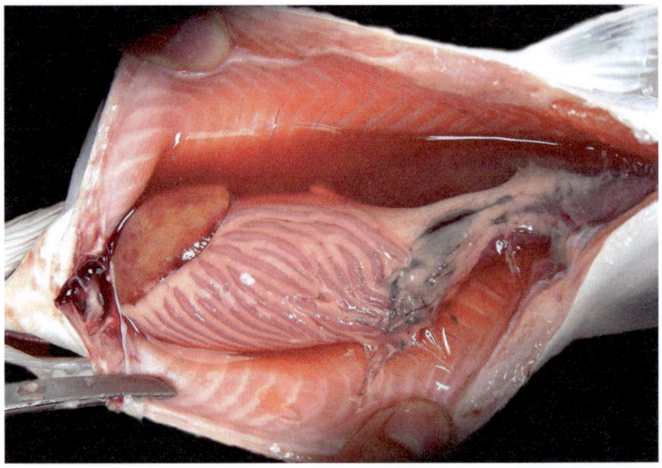

Fig. 4.25 Ascites between the tissues lining the abdomen and the peritoneal cavity

An excessive amount of fluid in or around cells, tissues or serous cavities is termed oedema and may follow changes in hydrostatic or osmotic pressure and increases in vascular permeability that accompany inflammation, as seen in fish infected with *Aliivibrio salmonicida*. For example, eye oedema can occur in the cornea, retina and choroid rete mirabile. Corneal oedema typically accompanies eye disease or wounding and follows separation of the collagen fibres and Malpighian cells of the stroma resulting in a hazy or cloudy appearance. Extensive fibrinous exudates, are often infiltrated with neutrophils, however these are common to a number of pathological conditions. Hepatocellular hydropic degeneration can be associated with bacterial infection and toxicity. Ascites occurs when excess fluid is observed in the space between the tissues lining the abdomen and abdominal organs (the peritoneal cavity) (Fig. 4.25).

4.3 Cell Injury and Death (Necrosis and Apoptosis)

Histologically, necrosis or cell death take on a different colour and consistency. Such metabolically inactive cells can be distinguished from those which are active and synthesising protein, as the nucleus within an inactive cell is round and compact with intense staining, whereas a synthesising cell has a large pale staining nucleus with large or multiple nucleoli. A coagulative necrosis is characterized by retention of tissue architecture and connected with infectious disease, ischemia, trauma and toxic damage. Caseous necrosis is observed with some bacterial infections, whereas liquefactive necrosis results in

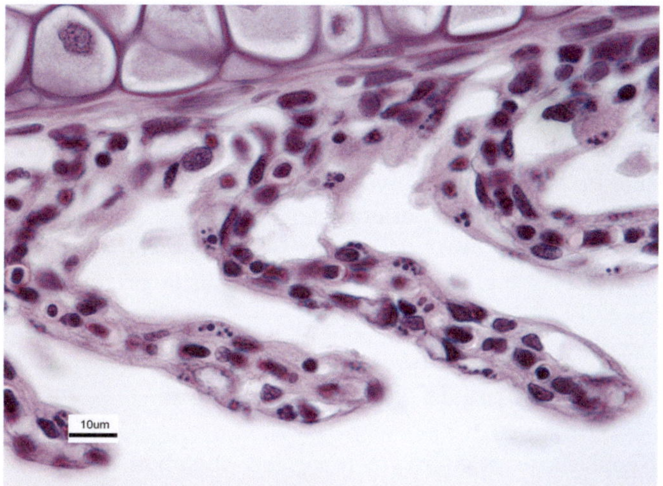

Fig. 4.26 Epithelial necrosis with karryorhectic nuclei in lamellae of adult sea water farmed Atlantic salmon

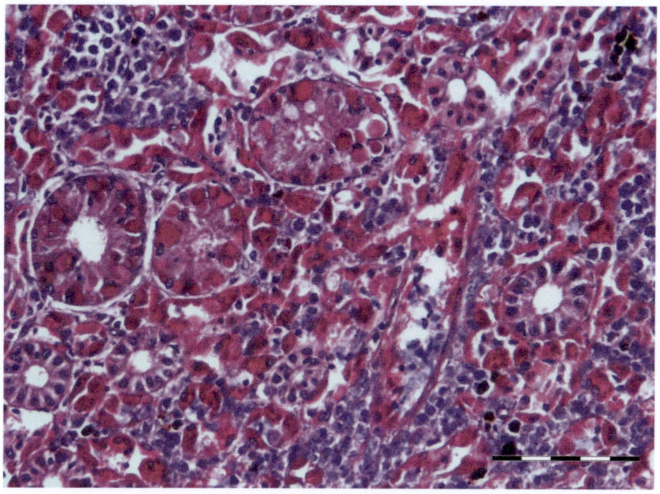

Fig. 4.28 Increase in protein deposition in kidney of rainbow trout. Bar = 100 μm

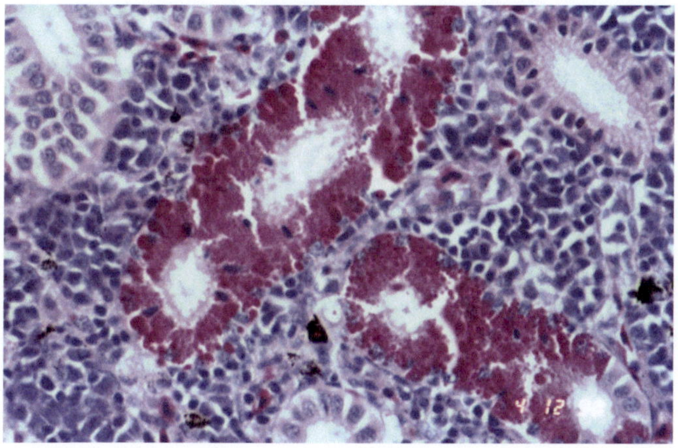

Fig. 4.27 Hyaline droplet degeneration in kidney tubules of rainbow trout. Medium power

complete loss of all histological features. Overall, necrotic changes in affected nuclei are easily recognised, for example pyknosis is associated with cell death and apparent as an amorphous, compact mass of darkly stained material. Both karyolysis and karyorrhexis may follow leaving the cell devoid of any discernible nucleus (Fig. 4.26). Irreversible injury to muscle cells results in swelling and fragmentation of myocytes, loss of cross-striation, intracellular vacuolation and pale staining. With cessation of the damaging stimuli with time there can be regeneration, with increased basophilia and fibre hypertrophy. Segmental areas of scar tissue maybe interspersed with regenerated tissue.

Apoptosis is the process of programmed cell death. This is a genetically controlled and evolutionarily conserved biological process of widespread biological significance. The mechanism of cell death is complex and results in cells with condensed chromatin and cytoplasm that fragment into membrane-bound particles, those fragments being engulfed by phagocytic cells (see Fig. 5.6). In this case the organelles are still functional, which is not the case with a necrotic cell. Apoptosis can be initiated by intrinsic and extrinsic signals linked to normal physiology, damage to mitochondrial membranes and response to pathological conditions such as a protective response to bacteria or virus infected cells i.e. it is a pathologic cell death. For example, studies on infectious pancreatic necrosis virus have shown that the associated liver pathology is characterised by progressive changes of increasing severity, leading to apoptosis preceding necrosis of the tissue.

Responses to injury are adaptive and often provoke changes in cellular structure which are not lethal but seen as reversible, and include acute cellular lamellar telangiectasis, swelling, hydropic change and lipidosis. Acute cellular swelling represents an early and completely reversible manifestation of injury which occurs when cells swell due to increased water uptake, following alterations in membrane permeability. Hydropic changes represent a pronounced form of swelling with large distinct water vacuoles forming within the cell cytoplasm. In both cases these typically occur in epithelial cells. Hyaline droplet degeneration refers to a particular histological appearance of cells or tissues when stained with H&E (Fig. 4.27) and represent an accumulation reabsorbed protein from glomerular filtrate, or arising as a result of cell degeneration (Fig. 4.28).

Lipidosis can be severe with disruption of cell function and is commonly seen in the liver. Distinct vacuoles of fat lie in the cell cytoplasm displacing and compressing the nucleus; they appear as non-stained as the content is dissolved during tissue processing (Fig. 4.29).

4.4 Pigments and Mineralization

Melanomacrophages usually contain a variety of pigments, including melanin, which are known to increase in number in older fish (Fig. 4.30). Melanin is capable of neutralising free radicals and cation activity associated with oxidising conditions, which partly explains their increased accumulation in the presence of cachectic disease and injury. These cells may be the forerunners of the germinal centres present in the spleen and lymph nodes of birds and mammals. They often cluster in chronic, granulomatous inflammation, especially as a response to encysted parasites or foreign material, and their number may escalate in fish with chronic infections, however their role in modulating infections is speculative.

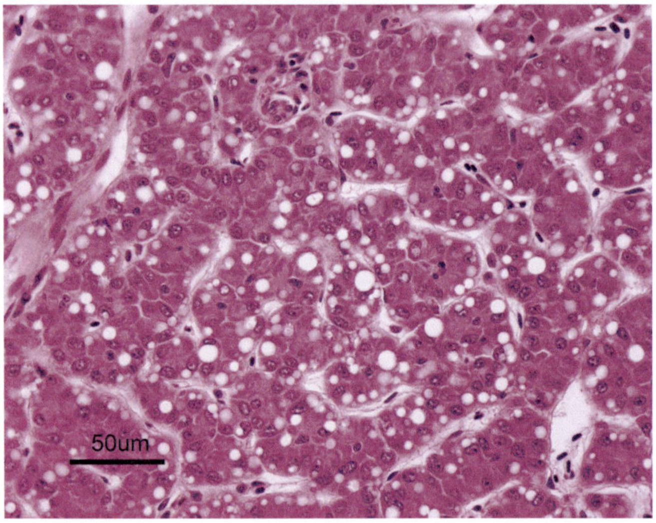

Fig. 4.29 Vacoluation in liver of rainbow trout

Dystrophic mineralization is associated with degenerative changes in cells and tissues and subsequently by the deposition of the mineral. This has been observed as a prominent and diffuse mineralization of the bulbous arteriosus in clinically normal adult female rainbow trout, and in the compact myocardium and cardiac valves of salmon with coronary arteriosclerosis (see Fig. 4.24). Calcification may also occur in old and well-organized granulomas, e.g. of mycobacterial or parasitic origin.

4.5 Neoplasia

Neoplasia or new growth represents abnormal tissue that starts from the autonomous, progressive and excessive proliferation of the animals own cells. This usually develops into a distinct mass of tissue which may be either benign or malignant, incited by a number of mechanisms ranging from genetic events, to toxin exposure, to some types of infections. Neoplasia may be observed grossly from virtually any tissue and are classified partly from the recognition of the parent tissue, but largely from their microscopical structure. Histological features are extremely useful in determining whether the lesion is benign or malignant. Benign neoplasia shows a homologous tissue type, uniform cut surface, lack metastases and cell atypia. Malignant neoplasia conversely shows infiltration with a heterologous tissue type compared to parent tissue, often with a variable cut surface including haemorrhage and necrosis. Metastases and atypia are common with a prominent nucleolus, irregular cell size and shape. Some neoplastic tissue may show characteristics of both benignancy and malignancy. For more detail please refer to Chap. 12.

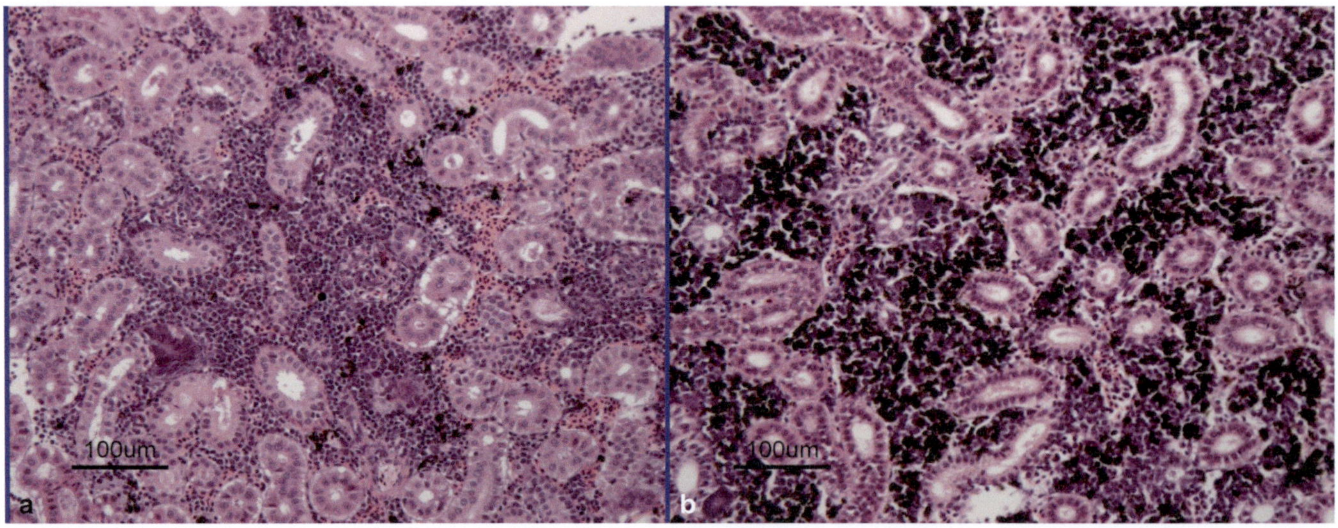

Fig. 4.30 (**a**) Normal distribution of melanomacrophages in farmed Atlantic salmon. (**b**) Increased number of melanomacrophages

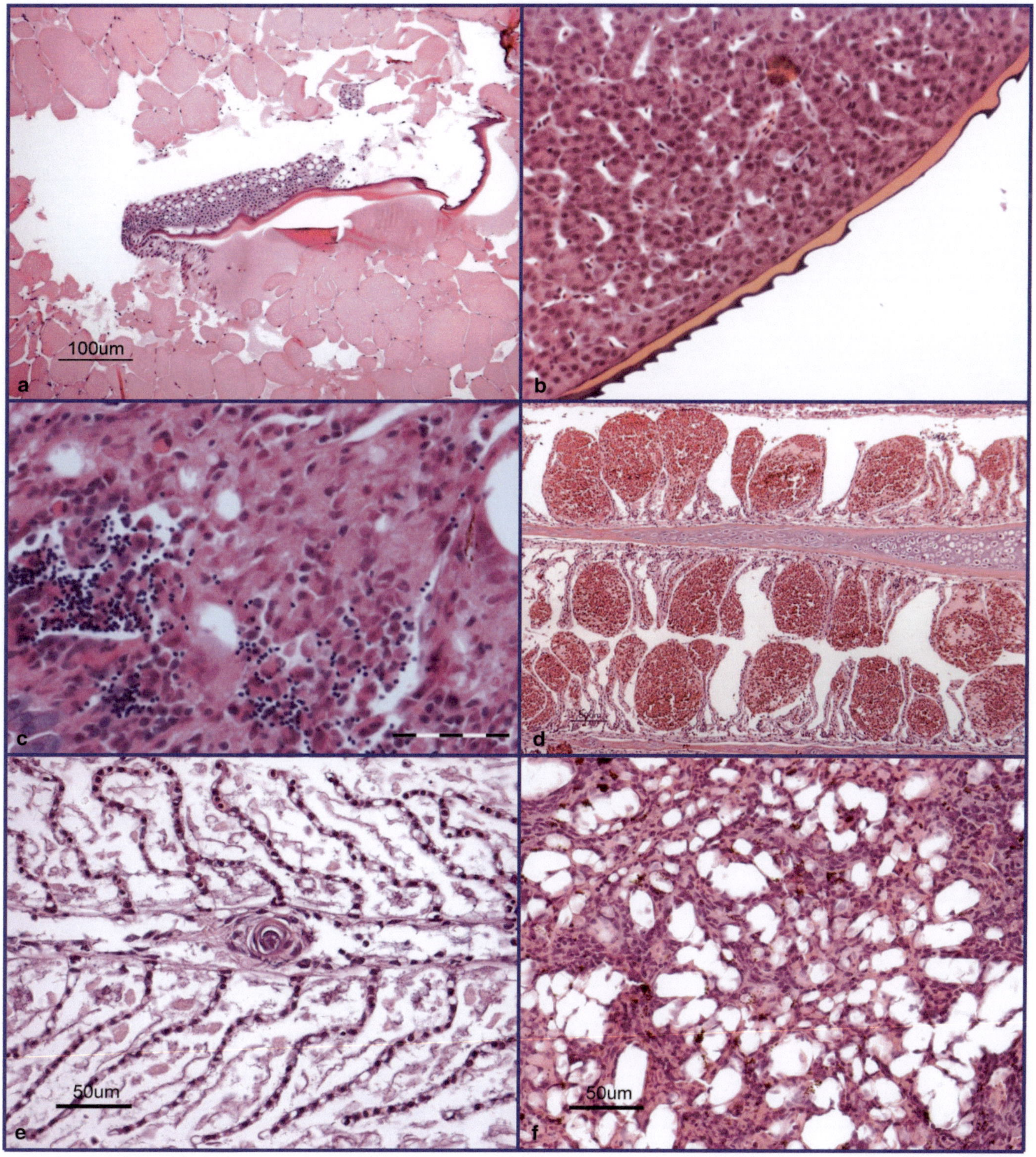

Fig. 4.31 (a) Scale with epidermis within muscle. (b) Scale on the surface of the liver. (c) Spermatozoa in liver parenchyma. (d) Gill aneurisms resulting from blow to the head. (e) Post mortem changes in gills sampled from a fish that was dead prior to sampling. (f) Freezing artifact (spleen). (g) Post mortem changes in pancreas; rounding of acinar cells and pyknotic nuclei. (h) Mountant artifact in liver. (i) Bile from the gall bladder has caused damage to the surface of the liver. (j) Staining artifact in the liver (normal staining to the *left*). (k) Yeast growing in stain. (l) Necropsy artifact in the liver caused by forceps

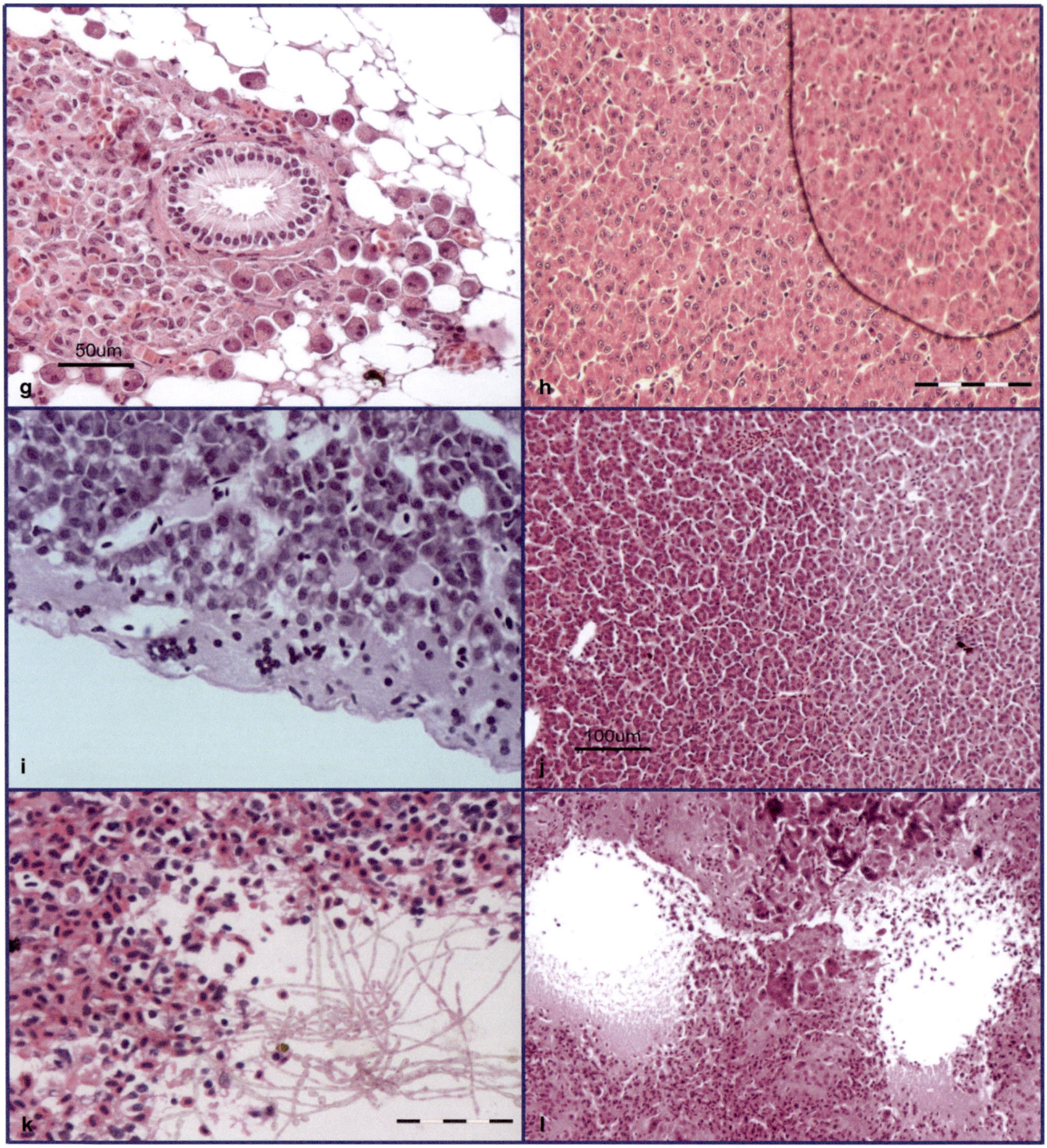

Fig. 4.31 (continued)

4.6 Artifacts

Many factors can affect the outcome of histological sections or its image capture when slides are viewed under light microscopy. Inadequate fixation, processing or staining, low-quality optics and improper adjustment of the microscope illumination are a few examples of interfering factors. In general, artefacts represent changes that were not originally present in the living tissue. These changes are accidentally or artificially produced at any of the steps from the tissue sampling, processing, to section interpretation, and it is important that they are properly recognised and steps taken to reduce them as their presence can compromise an accurate diagnosis. A range of artefacts are shown in Fig. 4.31.

Further Reading

Åsbakk K (2001) Elimination of foreign material by epidermal malpighian cells during wound healing in fish skin. J Fish Biol 4:935–966

Mitchill SO, Baxter EJ, Holland C, Rodger HD (2012) Development of a novel histopathological gill scoring protocol for assessment of gill health during a longitudinal study in marine-farmed Atlantic salmon (*Salmo salar*). Aquac Int 20:813–825

Plumb JA, Hanson LA (2010) Pathology and disease diagnosis, in health maintenance and principal microbial diseases of cultured fishes, 3rd edn. Wiley-Blackwell, Oxford. doi:10.1002/9780470958353.ch3

Reavill DR (2006) Common diagnostic and clinical techniques for fish. Vet Clin N Am Exot Anim Pract 9:223–235

5 Viral Diseases

Abstract

Some viral diseases are prevented through vaccination but for many no vaccines are available hence infection is a major concern for farmed fish. Among salmonids, RNA viruses have the greatest ecological and socio-economic impact. Acute infections may lead to host death, or induce a more chronic condition. Stress including overcrowding, sexual maturation and handling may also reactivate a latent infection resulting in clinical disease. This chapter covers a range of viral diseases in wild and farmed salmonids.

Keywords

Fish virus • Salmon • Trout

Infectious viral diseases continue to be a major concern in wild and farmed fish and responsible of important losses, particularly under farming conditions. These infections can manifest as a septicaemia-like disease with a characteristic acute presentation, but some viral agents may induce neoplasia or a more chronic condition. Overall, the consequences of viral infections depend on the outcome of the interaction between a number of factors both from the virus and the host. Acute infections may lead to host death, recovery with no long term effects, or develop into a chronic infection. The later can remain subclinical for life, have a long silent period before it manifests or can have periods of reactivation with relapses and exacerbation causing acute disease. Different types of stress including overcrowding, sexual maturation and handling may reactivate a latent infection resulting in clinical disease. Avoidance of viral infection is difficult, unless spring or disinfected water is used. This is possible in the early developmental fresh water stage, but very costly or impractical when farming in open environments e.g. sea water cage culture, or when high volumes are required for growers in land based systems.

Among salmonids, RNA viruses have the greatest ecological and socio-economic impact. The occurrence of viral diseases has seen an increment over the last 10 years and it is likely that more will be identified in the future. The most important conditions and host species are listed in Table 5.1.

5.1 Infectious Pancreatic Necrosis

Infectious pancreatic necrosis virus (IPNV) is the aetiological agent of the highly contagious and acute catarrhal enteritis 'infectious pancreatic necrosis', primarily affecting cultured fish including rainbow trout and Atlantic salmon worldwide. The virus has a widespread distribution in many wild fish species but there is little evidence of transmission to farmed stock.

Significant mortalities can occur from clinical outbreaks in fry, with a relative reduced mortality in older fish. However, outbreaks in post-smolts in sea water are common and fish which survive infection can become carriers and continue to shed infective material into the water. Carrier fish are a particular hazard if used as broodstock since the virus has a germ-line associated vertical transmission via ova or milt, which occur both extra- or intra-ovum. Surface extra-ovum infection of the gametes can be managed through proper biosecurity procedures and disinfection, but the intra-ovum transmission imposes an important risk and can only be controlled through rigorous testing of the broodstock.

Salmonids may show clinical signs with high mortality or become asymptomatic carriers, thus establishing covert infections. Clinical signs include moribund fish which are darker in appearance, slightly emaciated and lethargic.

D.W. Bruno et al., *A Colour Atlas of Salmonid Diseases*,
DOI 10.1007/978-94-007-2010-7_5, © Springer Science+Business Media B.V. 2013

Table 5.1 Principal viral diseases of salmonids

Virus name	Family	Nucleic acid	Principal salmonid host (s)	Environment
Infectious pancreatic necrosis virus	Birnaviridae	ssRNA	Rainbow trout, salmon	FW, SW
Infectious salmon anaemia virus	Orthomyxoviridae	ssRNA	Atlantic salmon	SW
Oncorhynchus masou virus	Herpesviridae	ssRNA	Pacific salmon (e.g. masu)	FW?
Piscine reovirus (Heart and skeletal muscle inflammation)	Reoviridae	dsRNA	Atlantic salmon	SW
Salmon leukaemia virus	Retroviridae	ssRNA	Chinook salmon	SW, FW reared salmon in United States
Viral haemorrhagic septicaemia virus	Rhabdoviridae	ssRNA	Rainbow trout	Mainly FW
Infectious haematopoietic necrosis virus	Rhabdoviridae	ssRNA	Salmon, trout	FW
Salmonid alphavirus	Togaviridae	ssRNA	Atlantic salmon, rainbow trout	FW, SW
Piscine myocarditis virus (Cardiomyopathy syndrome)	Totiviridae	dsRNA	Atlantic salmon	SW
Erythrocytic inclusion body syndrome	Iridovirus	dsDNA	Pacific, Atlantic salmon	FW, SW

SS, single-stranded ribonucleic acid, DS double-stranded ribonucleic acid
FW freshwater, SW sea water

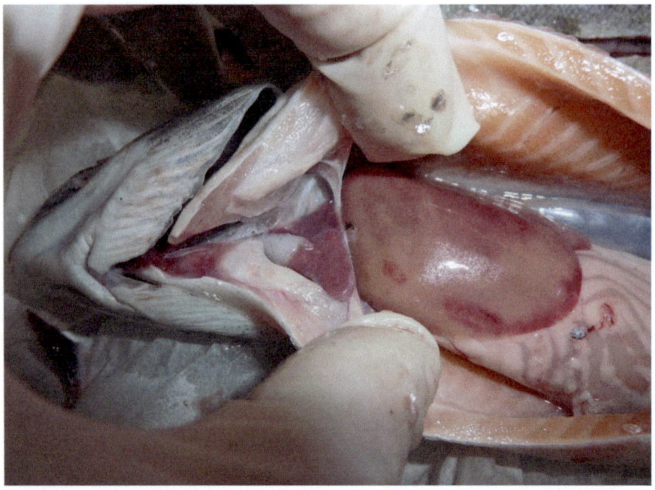

Fig. 5.1 Pale heart and liver with haemorrhage in farmed salmon post smolt with infectious pancreatic necrosis

Abdominal distension, mild to moderate exophthalmia, haemorrhage in ventral areas, oedema and swelling at the vent are typical macroscopic findings. The liver and spleen appear pale with the stomach and intestine devoid of food (Fig. 5.1). Some petechial haemorrhage, particularly on the peri-pancreatic fat among the pyloric caeca may be recorded and similar gross signs are reported in infected salmonids reared in sea water (Fig. 5.2).

Histologically, IPN is a subacute to acute infection with pyknotic nuclei and associated focal necrosis of the pancreatic acinar tissue (Fig. 5.3). Areas of focal necrosis are replaced by a loose fibrous network with fat degeneration and consequently the tissue has a loose appearance in stained sections. In the gut, some apoptotic cells are noted within the mucosal intestinal epithelium ('McKnight' cells),with sloughing combined with excess mucous to form a haemorrhagic exudate (Fig. 5.4). An increase in eosinophilic granular cells in the granulosum layer of the intestinal wall can be recorded. Pancreatic and hepatic tissues may be infiltrated by macrophages and polymorphonuclear leucocytes. Involvement of the liver has been documented as a feature of IPN, with early changes described as fine vacuolation or vesicles within individual hepatocytes. Progressively the vacuolation becomes widespread, particularly towards the external edge of the hepatic cords and individual lobules, and mirrored by an increase in apoptotic figures (Fig. 5.5). Apoptotic bodies are phagocytised by neighbouring cells with only a mild or generally absent inflammatory response (Fig. 5.6). Post apoptotic necrosis is observed in severely damaged livers, but extensive apoptosis can also take place without necrosis. Pyknotic nuclei with loss of cell integrity are found concurrently with the occurrence of apoptosis, becoming a prominent feature in advanced stages (Fig. 5.7). Occasionally the entire tissue becomes affected and the number of cells undergoing necrosis can be significant and outnumber those cells that are specifically apoptotic. Differential diagnosis of IPN includes other viral infections that can occur concurrently, such as salmonid alphavirus in Atlantic salmon post-smolts.

Both clinical signs and histopathology are used to provide a presumptive diagnosis. A confirmed diagnosis of IPN can be

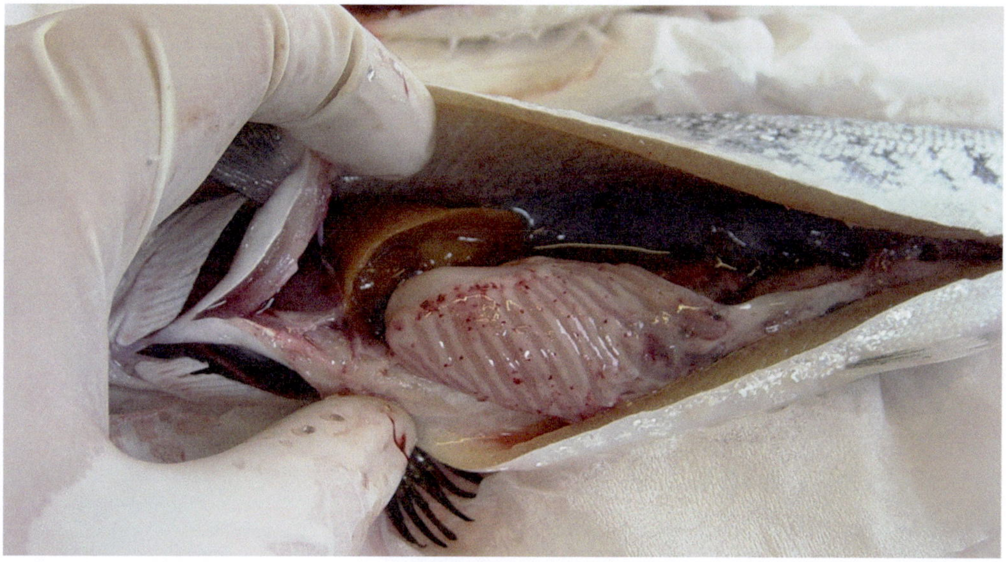

Fig. 5.2 Petechiae in pancreatic tissue and peripancreatic fat in farmed salmon smolt with infectious pancreatic necrosis

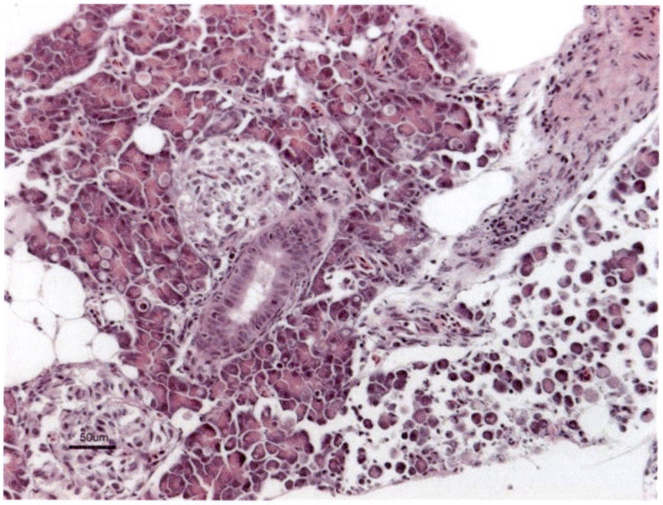

Fig. 5.3 Infectious pancreatic necrosis in farmed Atlantic salmon smolt. Necrosis of exocrine pancreatic cells (*lower right*) and normal tissue (*upper left*)

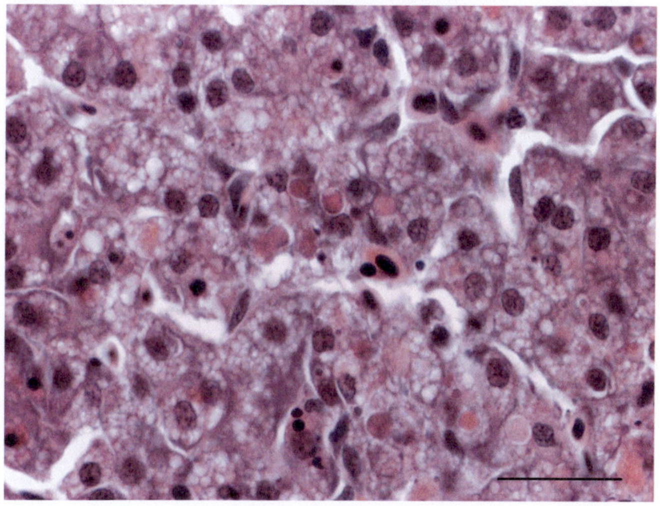

Fig. 5.5 Fine vacuolation or vesicles within individual hepatocytes in Atlantic salmon with infectious pancreatic necrosis. Medium power

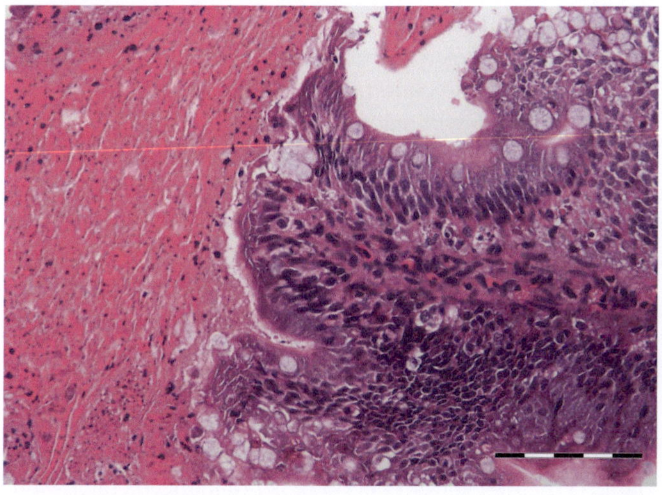

Fig. 5.4 Exudate in the intestine of farmed Atlantic salmon with infectious pancreatic necrosis. Bar = 100 μm

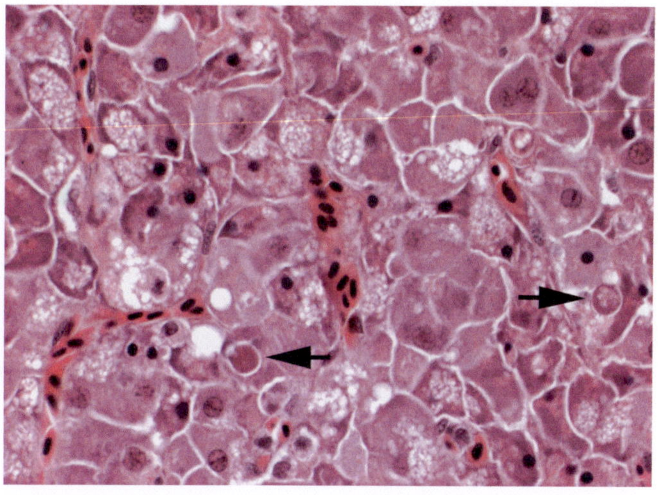

Fig. 5.6 Liver of Atlantic salmon infected with infectious pancreatic necrosis showing apoptosis (*arrows*). High power

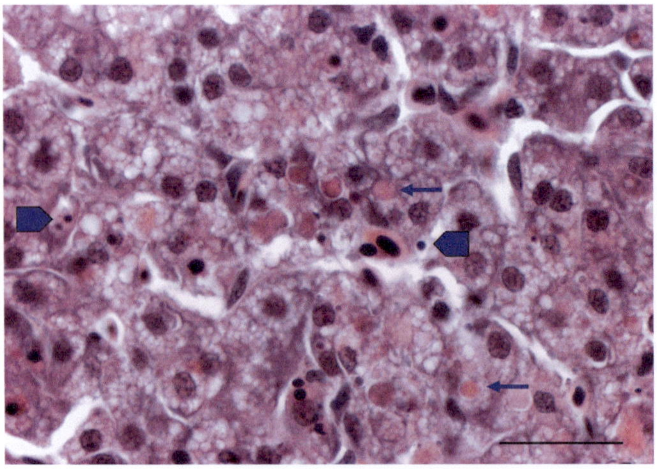

Fig. 5.7 Liver of Atlantic salmon infected with infectious pancreatic necrosis. Pyknotic nuclei (*block arrows*) with loss of cell integrity are found concurrently with apoptosis (*arrows*) and necrosis. Bar = 20 µm

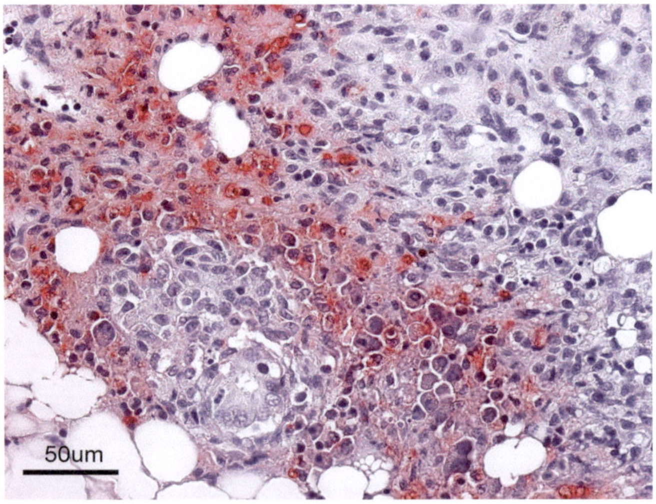

Fig. 5.8 Farmed Atlantic salmon with infectious pancreatic necrosis. Strong positive immunohistochemical reaction in post-vaccination granulomatous tissue in pancreatic area

achieved through tissue culture using a variety of salmonid and non-salmonid cell lines including CHSE-214 (Chinook salmon embryo), BF-2 (bluegill fry) RTG-2 (rainbow trout gonad), a neutralisation test, or a reverse transcription-polymerase chain reaction (RT-PCR) and sequence analysis. In addition, the detection in adherent leucocytes using immunofluorescence labelled specific antibodies is useful (Fig. 5.8). Infectious pancreatic necrosis virus is a bisegmented double-stranded RNA virus of the family Birnaviridae and at least nine serotypes exist. Commercially, both oral and injectable vaccines are available.

5.2 Infectious Salmon Anaemia

Infectious salmon anaemia (ISA) is a highly infectious viral disease which has only been observed naturally in Atlantic salmon. ISA was first diagnosed in Norwegian aquaculture

Fig. 5.9 Characteristic dark liver in Atlantic salmon with infectious salmon anaemia

in the mid 1980s, but also reported from farmed fish in Canada, Faroe Islands, Scotland, USA, Ireland and Chile. Outbreaks of ISA occur in sea water or in hatcheries where sea water is mixed with fresh water for pH or adjustment to enhance smoltification. Under experimental conditions, sea-run brown trout can harbour and transmit the virus. Most new outbreaks have occurred in connection with rapid changes in temperature during the spring and to some extent in the autumn. Importantly, salmon stocks show large variation in their susceptibility to the virus.

Diseased fish are lethargic and listless and tend to sink to the bottom of the cages. During the acute stage, mortality is usually high but in the pre-acute stages mortality may not be observed. Externally, fish show abdominal distension, haemorrhagic spots surrounding the lens, pale gills and petechiae. Occasionally, some fish are observed with hyperactive behaviour and presumably nervous movements, spinning around the longitudinal axis. Internally, ascites, oedema of the swim bladder, splenomegaly and a homogeneous dark liver are common (Fig. 5.9).

The virus is described as non-cytolytic endotheliotropic and infection of the endothelium lining the circulatory system is observed without vasculitis and with virus absent from necrotic parenchymal cells, e.g. in liver, heart or kidney. The pattern of virus attachment mirrors the distribution of infection, showing that the virus receptor is important for cell tropism as well as for adsorption to erythrocytes. During the final stages, a confluent haemorrhagic focus in the kidney is characteristic, but more frequent in North American stock. The liver lesions in affected salmon from Norway and Scotland show dilation of the hepatic sinusoids and zonal haemorrhagic necrosis with a bridge-like pattern that leaves the hepatic tissue surrounding small and medium-sized veins viable (Fig. 5.10 and 5.11). There is pronounced congestion, often combined with extensive haemorrhage in the lamina propria of the foregut and congestion of the splenic parenchyma with occasional erythrophagocytosis (see Fig. 4.11).

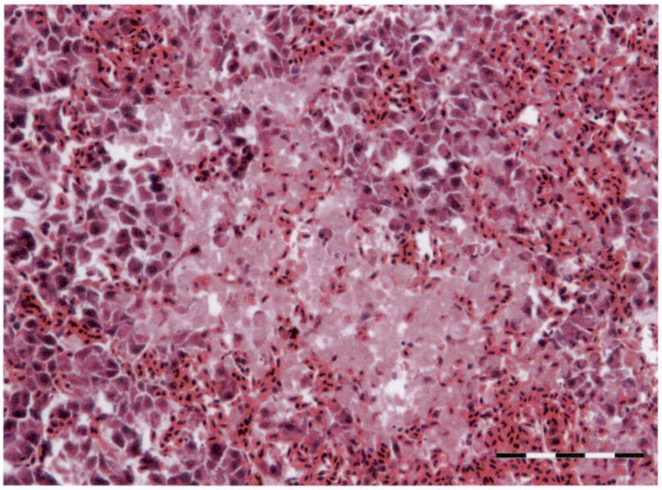

Fig. 5.10 Localised haemorrhage in head kidney from Atlantic salmon with infectious salmon anaemia. Low power

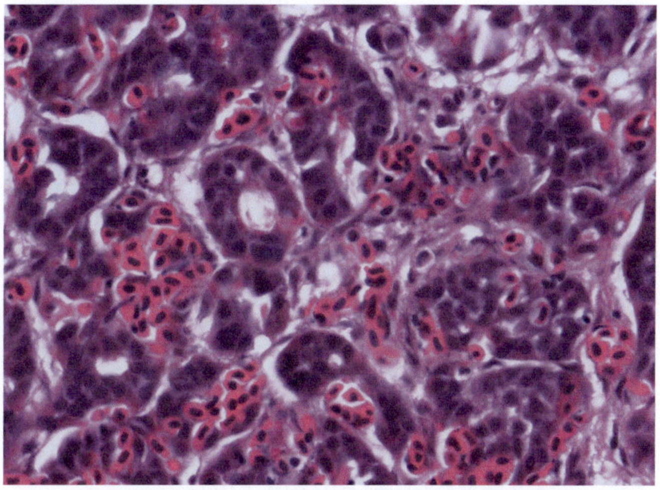

Fig. 5.12 Corpuscle of Stannius with diffuse haemorrhage in Atlantic salmon infected with infectious salmon anaemia virus. Medium power

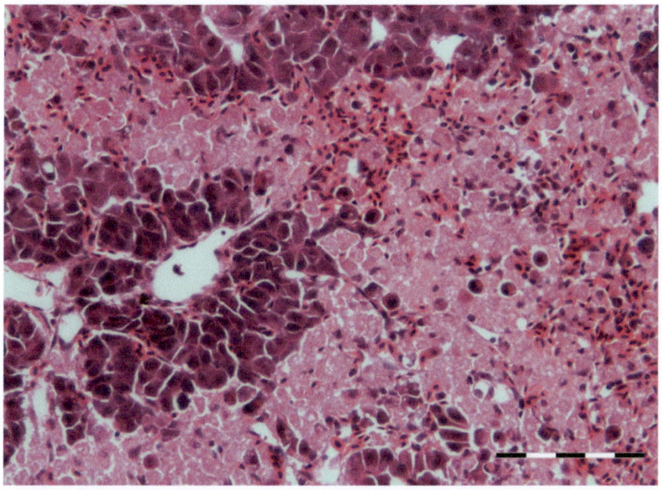

Fig. 5.11 Characteristic anastomosing liver necrosis and haemorrhage in farmed Atlantic salmon with infectious salmon anaemia. Bar scale = 100 μm

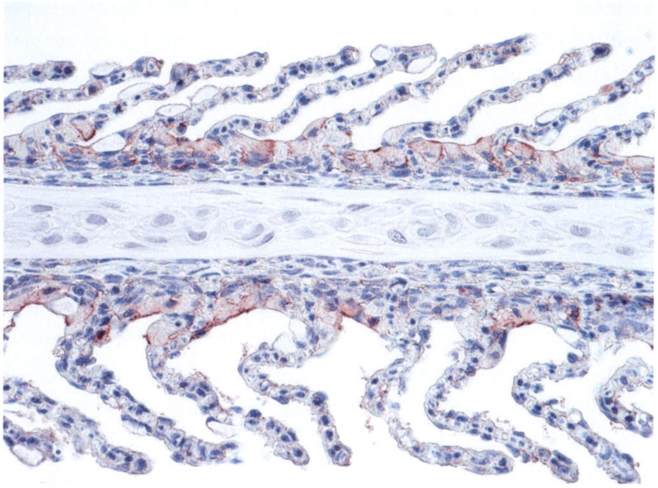

Fig. 5.13 Gill chloride cells show a strong positive reaction in farmed salmon with infectious salmon anaemia. Medium power

Similarly, there can be extensive haemorrhage in the corpuscles of Stannius (Fig. 5.12). Haematological changes include vacuolation of erythrocytic cytoplasm, leucopenia and decline in haematocrit to less than 5 in severely affected fish. A differential diagnosis would be viral haemorrhagic septicaemia.

ISA virus is a member of the Orthomyxoviridae family. Although the natural reservoir of the virus remains unknown, a marine source is considered likely. Subclinical infection has been recorded in several species including trout, rainbow trout, Arctic char, chum, Chinook and coho salmon plus marine species such as herring and pollock. Experimental data indicate that transmission may occur effectively through blood, skin mucous and faeces, with the gill capillary network being the most important route. Spread of the virus also occurs via movement of latent carriers or diseased fish, and historically through water and blood discharged from slaughtering facilities. Sea lice (*Lepeophtheirus* spp. and *Caligus* spp.) may act as vectors of the virus and the stress associated to lice infestation may make the fish more susceptible to infection. Vertical transmission of the virus has been suggested, but this is still controversial and not fully resolved.

Diagnosis is based upon characteristic clinical changes, histopathology and isolation of the virus in cell lines such as salmon head kidney (SHK-1), Atlantic salmon kidney (ASK) and CHSE. Immunohistochemistry and PCR tools are also available to demonstrate the presence of the virus, particularly in haematopoietic tissue, endocardial and chloride cells (Fig. 5.13). Variants of ISAV have been genetically differentiated on the basis of the sequence of a highly polymorphic region (HPR of genomic segment 6 which encodes the Haemagglutinin-Esterase (HE) protein). A deletion within the HPR region (named HPRΔ ISAV) in certain ISAV variants appears to be a dependable indicator of pathogenicity. ISAV without any deletions in the HPR region (HPR0 ISAV) has been reported only in apparently healthy fish and to date have not been associated with ISA disease. A reverse transcriptase-polymerase chain reaction has been developed as a sensitive method to detect carrier fish. A commercial vaccine is also available.

5.3 *Oncorhynchus masou* Virus

Oncorhynchus masou virus (OMV) is a virulent and economical significant disease that was originally isolated from ovarian fluids of a landlocked population of adult masou salmon in Hokkaido, Japan, but now also occurring in wild stocks. Other salmonid species are susceptible to OMV including coho, chum, kokanee and rainbow trout with high mortality in young fish.

Affected fish are dark, frequently showing severe exophthalmia and petechial haemorrhage under the lower jaw and along the ventral surface. Epithelioma around the mouth (upper and lower jaw), and, to a lesser extent, on the caudal fin, operculum and body surface occur progressively (Fig. 5.14). A white mottled appearance of the liver, progressing to a pearly white colour of the whole organ is recorded. A pale kidney and a multifocal, severe necrosis of the liver are also common. Gill epithelial cells become swollen and slough. There is a marked splenomegaly with associated necrosis of the ellipsoids and the digestive tract is generally devoid of food.

Studies involving experimental infection with OMV have shown that there is some variation in histopathological findings between species of juvenile salmon. In chum salmon, the apparent target organ is the kidney with necrosis of haematopoietic tissue, hyaline droplet degeneration and pyknotic nuclei. Partial necrosis occurs in the spleen, liver, pancreas and stomach. However, in masou salmon haematopoietic necrosis has been reported without the glomeruli or tubules being affected. OMV has oncogenic potential and induces a mandibular epithelial neoplasm in surviving fish and other neoplasms of the fins, body surface and cornea. These growths are characterised as papillomatous (Fig. 5.15). Multiple mitotic figures confirm the proliferative nature of the swelling.

OMV is a salmonid herpesvirus type 2 (SalHV-2) and transmitted by diseased fish and asymptomatic carriers. This virus is shed in the faeces, urine, sexual products at spawning, and probably with skin mucus. Transmission is by direct contact or through the water, but 'egg-surface associated' transmission probably also occurs. Symptomatic and asymptomatic carriers can spread the virus to uninfected stocks.

Diagnosis involves virus isolation using cell lines such as CHSE-214 or RTG-2 and a serum neutralisation test using a specific OMV antiserum. Viral antigens can be identified directly in tissues by immunofluorescence or ELISA. The differential diagnosis includes infectious haematopoietic necrosis, whirling disease and viral haemorrhagic septicaemia.

Fig. 5.14 Papillomatous neoplasia in the mandible in coho salmon due to *Oncorhynchus masou* virus. Bar = 100 μm

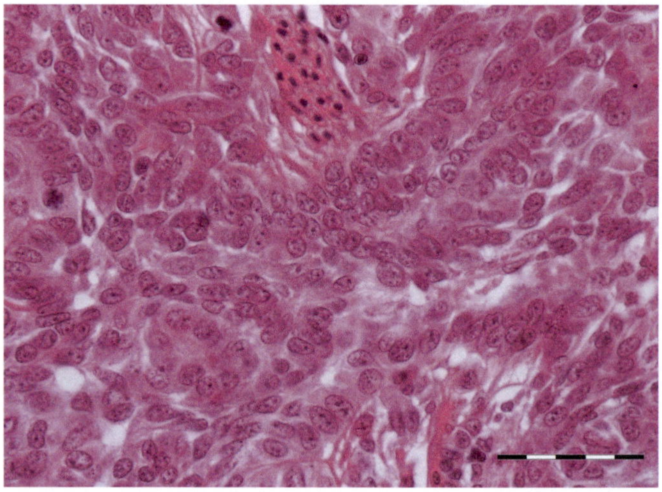

Fig. 5.15 Transverse section of papillomatous neoplasia showing proliferating epithelial cells supported by thin connective tissue in chum salmon with *Oncorhynchus masou* virus. Bar = 50 μm

5.4 Piscine Reovirus (Heart and Skeletal Muscle Inflammation, HSMI)

Piscine reovirus (PRV) has been recently reported as the aetiological agent of HSMI, a systemic viral disease of seawater farmed Atlantic salmon. The first cases were identified in Norway in 1999 and the disease is currently widespread in Norwegian aquaculture where it causes substantial losses. This condition has also been described from farmed salmon in Scotland. The PCR-screening of marine fish caught along the Norwegian coast has revealed PRV in great silver smelt, capelin, Atlantic herring and horse mackerel.

Clinical outbreaks typically occur 5–9 months after transfer to sea water. Morbidity may be very high in affected cages, while mortality may reach 20 %. Clinical signs include anorexia and abnormal swimming behaviour and internally, pale heart, yellow-orange liver, ascites, splenomegaly and visceral petechiae.

Characteristic histopathological changes are found in heart and skeletal red muscle. Red skeletal muscle is usually heavily affected with myocyte degeneration and infiltration of inflammatory cells (Figs. 5.16 and 5.17). In the heart, early lesions in the ventricular compactum typically include perivasculitis associated with branches of the coronary vessels, endocarditis and focal myocarditis (Fig. 5.18). A highly cellular epicarditis can also be observed (Fig. 5.19). Cardiac lesions subsequently spread to the entire myocardium developing an extensive panmyocarditis, multifocal necrosis and inflammation dominated by neutrophils and macrophages in both spongy and compact myocardium, within and between muscle fibres, and aggregates or 'nests' of small nuclei may be seen in affected myocardium. Additional cardiac lesions are compensatory karyomegaly and show elongated Anitschkow-like nuclei. Atrial lesions are similar to those seen in the spongy myocardium, but often milder.

Lesions in other organs are few but general congestion and multifocal liver necrosis with vacuolated and pyknotic or karyolytic cells may be seen. In addition, haemorrhage and accumulation of erythrocytes can be recorded in gills, kidney and spleen.

PRV belongs to the reovirus group and appears to be widespread in farmed salmon. The route of infection is presently unknown. However, a low prevalence in several non-salmonid species from Norwegian waters using a real-time PCR suggests there is a complex relationship that involves

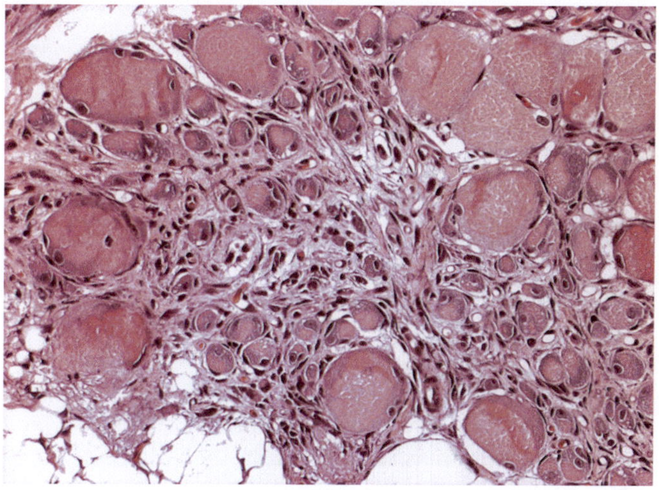

Fig. 5.17 Transverse section showing degeneration and inflammation of red muscle in farmed Atlantic salmon with heart and skeletal muscle inflammation. Medium power

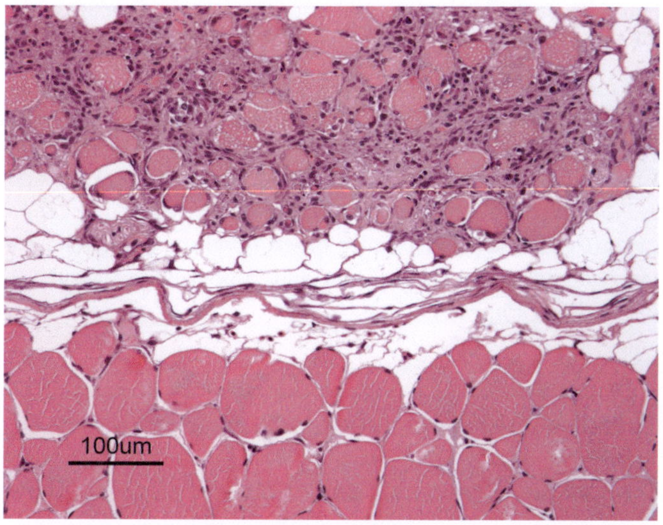

Fig. 5.16 Degeneration and inflammation of red muscle (*top*) in Atlantic salmon with heart and skeletal muscle inflammation

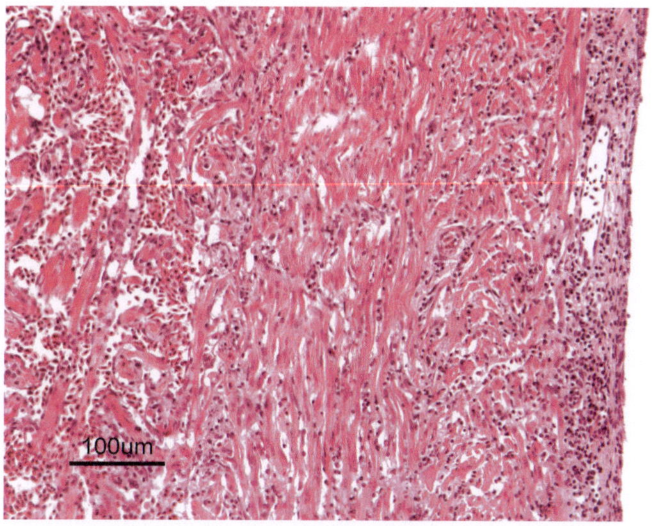

Fig. 5.18 Heart and skeletal muscle inflammation in farmed Atlantic salmon. Severe inflammation in both myocardial layers and epicardium

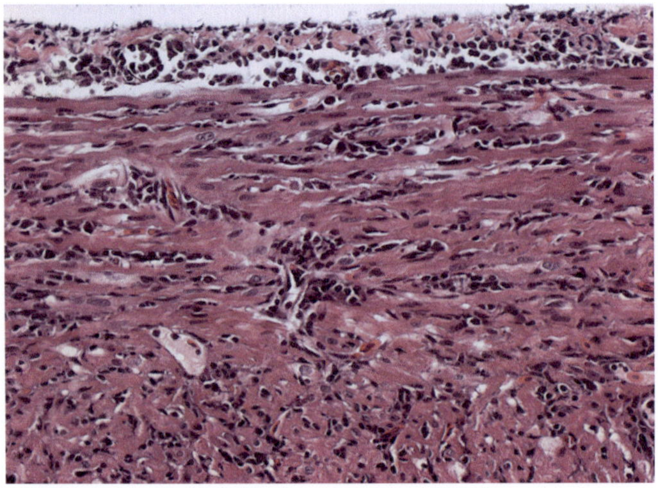

Fig. 5.19 Cellular epicarditis and inflammation of compact myocardium in farmed Atlantic salmon with heart and skeletal muscle inflammation. Medium power

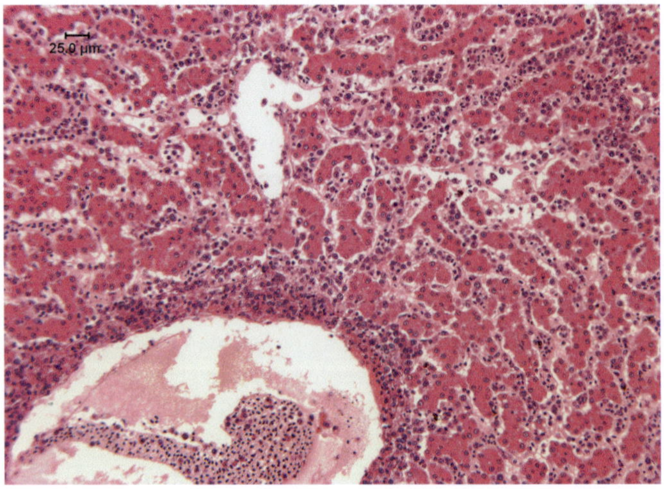

Fig. 5.20 Plasmacytoid leukaemia in farmed Chinook salmon. The liver sinusoids are infiltrated with proliferating plasmablasts. Medium power

carriers and virus reservoirs. Diagnosis comes from the characteristic histopathological lesions in heart and red muscle, immunohistochemistry and the presence of PRV demonstrated by PCR. Pancreas disease (PD) and cardiomyopathy syndrome (CMS) are important differential diagnoses, but may be differentiated from HSMI by the type and distribution of cardiac lesions, pancreatic lesions and pathological changes in red muscle. There is no treatment or vaccines available. Prophylactic measures include reduction of stress and restrictions on the transport of live fish.

5.5 Salmon Leukaemia Virus

Plasmacytoid leukaemia (PL) of farmed Chinook on the west coast of North America is attributed to salmon leukaemia virus (SLC). Evidence of PL has also been recorded in freshwater-reared salmon in the United States and in salmon from net-pens in Chile. Other salmonids such as coho, sockeye and Atlantic salmon are considered susceptible under experimental conditions. Affected fish are dark, lethargic and often show severe bilateral exophthalmia. Pale gills are a regular finding and affected fish usually stay near the water surface. Spleen, kidney and retrobulbar tissue are enlarged and mottled, with petechial haemorrhage in several organs including the liver, mesenteric fat, pancreas and skeletal muscle. Histologically, infection is characterised by proliferation and infiltration of plasmablasts and other lymphoid cells into the visceral organs and retrobulbar tissues (Figs. 5.20 and 5.21). The plasmablasts have lobate nuclei and prominent nucleoli. The kidney shows mild to moderate hyperplasia of the haematopoietic interstitium. Evidence suggests this is a neoplastic condition rather than a reactive plasmacytosis. There

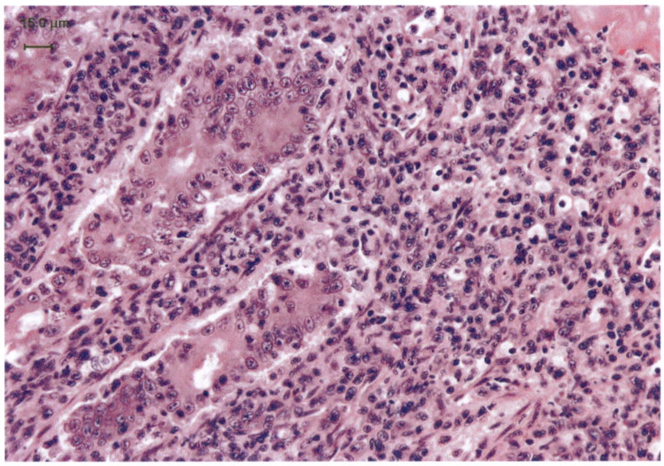

Fig. 5.21 Infiltration of plasmablasts in the stratum proprium of the intestine of Chinook salmon. Medium power

is some indication that the presence of an intranuclear microsporidian, *Nucleospora salmonis* (previously *Enterocytozoon salmonis*) is associated with PL, and maybe a cofactor. The diagnosis of plasmacytoid leukaemia (a retrovirus) is provisionally based upon the detection of a large number of plasmablasts and supported by an IFAT using tissue imprints.

5.6 Viral Haemorrhagic Septicaemia

Viral haemorrhagic septicaemia (VHS) is a serious contagious disease that particularly affects rainbow and brown trout, but also Japanese flounder, turbot and whitefish. Overall, rainbow trout reared in fresh water are the main susceptible group but other salmonids and non-salmonid species

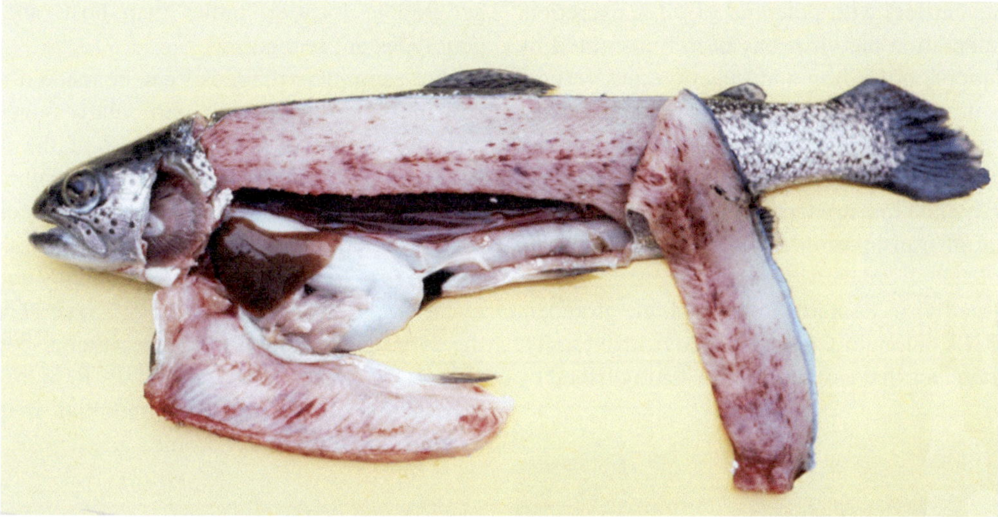

Fig. 5.22 Viral haemorrhagic septicaemia in rainbow trout showing widespread petechiae of the musculature and pale gills

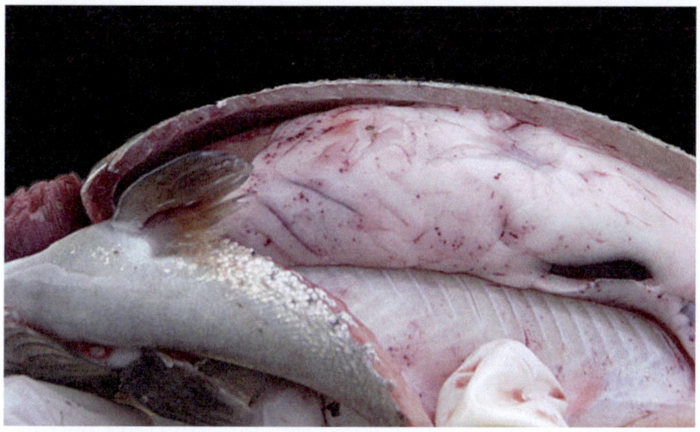

Fig. 5.23 Petecchia in the pyloric region in farmed rainbow trout with viral haemorrhagic septicaemia

may become infected in fresh and marine environments. The emergence of VHSV in the Great Lakes Basin of North America during 2003 highlighted this virus as a cause or mortality in a range of freshwater fish species. Outbreaks typically result in an acute to chronic disease when temperatures are fluctuating and generally below 14 °C.

A wide range of disease signs are recorded. Clinically, fish show lethargy with dark skin colour and exophthalmia, a severe haemorrhagic anaemia and a marked distension of the abdomen due to oedema in the liver, spleen and kidney with darkening of the body (Figs. 5.22 and 5.23). However, in many fish few pathological changes are noted. VHSV is characterised by destruction of the endothelial lining causing haemorrhage in the skeletal muscle, meninges, intestinal mucosa and in the eye. This may be accompanied by ataxia. Ascites can be recorded and there is an absence of food in the gastrointestinal tract. These acute signs are usually associated with a rapid onset of heavy mortality and linked to the age of the fish with up to 100 % in fry, often less in older fish, typically from 30 to 70 %. In sea water, a mortality of 80 % has been reported within a month following transfer of fish from fresh water.

The main histopathological findings include hepatitis with multifocal and sometimes haemorrhagic necrosis in the liver, with endocarditis and marked haematopoietic necrosis in kidney and spleen. At a later stage of infection the kidney tubules are also necrotic. Severe glomerular changes resembling a membranous glomerulonephritis occur with focal necrosis and degeneration, and associated with leukocyte infiltration and cell debris. The liver sinusoids become congested, together with a widespread necrosis with numerous pyknotic and karyolytic nuclei. The spleen can show a severe vasculitis and the brain haemorrhage occurs in conjunction with necrotic foci. Immunostaining of the cerebellum shows the positive staining in the Purkinje cell and inner granular layer (Fig. 5.24). The muscle fibres and bundles commonly show intermuscular haemorrhage (Fig. 5.25). A chronic stage is correlated with a lower mortality over an extended period. At this stage the liver sinusoids show congestion with hyperplasia. During the latent infection or the ‘nervous stage’, mortality is low and fish often appear normal. However, some fish are hyperactive with poor balance which is conspicuous, as an erratic and often spiralling, swimming behaviour. No remarkable histopathological changes occur in these carriers.

The transmission of this virus and the outbreak of disease in susceptible fish are related to stage of development and water temperature. VHSV can be transmitted to fish and survivors become carriers with potential of virus excretion. Through experimentation the virus has been transmitted by cohabitation, immersion, feeding and injection, but vertical transmission has not been reported. VHSV can persist in the water for several days after release from infected fish and this represents a significant hazard to other farmed fish or wild stocks which come into contact with the effluent. There is some evidence of transmission from wild marine fish to farmed fish.

VHSV or Egtved virus contains a linear single stranded, negative-sense RNA belonging to the Rhabdoviridae family. Sequence data indicates that isolates can be distinguished by geographic location rather than host; and currently four genotypes are recognised.

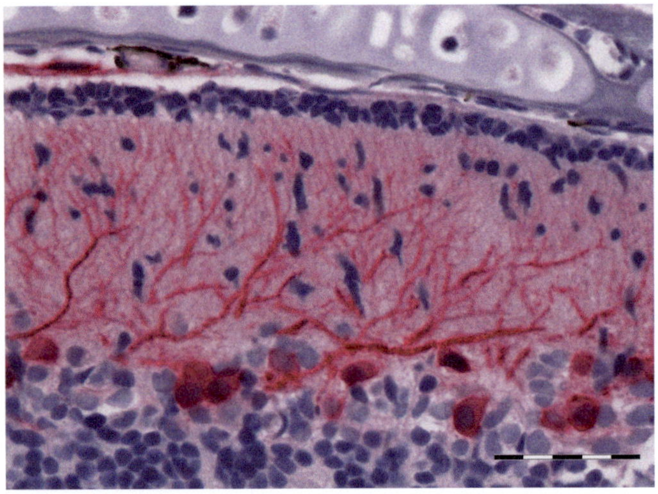

Fig. 5.24 Cerebellum from rainbow trout fry with viral haemorrhagic septicaemia. Positive immunostaining shows virus location in the Purkinje cell and inner granular layer. Bar = 50 μm

A presumptive diagnosis can be made if mortality among susceptible groups is high, particularly fingerling or yearling rainbow trout which exhibit most of the major signs and behavioural changes described. Histopathological changes also provide additional evidence of infection. Generally, diagnosis requires the isolation of the viral agent in tissue culture from organs including the kidney and spleen, and by reactivity with a specific antibody. Several cell lines support the virus growth and include epithelial papilloma of carp (EPC), BF-2 and RTG-2and RT-PCR followed by sequencing of the amplicon. A differential diagnosis includes haemorrhagic smolt syndrome.

5.7 Infectious Haematopoietic Necrosis

Infectious haematopoietic necrosis (IHN) is a virulent, generally lethal, systemic disease primarily of young fish. Wild Pacific salmon have a natural resistance to the virus, however, in farmed Atlantic salmon on the Pacific coast of Canada where they are not native and don't have natural immunity and consequently outbreaks have resulted in high mortality. The appearance of IHNV in farmed rainbow trout has been accompanied by genetic changes that appear to be related to a shift in host specificity and virulence.

The virus is enzootic throughout North America, and movement of infected animals believed to have been a significant factor in the spread to Asia and Europe. The IHN virus is carried by Pacific salmon, trout and herring.

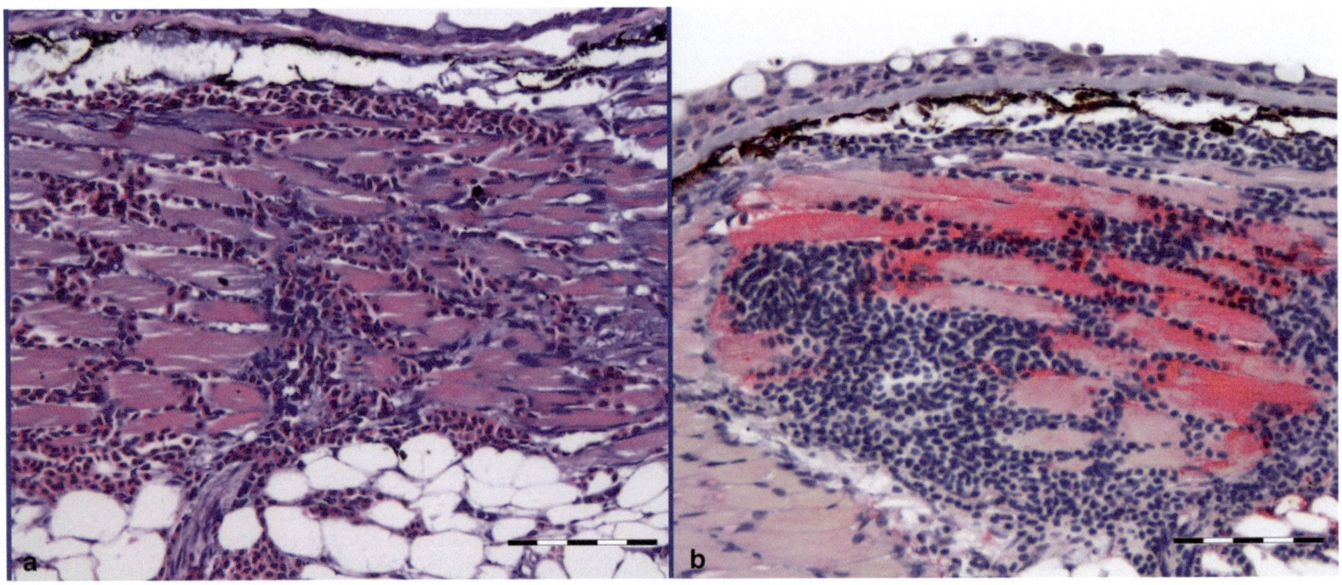

Fig. 5.25 (a) Intramuscular haemorrhage of rainbow trout fry with viral haemorrhagic septicaemia. (b) Immunostaining (*red staining*) of the same area to show location of viral haematopoietic necrosis virus. Bar = 100 μm

Fig. 5.26 Exopthalmia with haemorrhage in rainbow trout fry with infectious haematopoietic necrosis

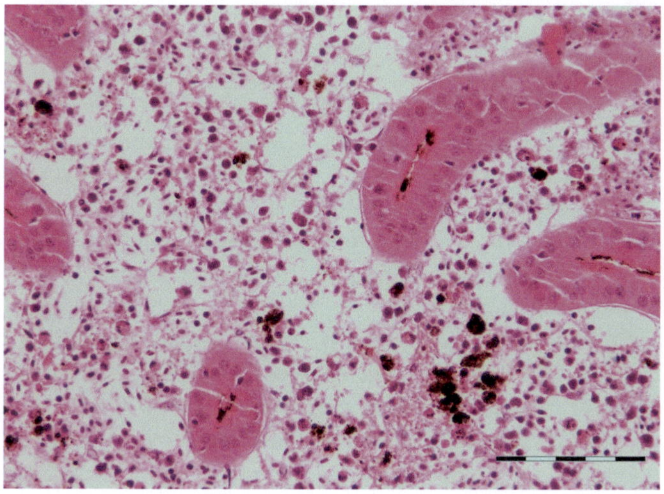

Fig. 5.27 Focal necrosis of kidney interstitium of rainbow trout with infectious haematopoietic necrosis. Bar = 100 μm

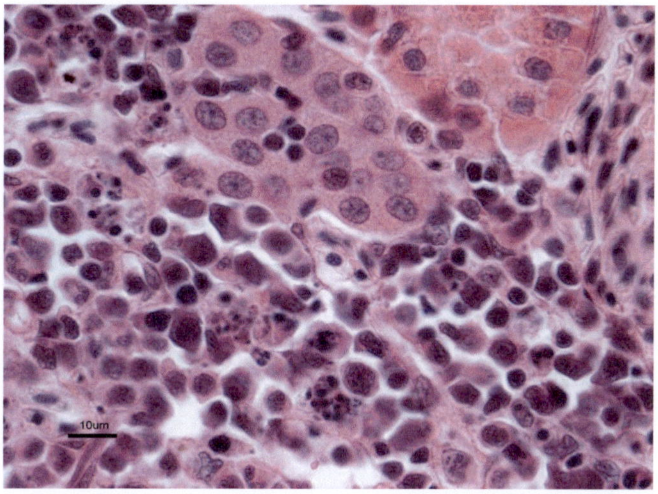

Fig. 5.28 Pyknotic and karryorrhectic nuclei in haematopoietic tissue in head kidney in sockeye salmon fry with infectious haematopoietic necrosis. Note pale eosinophilic adrenal cells with large nuclei

Natural outbreaks of IHN are rare above 15 °C and temperature is a factor which strongly influences disease progression. In acute infection mortality increases rapidly and fry show lethargy and move to the edges of tanks. Abnormal swimming patterns such as whirling and flashing may be observed. Older fish rarely show behavioural changes. Pale gills, skin darkening, exophthalmia (Fig. 5.26), distended abdomen and prominent sub-dermal haemorrhage can occur between the head and dorsal fin. Internally the intestine often contains a watery, yellow-coloured fluid with haemorrhage in the visceral mesenteries.

Significant histopathological lesions occur in the haematopoietic tissues, posterior kidney, spleen, pancreas and digestive tract (Figs. 5.27 and 5.28). Infection in the kidney progresses from degenerative necrotic changes, followed by increased abundance of macrophages, and then vacuolation and nuclear chromatin margination. Later tissues show necrotic lymphoid cells with extensive pyknosis, karyorrhexis and karyolysis in all organs. In addition affected fish show a severe damage to vessels and a multifocal myocarditis. Areas of necrosis may be present in the liver and pancreas. Necrosis of the stratum granulosum and stratum compactum of the digestive tract is considered pathognomonic, and the sloughing gives rise to faecal casts.

IHN-infected broodstock represent a source of infection and the most important route of transmission is considered to be via the gills. Although horizontal spread has been demonstrated, there is also strong evidence for vertical transmission.

The causative virus, IHNV is placed as the type species in the *Novirhabdovirus* genus of the family Rhabdoviridae. Diagnosis of IHN is based on the characteristic histopathological findings, immunostaining (Figs. 5.29 and 5.30) and the isolation and characterisation of virus in tissue culture using cell lines such as CHSE-214 followed by a serum neutralisation. A differential diagnosis would include infectious pancreatic necrosis. There is evidence that an orally delivered DNA vaccine for IHN is achievable, but this approach requires refinement.

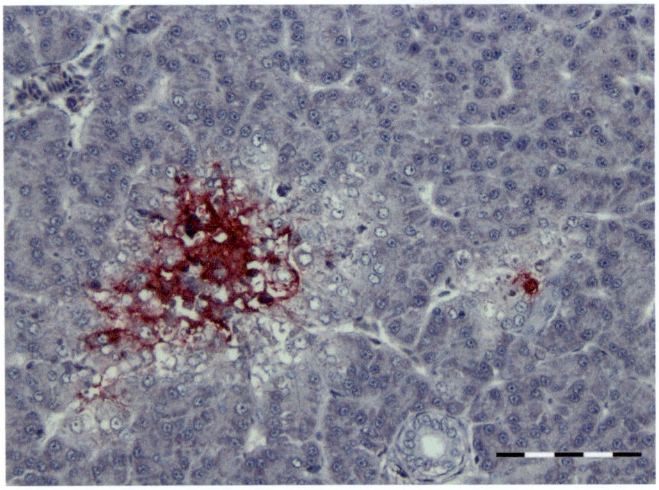

Fig. 5.29 Liver of sockeye salmon fry with infectious haematopoietic necrosis. Hepatic necrosis with strong positive immunohistochemical reaction. Bar = 100 μm

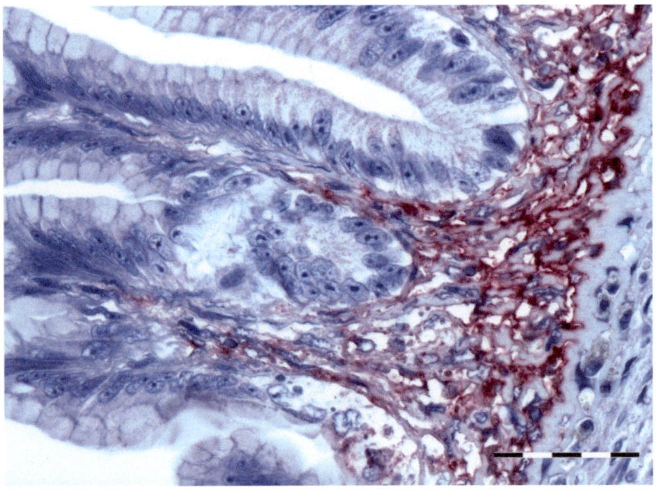

Fig. 5.30 Intestine of sockeye salmon fry with infectious haematopoietic necrosis. Strong positive immunohistochemistry reaction in stratum proprium. Bar = 50 μm

5.8 Hirame Rhabdovirus

Hirame rhabdovirus (HRV) was first reported from moribund Japanese flounder in 1984, and subsequently the virus has been reported as pathogenic to species of bream. The first description of HRV in Europe occurred in grayling and brown trout in a farm in Poland, raising concerns that spread may have occurred from Asiatic countries and there was potential for emergence in freshwater fish.

Experimental studies involving salmonid fish (i.e. rainbow trout, chum, coho, masu and pink salmon) have been carried out. The highest virus titre was obtained from rainbow trout infected following direct transmission through the water. Histologically, the experimentally HRV infected rainbow trout showed necrosis and haemorrhage in the kidney and spleen and the skeletal muscle revealed hyperaemia and haemorrhage. Due to similarities with VHS, this becomes an important differential diagnosis for farmed rainbow trout. A qRT-PCR assay represents a reliable, specific and sensitive tool for the quantitative diagnosis of HRV in fish samples. Further work is required to assess if the virus is spreading and its potential impact on farmed fish.

5.9 Salmonid Alphavirus (Pancreas and Sleeping Disease)

Salmonid alphavirus is responsible for a serious infectious disease affecting farmed Atlantic salmon (Pancreas disease, PD), and rainbow trout (Sleeping disease, SD). It was first reported in Scotland in 1976 in salmon followed by observations in Norway and Ireland. The rainbow trout condition was recognised by 1994 in France as a similar disease and the confirmation of a viral aetiologies came soon after with the isolation of the first alphavirus from fish (PD) in Ireland (1995) and of SD, in France (1997). Evidence shows that SAV represents an atypical and new member of the genus Alphavirus within the family Togaviridae. Currently six subtypes have been described, and all of them affects sea water salmon (SAV 1, 2, 3, 4, 5 and 6), however only SAV 3 and recently SAV 2 have been recorded in farmed salmon in Norway. The rainbow trout fresh water virus is categorised as SAV-2, and SAV-2 marine strains have been reported from salmon in Scotland. SAV infections are recognised as a major disease problem in farmed salmonids in the UK, Ireland and Norway, while in continental Europe SD has been reported in France, Spain, Italy and Germany, and on one occasion in Atlantic salmon from North America in 1987.

Clinical signs of both diseases are similar with the exception of the 'sleeping behaviour' described in rainbow trout. Other shared clinical signs include inappetence, cachectic appearance (Fig. 5.31), lethargy, increased mortality, cease of shoaling and a disrupted swimming pattern. During the latter stages of PD, fish that have a normal external appearance and may display spiralling swimming, lying on the bottom of the cage (i.e. similar to SD), with an increased sensitivity to handling which can lead to 'sudden death'. This appears to be frequent in older fish as seen in Scottish and Norwegian salmon farming. Necropsy shows an empty gut and petechial haemorrhage around the depleted fat. Concurrent signs of parasitic or bacterial infections can be frequent in those fish that have become 'runts'.

The histopathological changes in naturally occurring PD/SD primarily occur in the pancreas, heart and skeletal

Fig. 5.31 Chronic pancreas disease in farmed Atlantic salmon

muscle, although the severity and the distribution of the lesions depend on the time after initial infection. As infection may occur asynchronously within the same population in a cage, it is common to find fish from the same unit displaying different levels and severity of pathological changes. The first lesions are an acute phase of pancreatic acinar cell necrosis, with variable inflammatory response spanning from virtually none to a moderate mononuclear cell infiltration and/or fibrosis of the periacinar tissue, with loss of exocrine tissue (Fig. 5.32). The endocrine pancreas is not a target tissue. Almost simultaneously or slightly delayed, heart lesions can be observed. A severe degeneration with multifocal cardiomyocytic necrosis affecting the spongy and compact myocardium is described. Lesions are characterised by individual or clusters of shrunken cells becoming strongly eosinophilic with pyknotic nuclei. Increased cellularity at the junction of the ventricle compact –spongy layer can be observed and hypertrophy of cardiomyocytic nuclei also described, the latter being evident in the recovery phase (Figs. 5.33 and 5.34). A distinct difference in the response of affected fish in relation to their life stage (smolt versus growers) has been observed, with mitotic figures being reported as a consistent feature in the heart of affected smolts, but less frequently or absent in older fish. Finally, the skeletal muscle becomes involved at least 3–4 weeks after the lesions in the pancreas and heart are observed. Both red and white muscle are affected and characterised by myofibre hyaline degeneration with swollen and fragmented sarcoplasm (Fig. 5.35). In the white muscle individual fibres are affected and can be seen as highly eosinophilic with central migration of nuclei. Sarcoplasm infiltration by phagocytic macrophages and at later stage, a variable inflammation and fibrosis can be noted. The red muscle layer shows similar sarcoplasmic pathological changes, although frequently the proportion of damaged tissue is greater than in the white muscle. A variable degree

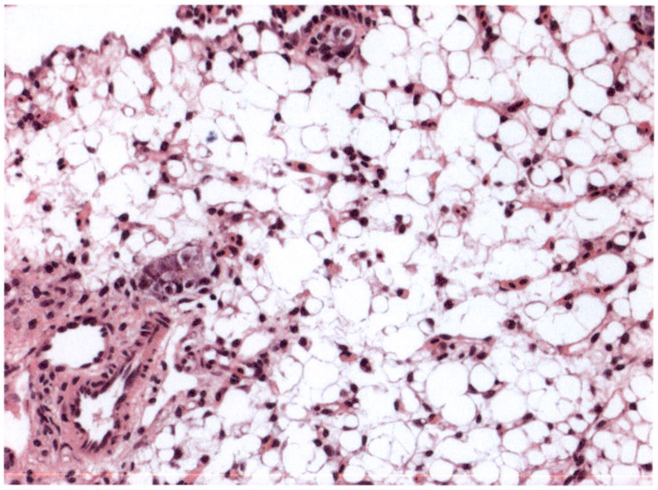

Fig. 5.32 Slight haemorrhage and absence of pancreatic tissue in farmed Atlantic salmon with pancreas disease. Medium power

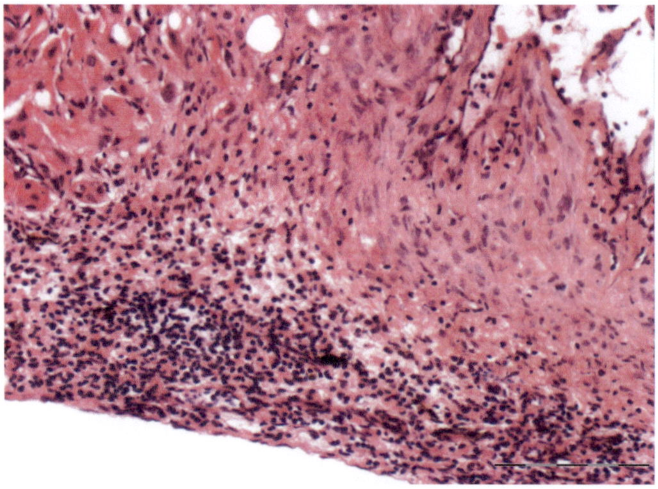

Fig. 5.33 Increased cellularity at the junction of the ventricle compact-spongy layer in Atlantic salmon with pancreas disease. Medium power

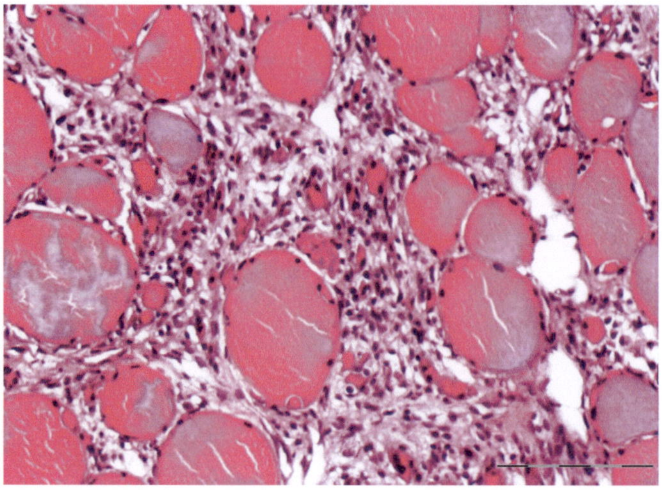

Fig. 5.35 Myofibre degeneration with swollen and fragmented sarcoplasm in Atlantic salmon with pancreas disease. Medium power

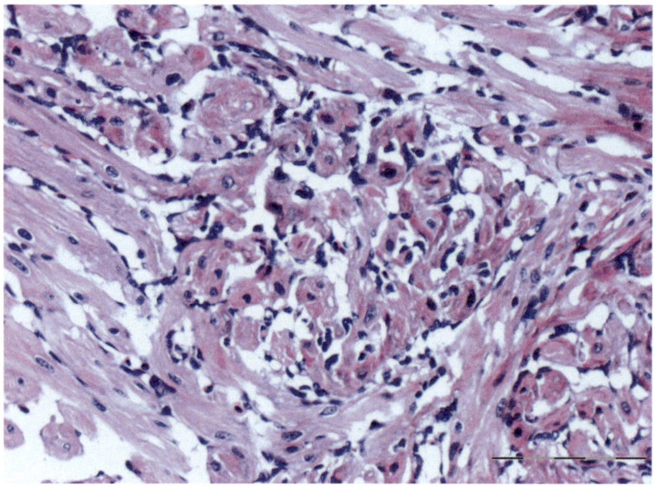

Fig. 5.34 Focal necrosis and inflammation of cardiomyocytes in spongy myocardium in farmed Atlantic salmon with pancreas disease. Medium power

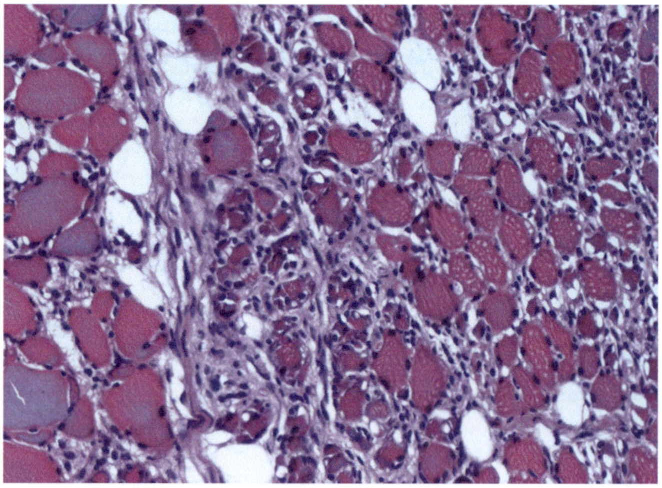

Fig. 5.36 Inflammation and fibrosis of endomysium with recovering red muscle fibres in Atlantic salmon with pancreas disease. Medium power

of inflammation and fibrosis of endomysium can be observed with recovering basophilic red muscle fibres (Fig. 5.36). Creatine phosphokinase (CPK) levels are initially low but increase during the chronic stages.

In surviving fish the pancreas can recover, with regeneration of the exocrine tissues occurring as early as 4 weeks after infection. However, some fish can develop chronic pancreatic lesions with significant loss of acinar cells with or without fibroplasia of the surrounding tissue; a situation more usually found associated with fish that become runts.

PD in salmon primarily occurs during the first year at sea, but other year classes are also susceptible. Outbreaks are common between late June and November, although the disease can recur at any time in sea water. In SD all ages are susceptible to infection but natural outbreaks appear almost exclusively when water temperatures are around 10 °C. The disease is serious in fingerlings (10–15 g) with mortality reaching up to 50 %. As with PD, older fish can be infected without showing any clinical signs and may become long term carriers of the virus.

Sub-clinical SAV infection with negligible clinical signs, no significant mortalities and relatively mild histopathological lesions have been described, suggesting a complex situation where several aspects influence the severity of the outbreaks in a given population, including environmental, host and/or pathogen. Furthermore, disease outbreaks tend to reappear in consecutive fish groups introduced onto historically infected sites, irrespective of the length of the fallowing period implying there is a marine SAV reservoir or a freshwater carrier state. Common dab have tested positive for SAV by PCR and phylogenetic analysis of an E2 dataset has confirmed a subtype V-like sequence.

The diagnosis of PD is currently based on a combination of histopathological examination, virus culture and PCR detection using heart tissue samples rather than kidney. The differential diagnosis includes CMS and HSMI. There is evidence that outbreaks are declining and attributed to improved management practices and the wider application of available vaccines.

5.10 Piscine Myocarditis Virus (Cardiomyopathy Syndrome)

Piscine myocarditis virus (PMV) has recently been recognised as the aetiology of cardiomyopathy syndrome (CMS), a chronic heart disease primarily affecting marine farmed Atlantic salmon. Lesions identical to those seen in fish with CMS have also been recorded in wild salmon but the disease has not been diagnosed in other salmonids. The first cases were reported from Norway in 1985, but subsequently described from other salmon-farming countries including Scotland, The Faroe Islands and Canada. CMS is a chronic progressive disease that develops over several months, with mortality typically occurring in large fish 12–18 months after transfer to sea water and in fish close to slaughter (Fig. 5.37). The economic losses are not due to high mortality during production, but to the effect of sudden death on large and valuable market sized fish. Widespread ventral skin scale-pocket oedema and haemorrhage can be observed in these fish (Fig. 5.38). At necropsy, haemopericardium and/or blood clots in the pericardial cavity are a typical finding. The haemorrhage is the result of a small or larger rupture of the atrium or sinus venosus resulting from severe congestion with clots following long-lasting cardiac insufficiency. Severe haemorrhage may also occur in the anterior part of the abdominal cavity when the sinus venosus ruptures caudally to the septum transversum. Other common findings are ascites, a mottled liver with a fibrinous coat and general congestion. Acute death is the result from cardiac tamponade and blood loss (Fig. 5.39). The atrium is typically considerably enlarged and sometimes filled with blood clots. The entire ventricle may be hidden within one large blood clot (Fig. 5.40).

Histologically, lesions initially occur in the atrium with scattered pleomorphic nuclei and sub-endocardial infiltration. This progresses to the spongy part of the ventricular endocardium with marked thickening of myofibres

Fig. 5.37 Good quality, harvest-size farmed Atlantic salmon with cardiomyopathy syndrome

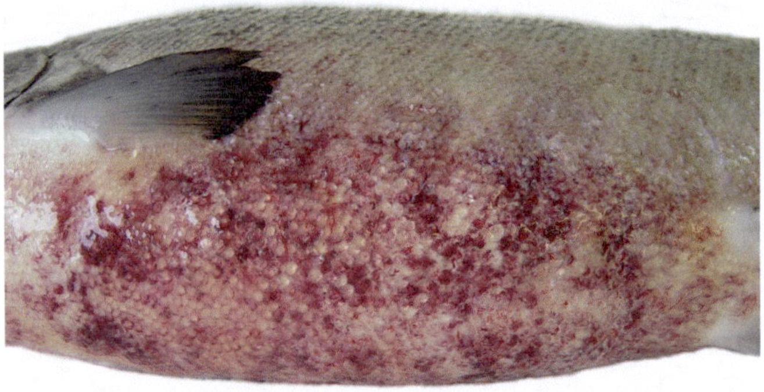

Fig. 5.38 Diffuse ventral scale-pocket oedema and haemorrhage in farmed Atlantic salmon with cardiomyopathy syndrome

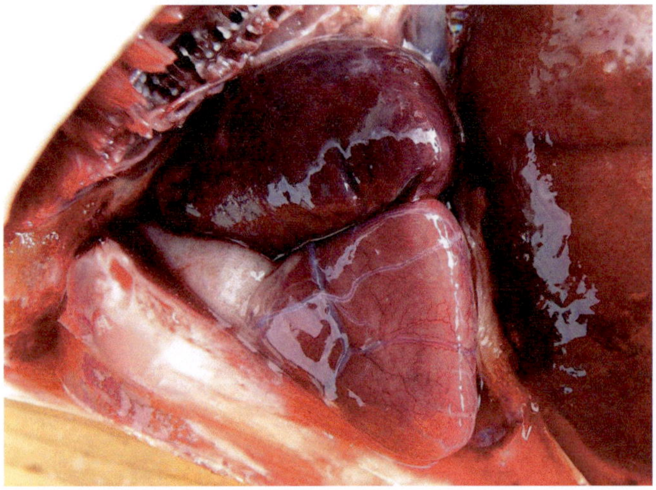

Fig. 5.39 Severely dilated atrium in farmed Atlantic salmon with terminal cardiomyopathy syndrome

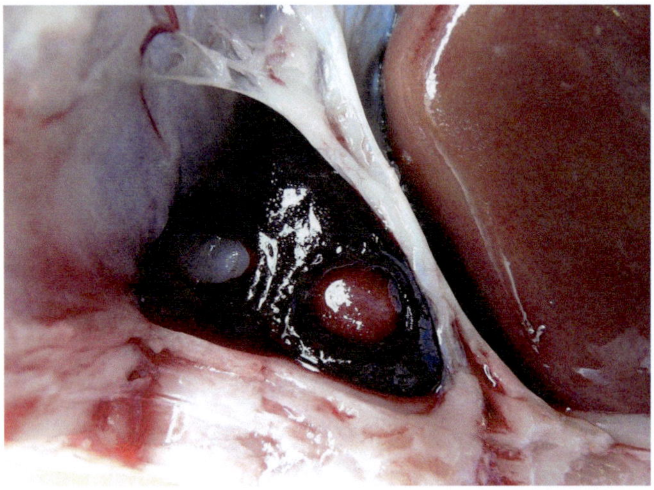

Fig. 5.40 Cardiac tamponade in farmed Atlantic salmon with cardiomyopathy syndrome. Parts of ventricle and bulbus are visible through the blood clot

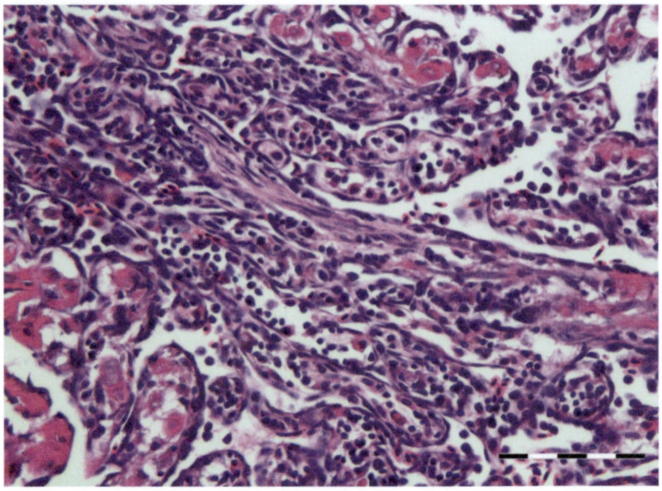

Fig. 5.41 Necrosis of spongy myocardium, mononuclear cell infiltration and marked endocarditis in farmed Atlantic salmon with cardiomyopathy syndrome. Bar = 50 μm

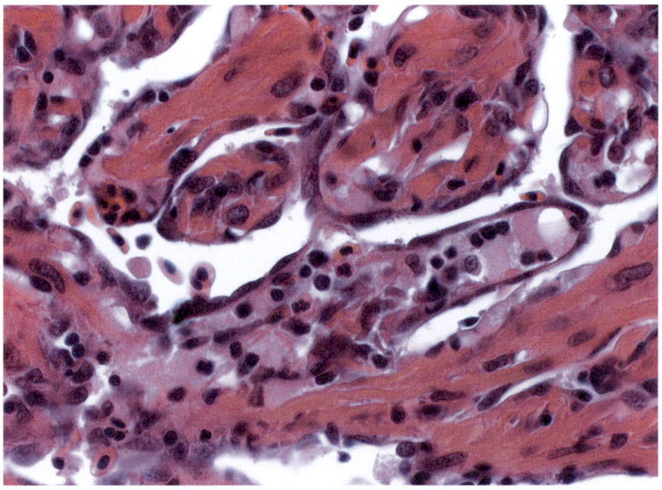

Fig. 5.42 Severe myocardial degeneration and inflammation in spongy ventricle of farmed Atlantic salmon with cardiomyopathy syndrome. High power

extending from the outer compact layer into the spongy layer with slight loss of striation. Lesions progress from focal to multifocal or diffuse degeneration of the trabecular myocardium and increased number of mononuclear inflammatory cells, lymphocytes and plasma cells, around multiple intramural coronary vessels and, intermittently, throughout the epicardium (Figs. 5.41 and 5.42). These cells frequently invade the sub-endocardial spaces (Fig. 5.43). Associated lesions may include mural thrombi (Fig. 5.44), hypertrophy of myocardial nuclei, which are believed to represent a compensatory reaction in a failing heart, and 'nest-like' aggregates of nuclei. The compact ventricular myocardium is usually unaffected, but a highly cellular epicarditis is

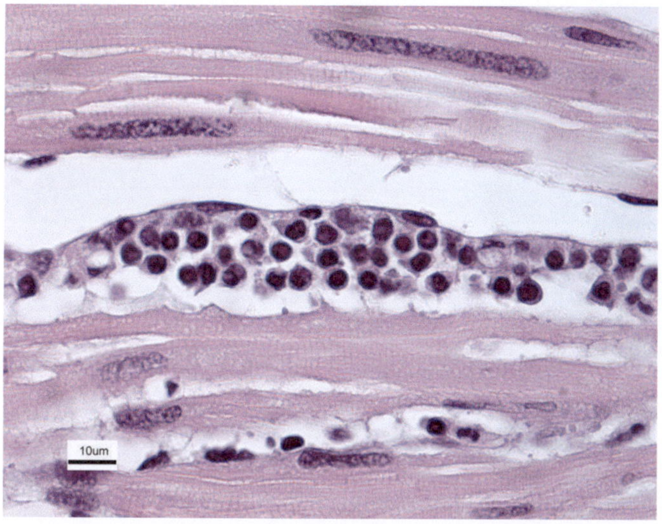

Fig. 5.43 Longitudinal section showing early subendocardial infiltration of mononuclear cells in the spongy myocardium of farmed Atlantic salmon with cardiomyopathy syndrome

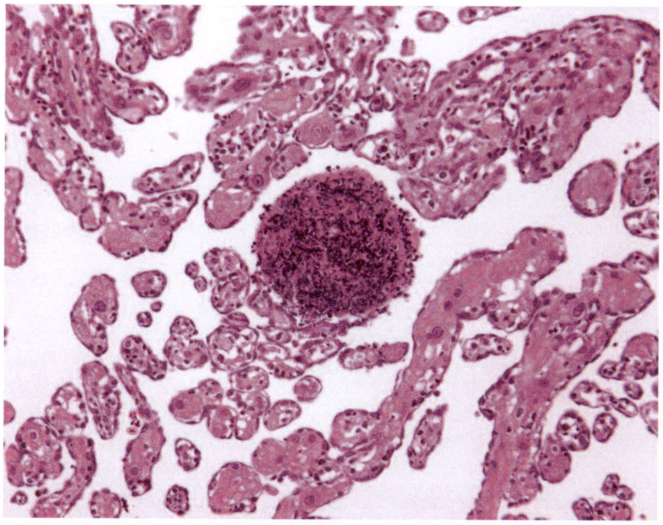

Fig. 5.44 Thrombus in spongy myocardium in farmed Atlantic salmon with cardiomyopathy syndrome. Medium power

common (Fig. 5.45). Nuclear enlargement is seen in some fish which resemble Anitschkow myocytes which are believed to be linked with myocardial repair. Within select sections of liver, a mild, circumferential, subintimal fibrosis of the central vein can be observed. HSMI and salmonid pancreas disease (PD) are important differential diagnosis and the potential overlap of more than one viral agent may obscure the histopathological evaluation.

PMV belongs to the Totiviridae family and the virus appears to be widespread in farmed fish. The recent identification of the causative virus will help in developing control strategies.

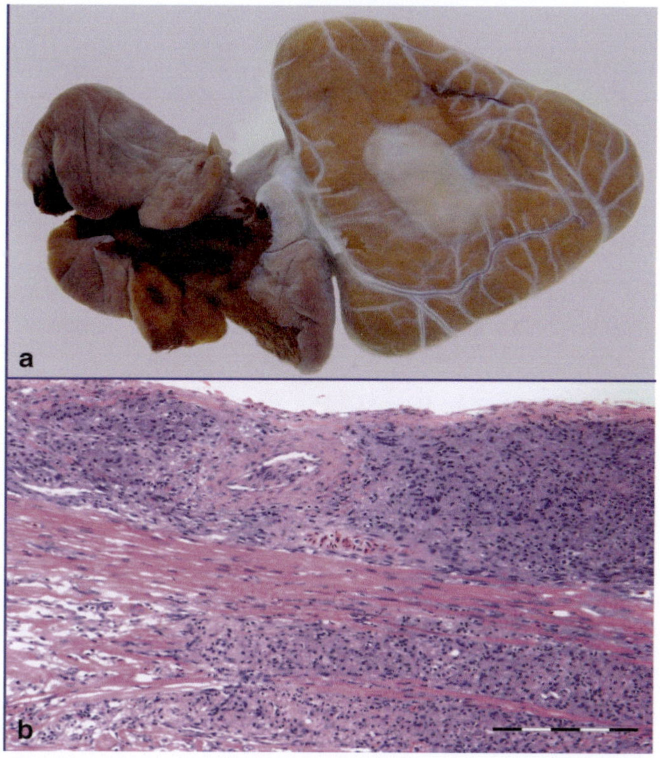

Fig. 5.45 (**a**) Focal fibrinous epicarditis on the caudal face of the ventricle of farmed Atlantic salmon broodstock. (**b**) Proliferative epicarditis in farmed Atlantic salmon. Bar =200 μm

5.11 Erythrocytic Inclusion Body Syndrome

Erythrocytic inclusion body syndrome (EIBS) is one of a number of intraerythrocytic inclusion viruses which has been reported from Atlantic, chum, Chinook, coho salmon and rainbow trout, both from wild and farmed origin. Most incidences are from sea water stock but also reported from freshwater fish.

Affected fish may be lethargic with pale gills and pigmentation abnormalities. Internally, fish may show splenomegaly and evidence of a progressive anaemia, but in many reports there is no specific correlation between infection and clinical signs. Laboratory studies show a decreased haematocrit. Each cycle of infection lasts around 45 days, with higher water temperatures reported as the factor that initiates infection, after which surviving fish appear to be resistant to re-infection. Losses are generally low but fish may become susceptible to secondary bacterial and oomycete infections, with occurrence of indirect mortalities. Histological assessment shows haemosiderin accumulation within the splenic ellipsoidal sheaths and less frequently, the kidney, although areas of haematopoietic necrosis are reported. Similarly, there is a limited hepatocyte necrosis.

EIBS is recognised by the finding of round to ovoid, randomly scattered, single or multiple basophilic bluish

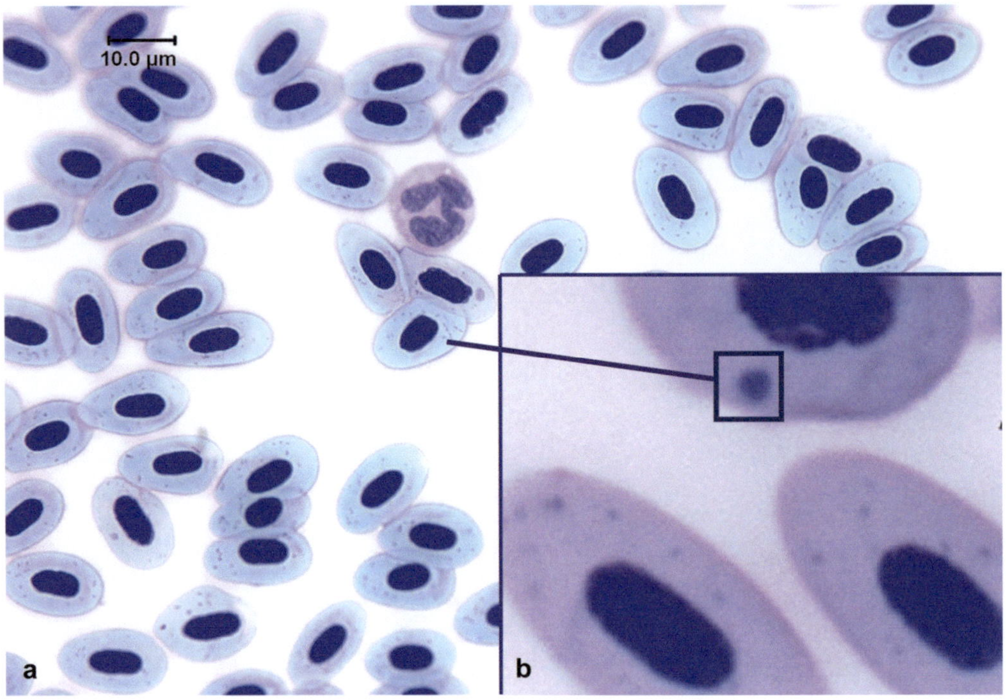

Fig. 5.46 (**a**) Blood smear from Chinook salmon with erythrocytic inclusion body syndrome. Basophilic inclusions in erythrocytes. Giemsa stain. (**b**) High power insert

cytoplasmic inclusions (0.8–3 μm) in Giemsa stained blood films (Fig. 5.46). At electron microscopy, spherical virion morphology with icosahedral core particles of 60–80 nm diameter can be seen. The viral genome is a single-stranded RNA with characteristics similar to a member of the family Togaviridae. There are no successful reports of virus isolation although a cohabitation model has been successful.

Further Reading

Aamelfot M, Dale OB, Weil SC, Koppang E, Falk K (2012) Expression of the infectious salmon anemia virus receptor on Atlantic salmon endothelial cells correlates with the cell tropism of the virus. J Virol 86:10571–10578

Ahmadi N, Oryan A, Aklaghi M, Hosseini A (2013) Tissue distribution of infectious pancreatic necrosis virus serotype Sp in naturally infected cultured rainbow trout, *Oncorhynchus mykiss* (Walbaum): an immunohistochemical and nested-PCR study. J Fish Dis. doi:10.1111/jfd.12072

Al-Hussinee L, Lord S, Stevenson RMW, Casey RN, Groocock GH, Britt KL, Kohler KH, Wooster GA, Getchell RG, Bowser PR, Lumsden JS (2011) Immunohistochemistry and pathology of multiple Great Lakes fish from mortality events associated with viral haemorrhagic septicemia virus type IVb. Dis Aquat Org 93:117–127

Aldrin M, Storvik B, Frigessi A, Viljugrein H, Jansen PA (2010) A stochastic model for the assessment of the transmission pathways of heart and skeletal muscle inflammation, pancreas disease and infectious salmon anaemia in marine fish farms in Norway. Prev Vet Med 93:51–61

Boucher P, Baudin Laurencin F (1994) Sleeping disease (SD) of salmonids. Bull Eur Assoc Fish Pathol 14:179–180

Brudeseth BE, Castric J, Evensen Ø (2002) Studies on pathogenesis following single and double infection with viral haemorrhagic septicemia virus and infectious hematopoietic necrosis virus in rainbow trout (*Oncorhynchus mykiss*). Vet Rec 39:180–189

Brun E, Poppe T, Skrudland A, Jarp J (2003) Cardiomyopathy syndrome in farmed Atlantic salmon *Salmo salar*: occurrence and direct financial losses for Norwegian aquaculture. Dis Aquat Org 56:241–247

Bruno DW (2004) Changes in prevalence of clinical infectious pancreatic necrosis among farmed Scottish Atlantic salmon, *Salmo salar* L. Aquaculture 235:13–26

Bruno DW, Noguera PA (2009) Experimental transmission of cardiomyopathy syndrome (CMS) in Atlantic salmon *Salmo salar*. Dis Aquat Org 87:235–242

Christiansen DH, Østergaard PS, Snow M, Dale OB, Falk K (2011) A low-pathogenic variant of infectious salmon anemia virus (ISAV-HPR0) is highly prevalent and causes a non-clinical transient infection in farmed Atlantic salmon (*Salmo salar* L.) in the Faroe Islands. J Gen Virol 92:909–918

Craig S, Kent ML, Dawe SC (1993) Hepatic megalocytosis in wild and farmed chinook salmon *Oncorhynchus tschawytscha* in British Columbia, Canada. Dis Aquat Org 16:35–39

Crane M, Hyatt A (2011) Examples of emerging virus diseases in salmonid aquaculture. Viruses 3:2025–2046

Eaton WD, Folkins B, Bagshaw J, Traxler G, Kent M (1993) Isolation of a retrovirus from two fish cell lines developed from chinook salmon (*Oncorhynchus tshawytscha*) with plasmacytoid leukaemia. J Gen Virol 74:2299–2302

Evensen Ø, Thorud KE, Olsen YA (1991) A morphological study of the gross and light microscopic lesions of infectious salmon anaemia in Atlantic salmon (*Salmo salar*). Res Vet Sci 51:215–222

Ferguson HW, Poppe T, Speare DJ (1990) Cardiomyopathy in farmed Norwegian salmon. Dis Aquat Org 8:225–231

Ferguson HW, Kongtorp RT, Taksdal T, Graham D, Falk K (2005) An outbreak of disease resembling heart and skeletal muscle inflammation in Scottish farmed salmon, *Salmo salar* L., with observations on myocardial regeneration. J Fish Dis 28:119–123

Finstad OW, Falk K, Løvoll M, Evensen Ø, Rimstad E (2012) Immunohistochemical detection of piscine reovirus (PRV) in hearts of Atlantic salmon coincide with the course of heart and skeletal muscle inflammation (HSMI). Vet Res 43:27–37

Fritsvold C, Kongtorp RT, Taksdal T, Ørpetveit I, Heum M, Poppe TT (2009) Experimental transmission of cardiomyopathy syndrome (CMS) in Atlantic salmon *Salmo salar*. Dis Aquat Org 87:225–234

Garseth ÅH, Fritsvold C, Opheim M, Skjerve E, Biering E (2012) Piscine reovirus (PRV) in wild Atlantic salmon, *Salmo salar* L and seatrout *Salmo trutta* L. in Norway. J Fish Dis. doi:10.1111/j.1365-2761.2012.01450.x

Godoy MG, Aedo A, Kibenge MJT, Groman DB, Yason CV, Grothusen H, Lisperguer A, Calbucura M, Avendaño F, Imilán M, Jarpa M, Kibenge FSB (2008) First detection, isolation and molecular characterization of infectious salmon anaemia virus associated with clinical disease in farmed Atlantic salmon (*Salmo salar*) in Chile. BMC Vet Res 4:28. doi:10.1186/1746-6148-4-28

Graham DA, Frost P, McLaughlin K, Rowley HM, Gabestad I, Gordon A, McLoughlin MF (2011) A comparative study of marine salmonid alphavirus subtypes 1–6 using an experimental cohabitation challenge model. J Fish Dis 34:273–286

Graham DA, Brown A, Savage P, Frost P (2012a) Detection of salmon pancreas disease virus in the faeces and mucus of Atlantic salmon, *Salmo salar* L., by real-time RT-PCR and cell culture following experimental challenge. J Fish Dis 35:949–951

Graham DA, Fringuelli E, Rowley HM, Cockerill D, Cox DI, Turnbull T, Rodger D, Morris D, McLoughlin MF (2012b) Geographical distribution of salmonid alphavirus subtypes in marine farmed Atlantic salmon, *Salmo salar* L., in Scotland and Ireland. J Fish Dis 35:755–765

Grammes F, Rørvik K-A, Takle H (2012) Tetradecylthioacetic acid modulates cardiac transcription in Atlantic salmon, *Salmo salar* L., suffering heart and skeletal muscle inflammation. J Fish Dis 35:109–117

Grove G, Austbø L, Hodneland K, Frost P, Løvoll M, McLoughlin M, Thim HL, Braaen S, König M, Syed M, Jørgensen JB, Rimstad E (2013) Immune parameters correlating with reduced susceptibility to pancreas disease in experimentally challenged Atlantic salmon (*Salmo salar*). Fish Shell Immunol 34:789–798

Haugland Ø, Mikalsen AB, Nilsen P, Lindmo K, Thu BJ, Eliassen TM, Roos N, Rode M, Evensen Ø (2011) Cardiomyopathy syndrome of Atlantic salmon (*Salmo salar* L.) is caused by a double-stranded RNA virus of the Totiviridae family. J Virol 85:5275–5286

Hjortaas MJ, Skjelstad HR, Taksdal T, Olsen AB, Johansen R, Bang-Jensen B, Ørpetveit I, Sindre H (2012) The first detections of subtype 2–related salmonid alphavirus (SAV2) in Atlantic salmon, *Salmo salar* L., in Norway. J Fish Dis 36:71–74

Hodneland K, Bratland A, Christie KE, Endressen C, Nylund A (2005) New subtype of salmonid alphavirus (SAV), Togaviridae, from Atlantic salmon, *Salmo salar* and rainbow trout *Oncorhynchus mykiss* in Norway. Dis Aquatic Org 66:113–120

Jensen BB, Kristofferesen AB, Myr C, Brun E (2012) Cohort study of effect of vaccination on pancreas diseases in Norwegian salmon aquaculture. Dis Aquat Org 102:23–31

Kent ML, Groff JM, Traxler GS, Zinkl JG, Bagshaw JW (1990) Plasmacytoid leukemia in salt water reared chinook salmon *Oncorhynchus tshawytscha*. Dis Aquat Org 8:199–209

Kim R, Faisal M (2011) Emergence and resurgence of the viral hemorrhagic septicemia virus (*Novirhabdovirus*, *Rhabdoviridae*, *Mononegavirales*). J Adv Res 2:9–23

Kongtorp RT, Taksdal T (2009) Studies with experimental transmission of heart and skeletal muscle inflammation in Atlantic salmon, *Salmo salar* L. J Fish Dis 32:253–262

Kongtorp RT, Kjerstad A, Taksdal T, Guttvik A, Falk K (2004a) Heart and skeletal muscle inflammation in Atlantic salmon, *Salmo salar* L.: a new infectious disease. J Fish Dis 27:351–358

Kongtorp RT, Taksdal T, Lyngøy A (2004b) Pathology of heart and skeletal muscle inflammation (HSMI) in farmed Atlantic salmon *Salmo salar*. Dis Aquat Org 59:217–244

Kongtorp RT, Halse M, Taksdal T, Falk K (2006) Longitudinal study of a natural outbreak of heart and skeletal muscle inflammation in Atlantic salmon, *Salmo salar* L. J Fish Dis 29:1–12

Koren CWR, Nylund A (1997) Morphology and morphogenesis of infectious salmon anaemia virus replicating in the endothelium of Atlantic salmon *Salmo salar*. Dis Aquat Org 29:99–109

Larsson T, Krasnov A, Lerfall J, Taksdal T, Pedersen M, Mørkøre T (2012) Fillet quality and gene transcriptome profiling of heart tissue of Atlantic salmon with pancreas disease (PD). Aquaculture 330–333:82–91

Leek SL (1987) Viral erythrocytic inclusion body syndrome (EIBS) occurring in juvenile spring chinook salmon (*Oncorhynchus tshawytscha*) reared in freshwater. Can J Fish Aquat Sci 44:685–688

Lester K, Black J, Bruno DW (2011) Prevalence and phylogenetic analysis of salmonid alphavirus in Scottish fish farms from 2000–2009. Bull Eur Assoc Fish Pathol 31:199–202

Løvoll M, Wiik-Nielsen J, Grove S, Wiik-Nielsen CR, Kristoffersen AB, Faller R, Poppe T, Jung J, Pedamallu CS, Nederbragt AJ, Meyerson M, Rimstad E, Tengs T (2010) A novel totivirus and piscine reovirus (PRV) in Atlantic salmon (*Salmo salar*) with cardiomyopathy syndrome (CMS). Virol J 7:309–315

Lunder T, Thorud K, Poppe TT, Holt RA, Rohovec JS (1990) Particles similar to the virus of erythrocytic inclusion body syndrome, EIBS, detected in Atlantic salmon (*Salmo salar*) in Norway. Bull Eur Assoc Fish Pathol 10:21–23

McKnight IJ, Roberts RJ (1976) The pathology of infectious pancreatic necrosis. 1. The sequential histopathology of the naturally occurring condition. Br Vet J 132:76–86

McLoughlin M, Graham AA (2007) Alphavirus infections in salmonids – a review. J Fish Dis 9:511–531

McVicar AH (1987) Pancreas disease of farmed Atlantic salmon, *Salmo salar*, in Scotland: epidemiology and early pathology. Aquaculture 67:71–78

Mikalsen AB, Haugland O, Rode M, Solbakk IT, Evensen O (2012) Atlantic salmon reovirus infection causes a CD8 T Cell myocarditis in Atlantic salmon (*Salmo salar* L.). PLoS One 7(6):e37269

Mladineo I, Zrnčić S, Loijkić I, Oraić D (2011) Molecular identification of a new strain of infectious pancreatic necrosis virus (IPNV) in a Croatian rainbow trout (*Oncorhynchus mykiss*) farm. J Appl Ichthyol 27:1165–1168

Mulcahy D, Klaybor D, Batts WN (1990) Isolation of infectious hematopoietic necrosis virus from a leech (*Piscicola salmositica*) and a copepod (*Salmincola sp.*), ectoparasites of sockeye salmon *Oncorhynchus nerka*. Dis Aquat Org 8:29–34

Munro ALS, Ellis AE, McVicar AH, McLay HA, Needham EA (1984) An exocrine pancreas disease of farmed Atlantic salmon in Scotland. Helgolander Meeresuntersuch 37:571–586

Murray AG, Busby CD, Bruno DW (2003) Infectious pancreatic necrosis virus in Scottish Atlantic salmon farms 1996–2001. Emerg Infect Dis 9:455–460

Newbound GC, Kent ML (1991) Experimental interspecies transmission of plasmacytoid leukemia in salmonid fishes. Dis Aquat Org 10:159–166

Noguera PA, Bruno DW (2010) Liver involvement in post smolt Atlantic salmon, *Salmo salar* infected with infectious pancreatic necrosis virus (IPNV) a retrospective histopathological study. J Fish Dis 33:819–832

Palacios G, Lovoll M, Tengs T, Hornig M, Hutchison S, Hui J, Kongtorp R-T, Savji N, Bussetti AV, Solovyov A, Kristoffersen AB, Celone C, Street C, Trifonov V, Hirschberg DL, Rabadan RR, Egholm M, Rimstad E, Lipkin WI (2010) Heart and skeletal muscle inflammation of farmed salmon is associated with infection with a novel reovirus. PLosOne 5:1–7. doi:10.1371/journal.pone.0011487

Poppe TT, Seierstad SL (2003) First description of cardiomyopathy syndrome (CMS)-related lesions in wild Atlantic salmon *Salmo salar* in Norway. Dis Aquat Org 56:87–88

Poppe T, Rimstad E, Hyllseth B (1989) Pancreas disease in Atlantic salmon (*Salmo salar*) postsmolts infected with infectious pancreatic necrosis virus (IPNV). Bull Eur Assoc Fish Pathol 9:83–85

Rimstad E (2011) Examples of emerging virus diseases in salmonid aquaculture. Aqua Res 42(suppl s1):86–89

Robertsen B (2011) Examples of emerging virus diseases in salmonid aquaculture. Aqua Res 42(suppl S1):125–131

Rodger HD (2007) Erythrocytic inclusion body syndrome virus in wild Atlantic salmon, *Salmo salar* L. J Fish Dis 30:411–418

Ronza P, Bermúdez R, Losada AP, Robles A, Quiroga MI (2011) Mucosal CD3ε + cell proliferation and gut epithelial apoptosis: implications in rainbow trout gastroenteritis (RTGE). J Fish Dis 34:433–443

Schönherz AA, Hansen MHH, Jørgensen HBH, Berg P, Lorenzen N, Einer-Jensen K (2012) Oral transmission as a route of infection for viral haemorrhagic septicaemia virus in rainbow trout, *Oncorhynchus mykiss* (Walbaum). J Fish Dis 35:395–406

Smail DA, McFarlane L, Bruno DW, McVicar AH (1995) The pathology of an IPN-Sp sub-type (Sh) in farmed Atlantic salmon, *Salmo salar* in the Shetland Isles, Scotland. J Fish Dis 18:631–638

Smail DA, Bain N, Bruno DW, King JA, Thompson F, Pendrey DJ, Morrice S, Cunningham CO (2006) Infectious pancreatic necrosis virus (IPNV) in Atlantic salmon, *Salmo salar* L. post-smolts in the in Shetland Isles, Scotland: virus identification, histopathology, immunohistochemistry and genetic comparison with Scottish mainland isolates. J Fish Dis 29:31–41

Snow M, Raynard RS, Bruno DW (2001) Comparative susceptibility of Arctic char (*Salvelinus alpinus*), rainbow trout (*Oncorhyncus mykiss*) and brown trout (*Salmo trutta*) to the Scottish isolate of infectious salmon anaemia virus (ISAV). Aquaculture 196:47–54

Snow M, Black J, Matejusova I, McIntosh R, Baretto E, Wallace IS, Bruno DW (2010) Evidence for the detection of salmonid alphavirus (SAV) RNA in wild marine fish caught in areas remote from aquaculture activity: implications for the origins of salmon pancreas disease (SPD) in aquaculture. Dis Aquat Org 91:177–188

Tengs T, Böckerman I (2012) A strain of piscine myocarditis virus infecting Atlantic argentine, *Argentina silus* (Ascanius). J Fish Dis 35:545–547

Watanabe K, Karlsen M, Devold M, Isdal E, Litlabø A, Nylund A (2006) Virus-like particles associated with heart and skeletal muscle inflammation (HSMI). Dis Aquat Org 70:183–192

Weli SC, Aamelfot M, Dale OB, Koppang EO, Falk K (2013) Infectious salmon anaemia virus infection of Atlantic salmon gill epithelial cells. Virol J 10:5. doi:10.1186/1743-422X-10-5

Wiik-Nielsen CR, Løvoll M, Fritsvold C, Kristoffersen AB, Haugland Ø, Hordvik I, Aamelfot M, Jirillo E, Koppang EO, Grove S (2012a) Characterization of myocardial lesions associated with cardiomyopathy syndrome in Atlantic salmon, *Salmo salar* L., using laser capture microdissection. J Fish Dis 35:907–916

Wiik-Nielsen CR, Løvoll M, Sandlund N, Faller R, Wiik-Nielsen J, Jensen BB (2012b) First detection of piscine reovirus (PRV) in marine fish species. Dis Aquat Org 97:255–258

Yousaf MN, Koppang EO, Skjødt K, Hordvik I, Zou J, Secombes C, Powell MD (2012) Cardiac pathological changes of Atlantic salmon (*Salmo salar* L.) affected with heart and skeletal muscle inflammation (HSMI), cardiomyopathy syndrome (CMS) and pancreas disease (PD). Vet Immunol Immunopathol 151:49–62

Bacterial Diseases

6

Abstract

Bacterial infections comprise true obligate and opportunistic facultative pathogens. The distinction between 'primary', 'facultative' or 'opportunistic' cannot be taken too strictly as the virulence of the agent will determine its ability to overcome the host defences. Bacterial diseases are largely represented by Gram negative organisms and affect salmonids in fresh, sea water or both. A few important diseases of salmon and trout, both in fresh and sea water are caused by Gram positive bacteria. This chapter covers a wide range of bacterial diseases representing acute, chronic, systemic or localised infections affecting salmonids in fresh and sea water.

Keywords

Bacteria • Infection • Salmon • Trout

Bacteria exist everywhere and the majority are capable of independent existence for varying periods without a 'host'. Specific bacteria cause disease, either because they are 'designed' to invade a host (true obligate pathogens) or simply because they are in the wrong place at the right time (opportunistic facultative pathogens). The distinction between 'primary', 'facultative' or 'opportunistic' cannot be taken too strictly, as the virulence of the agent will determine its ability to overcome the host defences. These vary between fish stock and circumstances including stress, water quality or other co-existing infections. When a bacterium is present in a host and associated with a disease condition, it will normally be deemed a 'primary pathogen', while those agents that can be found without compromising the fish health but are capable of inducing disease under certain conditions, will be addressed as 'facultative pathogens'. When bacteria survive and multiply in the host tissues without causing clinical disease, fish harbouring such bacteria are known as 'asymptomatic carriers'. The pathology and outcome of a bacterial infection may vary and dependant on factors linked to the bacterium, the host and/or the environmental conditions.

Bacteria gain entry to the fish through the gills, gut or via the skin, and then usually spread throughout the body. As infections become systemic they can induce acute changes that externally result in exophthalmia, hyperaemia and petechial haemorrhage, while internally, ascites, congestion and haemorrhage can be observed. However, some bacteria cause chronic infections with tissue proliferation and repairing processes, resulting in typical granulomatous responses.

Bacterial diseases are mainly represented by Gram negative organisms such as *Aeromonas salmonicida*, *Listonella anguillarum* and *Yersinia ruckeri*, affecting salmonids in fresh, sea water or both. Some important diseases however are caused by Gram positive bacteria, e.g. *Renibacterium salmoninarum* in salmon and trout, both in fresh and sea water. The Chlamydiales and Rickettsiales contain the genera *Chlamydia* and *Piscirickettsia* respectively, and are obligate intracellular pathogens which multiply within membrane-bound cytoplasmic vacuoles. Most of these infections occur in marine or anadromous hosts, but they have also been reported in freshwater fish. Measures to limit or control outbreaks in farmed fish include the use of vaccines and antimicrobial agents respectively.

Taxonomy has progressed dramatically in the last decade moving from culture-dependent techniques involving phenotype characterization and physiological data, to the application of molecular techniques, 16S rRNA and gene

D.W. Bruno et al., *A Colour Atlas of Salmonid Diseases*,
DOI 10.1007/978-94-007-2010-7_6,

Table 6.1 Principal and emerging bacterial and chlamydial diseases of salmonids

Gram negative bacteria	Disease	'Principal salmonid host	Environment
Aeromonas hydrophila		Rainbow trout	FW
Aeromonas salmonicida subsp. *salmonicida*	Furunculosis	Most salmonids	FW
Listonella anguillarum	Vibriosis	Salmonids	SW
Aliivibrio salmonicida, A. wodanis, A. logei	Cold-water vibriosis	Atlantic salmon	SW
Moritella viscosa	Winter ulcer disease	Atlantic salmon	SW
Yersinia ruckeri	Enteric redmouth	Rainbow trout	FW
Pseudomonas fluorescens		Rainbow trout	FW
Flavobacterium psychrophilum	Rainbow trout fry syndrome, peduncle disease	Rainbow trout	FW
Flavobacterium columnare, hydatis/johnsoniae	Columnaris disease	Rainbow trout	FW
Tenacibaculum maritimum	Black patch necrosis, marine flexibacteriosis	Atlantic salmon	SW
Hafnia alvei		Rainbow trout	FW
Chryseobacterium spp.		Rainbow trout	FW
Pasteurella skyensis		Atlantic salmon	SW
Francisella noatunensis supsp. *noatunensis*		Atlantic salmon	SW
Gram positive bacteria			
Renibacterium salmoninarum	Bacterial kidney disease	Rainbow trout/salmon	FW, SW
Carnobacterium maltaromaticum (synonym *C. piscicola*)	Pseudokidney disease	Rainbow trout/chinook salmon	FW
Streptococcus phocae		Atlantic salmon	SW
Acid fast bacteria			
Mycobacterium chelonae, M. fortuitum, M. marinum	Mycobacteriosis	Salmon	SW
Nocardia sp.	Nocardiosis		
Chlamydiaceae			
Piscirickettsia salmonis	Salmonid rickettsial septicaemia	Salmon	SW
Candidatus Piscichlamydia salmonis, *Candidatus* Clavochlamydia salmonicola	Epitheliocystis	Atlantic salmon, Arctic char, brown trout	FW, SW
Candidatus Branchiomonas cysticola	Epitheliocystis	Salmon	SW
Candidatus arthromitus	Rainbow trout gastroenteritis	Rainbow trout	FW

FW freshwater, SW saltwater

sequencing. While the traditional methods of culture remain relevant, new techniques are contributing to an accurate and rapid diagnosis of certain agents. Principal and emerging pathogens are listed in Table 6.1.

6.1 *Aeromonas hydrophila*

Aeromonas hydrophila is a widely distributed fresh water bacterium and the causative agent of a condition known as 'motile aeromonad septicaemia'. The bacteria affects wild and farmed fish species and aggravated by poor water quality principally when the water temperature is above 10 °C

A. hydrophila was significant for farmed fish during the 1970s until the emergence of successful vaccines during the 1980s. Nevertheless, it remains an important pathogen and clinical outbreaks are sporadically recorded. External signs include darkened skin, abdominal distension and exophthalmia, with erosion and apparent necrosis near the tail and other fins. Gills can be haemorrhagic or pale and swollen, with haemorrhage also seen in the vent and large areas of the skin, where oedema of superficial lesions may develop and ulcerate which then become prone to secondary infections e.g. *Saprolegnia*.

Internally, there is evidence of anaemia and the accumulation of clear or blood-tinged ascites. Splenomegaly and swollen kidney are common. Histopathological changes are characteristic of a generalised septicaemia with focal lesions involving several tissues such as gill, brain, heart, intestine, kidney and liver. A liquefactive necrosis and haemorrhage is detected in liver and kidney, with serous exudates in the intestine. The disease shows similar characteristics to those seen in a pseudomonad infection (see *Pseudomonas fluorescens*). Diseased fish transmit the infection horizontally through the water.

Aeromonas hydrophila is a motile, fermentative Gram negative rod which produces catalase and cytochrome oxidase. Material taken from the kidney and other organs onto a non-selective media is generally sufficient to obtain growth.

Diagnosis is based upon clinical disease signs, and the isolation and identification of the bacteria on medium such as trypticase soy agar (TSA) using morphological characteristics.

6.2 *Aeromonas salmonicida* subsp. *salmonicida*

Furunculosis is the name used to describe infections with *Aeromonas salmonicida* subsp. *salmonicida*. Infection occurs in wild and farmed salmonids, and has been recorded in Atlantic salmon, brown trout, Arctic char, brook trout and lake trout. Both chronic and acute furunculosis may occur depending on water temperature, fish age and virulence of the agent. Outbreaks are often triggered by stressors such as changes in temperature, poor handling, water quality and crowding.

In acute infections fish may die with few or no signs of disease or pathological changes. During the chronic stages the fish show lethargy, inappetence and darkening of the skin, which is similar to most bacterial septicaemias. Ventral haemorrhage is common, in particular near the base of the pectoral, pelvic and anal fins in addition to exophthalmia. At necropsy, ascites, splenomegaly, haemorrhagic enteritis, ascites, sub-capsular liver haemorrhage and pyloric caeca may be observed. Liquefactive, haemorrhagic 'boil' lesions that develop in the flank skeletal muscle are observed in chronic cases (Figs. 6.1, 6.2, 6.3 and 6.4). These 'furuncles' may rupture exposing open deep ulcers on the surface A large number of bacteria are released from these lesions and contribute to the spread of the infection. Although furuncles are characteristic, they are not always present in diseased fish and cannot be regarded as a diagnostic feature. A carrier state may be established after an infection.

Histopathological lesions are characterized by dense aggregates of bacteria in organs such as heart, kidney, spleen, muscle and gills (Figs. 6.5 and 6.6). In the latter, embolic spread with subsequent bacterial colonization may be seen in the gill capillaries. Mural thrombi with bacterial colonization are observed on the intimal aspect of vessels and ellipsoids. There is remarkably little tissue reaction around aggregates of bacteria in early stages of the disease, but tissue necrosis and liquefaction may become extensive in late and chronic stages.

Fig. 6.1 Furunculosis in farmed brook trout; mild ulcerated subcutaneous boil lesion

Fig. 6.2 Furunculosis in adult sockeye salmon; subcutaneous boil lesion in ventral muscle

Aeromonas salmonicida subspecies *salmonicida* is a Gram negative, non-motile, facultative anaerobic rod. Pathogenicity is dependent on the so-called A-layer, an external surface layer mainly composed of the A-protein that provides protection against the defence mechanisms of the host. At least 25 extracellular products are released during growth and responsible for the tissue damage-degeneration and ultimately, the death of the fish. Primary isolation of the pathogen can be achieved from the kidney and other organs on TSA or brain heart infusion agar (BHIA) at 22 °C. Most strains are oxidase positive and produce a water-soluble brown pigment on medium containing tryptone. A less common and atypical strain of this pathogen is the subspecies *achromogenes*, which does not produce pigment under standard incubation conditions and requires additional biochemical testing to differentiate. Diagnosis of *A. salmonicida* is based upon gross and histopathological lesions, isolation of the causative agent or its identification through immunohistochemistry, serology or molecular tools. In haematoxylin and eosin (H&E) sections the appearance of the bacteria is almost pathognomonic (see Fig. 6.5).

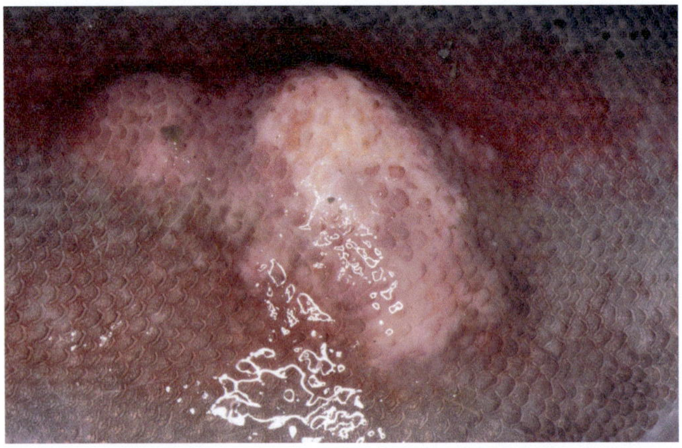

Fig. 6.3 Furunculosis in farmed rainbow trout broodstock; subcutaneous boil lesions

6.3 *Listonella anguillarum*

Listonella (*Vibrio*) *anguillarum* causes a haemorrhagic septicaemia, an economical important disease affecting salmonids in salt and brackish waters, particularly in the summer months at temperatures above 10 °C. However, infections with *Vibrio ordalii* and other *Vibrio* spp. are often collectively termed 'vibriosis', and may cause disease with similar clinical and pathological manifestation as *L. anguillarum*. Clinical signs and pathological lesions may be variable, but similar to several other Gram negative septicaemias, and dependant on water temperature, fish age and pathogen virulence. *L. anguillarum* may be present among the normal gut microflora of healthy fish and clinical outbreaks triggered only by stress activated virulent strains present in the gastrointestinal tract. Poor water quality and rapid temperature changes can also activate infection which then spreads horizontally. External signs of disease include dark skin coloration, anorexia, pale gills with increased mucous, periorbital oedema, swollen vent (Fig. 6.7) and haemorrhage near the base of the pectoral and pelvic fins. *V. ordalli* shows a predilection for muscle and skin with resulting haemorrhage. Extensive multifocal liquefactive muscle necrosis and haemorrhage with large numbers of bacteria is common (Fig. 6.8). Dermal or subdermal skin lesions are often coupled to hyperaemia and haemorrhage, and may be linked to occasional haemorrhagic 'boil' lesions in the muscle. These may rupture and release blood and bacteria to the surrounding water. Internally, petechiae may be present on the peritoneal surface and in internal organs. The liver is generally swollen and some fish show petechiae (Fig. 6.9). Colonies of the bacterium are also found throughout the

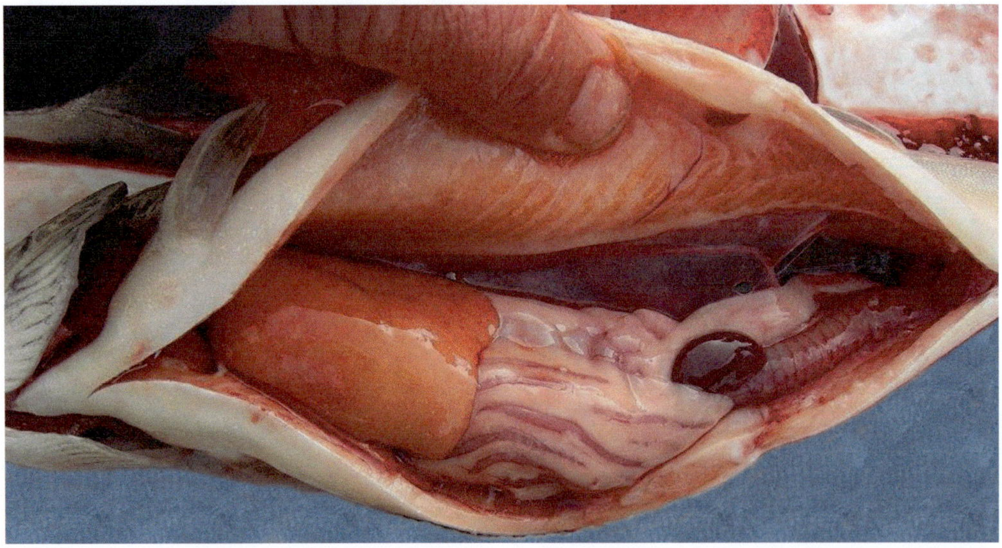

Fig. 6.4 Furunculosis in farmed Arctic char; ascites and swollen spleen

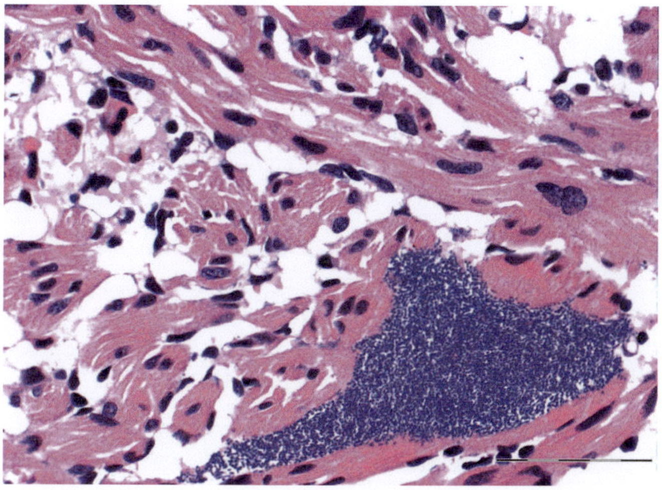

Fig. 6.5 Furunculosis in rainbow trout; microcolony in spongy myocardium. Medium power

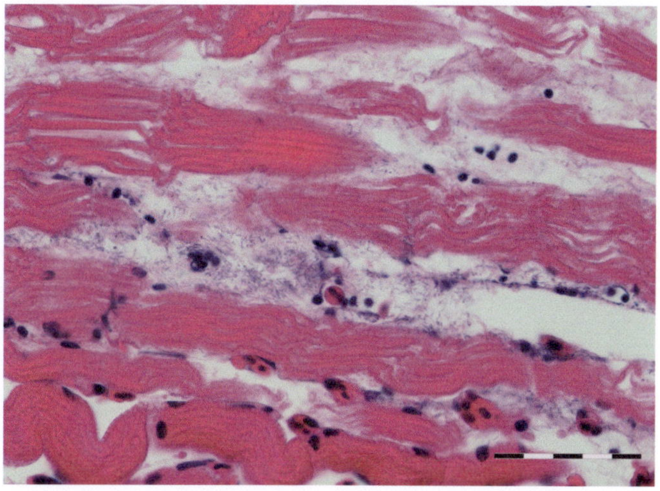

Fig. 6.8 *Listonella anguillarum* infection in muscle lesion in Atlantic salmon. Bar = 50 μm

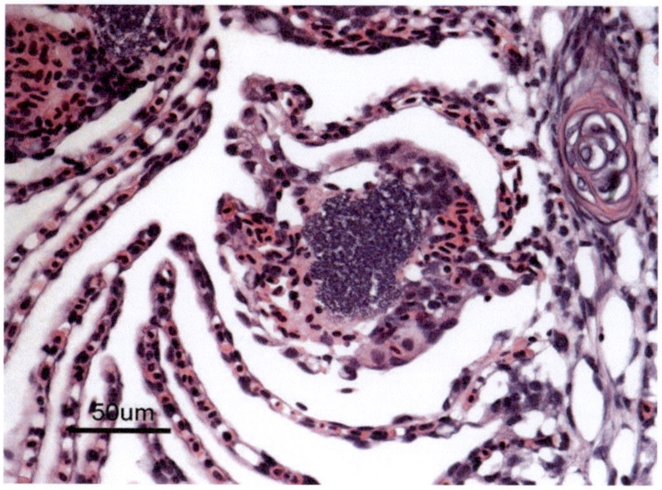

Fig. 6.6 Furunculosis in farmed Atlantic salmon, microcolony in lamella

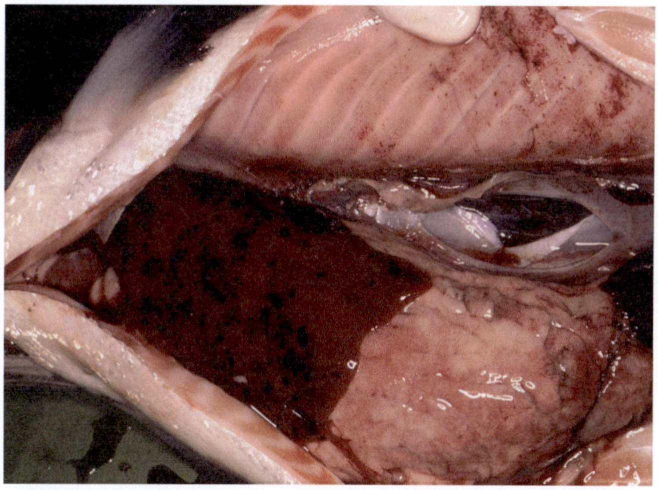

Fig. 6.9 Petecchia in liver and peritoneum of farmed Atlantic salmon resulting from *Listonella anguillarum* infection

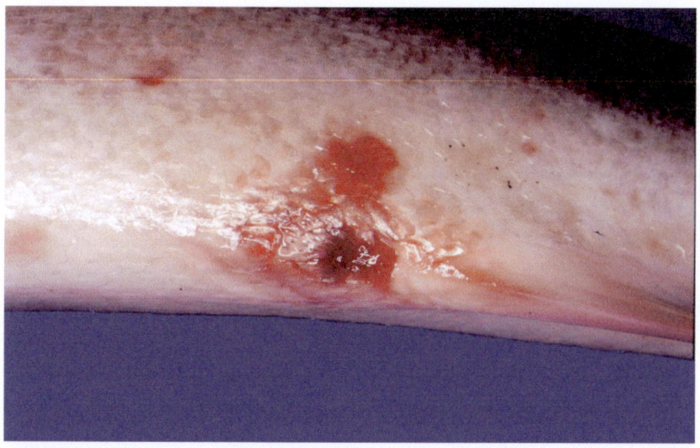

Fig. 6.7 Vent haemorrhage in Atlantic salmon with *Vibrio* sp. infection

digestive tract and in loose connective tissues such as in the gills. Bacteria can also be found at the back of the eye (Fig. 6.10), with opacity followed by corneal lesions, ulceration and avulsion of the orbital contents. Splenomegaly is common and even rupture of the organ may occur, particularly in rainbow trout. Histologically, the anterior part of the digestive tract may show vasodilatation and extensive necrosis of the mucosa and muscularis. Necrosis and oedema of haematopoietic tissue and spleen can also be found. A haemolytic anaemia occurs in chronic cases with resulting deposition of haemosiderin in the melanomacrophage centres of the splenic ellipsoids.

Listonella anguillarum is a halophilic with bipolar staining, Gram negative, slightly curved, flagellated motile rod. The bacterium grows well on standard medium at an optimum temperature of 22 °C and shows haemolytic

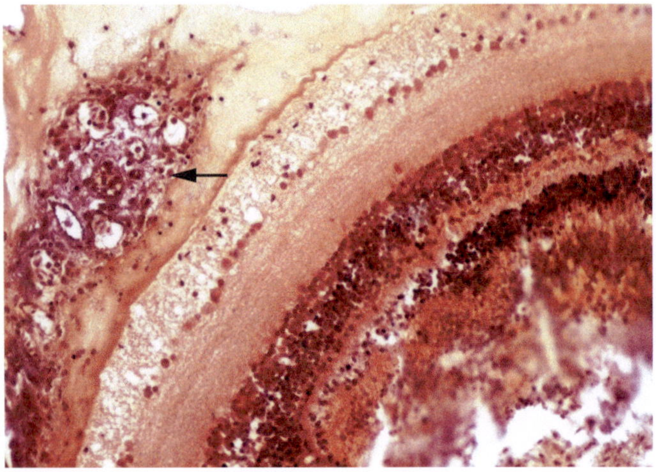

Fig. 6.10 *Listonella anguillarum* infection in rainbow trout. Retrobulbar proliferation of bacteria (*arrow*). Gram stain. Low power

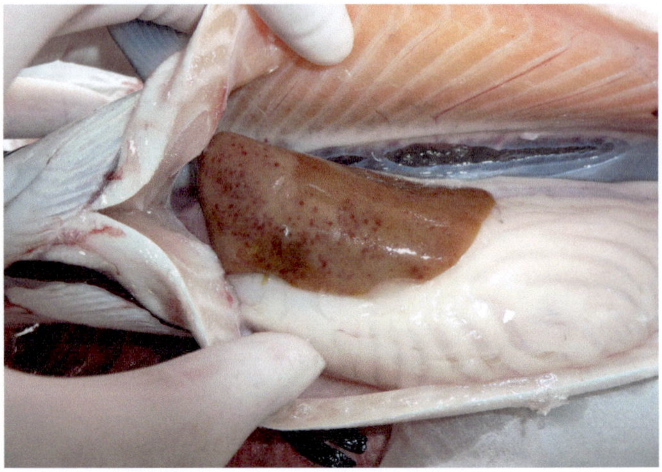

Fig. 6.11 *Aliivibrio wodanis* infection in farmed Atlantic salmon. Anaemic fish with pale liver and petecchia

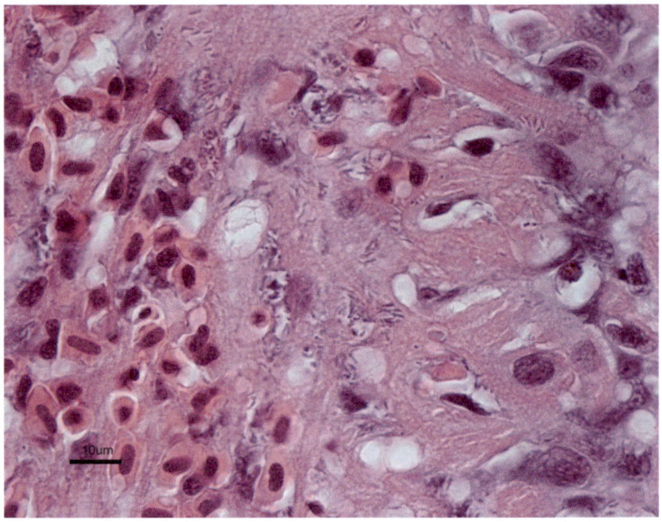

Fig. 6.12 Diffuse infiltration with *Aliivibrio salmonicida* in spongy myocardium of farmed Atlantic salmon with cold-water vibriosis. High power

activity on blood agar. Serovars O1 and O2 are the most common causing infection in salmonids, but more than 20 different serovars have been described.

Diagnosis is based upon classical pathology and isolation of *L. anguillarum* on NaCl-supplemented blood agar or TSA at room temperature and confirmation by ELISA. Serologically identification can be carried using a rapid agglutination test kit. Oil-adjuvant multivalent vaccines usually provide excellent protection against both O1 and O2 serovars of *L. anguillarum*.

6.4 *Aliivibrio* (*Vibrio*) *salmonicida*

Cold water vibriosis (Hitra disease or haemorrhagic syndrome) is a septicaemic condition of farmed Atlantic salmon caused by the motile bacterium *Aliivibrio* (*Vibrio*) *salmonicida*. The first observations were recorded in northern Norway in 1977 and the disease was a serious threat to the rapidly growing farming industry in the early 1980s. Outbreaks have also been recorded in Scotland, Faroe Islands, Iceland and the east coast of USA and Canada. Currently, oil-adjuvant multivalent vaccines give excellent protection against cold-water vibriosis and overall the impact of the disease has been greatly reduced.

Cold-water vibriosis typically occurs during the winter months. Affected fish go off the feed, are lethargic and dark coloured and often stay near the surface. As with some other diseases of farmed salmonids, apparently 'healthy fish' are often heavily affected. External lesions include exophthalmia, a swollen and haemorrhagic vent, petechial haemorrhage under the belly and at the base of the pectoral and pelvic fins. Necropsy may reveal a yellowish liver, sometimes with petechial or ecchymotic haemorrhage (Fig. 6.11) ascites, splenomegaly, haemorrhagic enteritis and a generalised oedema. Petechiae in the pyloric region are also common. Early histopathological lesions are characterized by large numbers of bacteria in blood vessels, followed by the heart, kidney, muscle and spleen, but with little tissue reaction (Fig. 6.12). Congestion with arteriole mural necrosis and thrombi are recorded, as well as kidney tubular necrosis and myolysis of skeletal muscle in the latter stages. Viral septicaemia may be an important differential diagnosis.

Diagnosis is based upon gross lesions, histopathology and the isolation of *Aliivibrio* (*Vibrio*) *salmonicida*. The bacterium is a psychrophilic, moderately halophilic Gram negative curved or straight rod, and is motile with up to 9 polar sheathed flagella. It can be grown at 15 °C on NaCl-supplemented blood agar where optimum salt concentration is 1.5 %. Growth shows small, greyish, non-haemolytic colonies on blood agar that are facultative anaerobic, oxidase positive and susceptible to the vibriostatic agent 0/129.

6.5 *Moritella viscosa*

Moritella viscosa is one of the causative agents of 'winter ulcers', a commonly occurring skin disease of farmed salmon and rainbow trout that has been diagnosed in Norway, Iceland, Faroe Islands and Scotland. *M. marina* has also been described. Infection results in increased mortality rates and major economic losses due to downgrade of the fish at slaughter, and typically occurs at a low but consistent prevalence in farmed Atlantic salmon during the coldest months of the year. The significance of the disease is associated to poor animal welfare, reduced osmoregulatory capacity, increased susceptibility to other infections and reduced marketability. Clinical observations show small raised skin lesions on the flank of the fish. These increase in size and gradually break the skin exposing the underlying muscle (Fig. 6.13). Lesions are characteristically rounded or oval with a white demarcation zone towards normal skin, which may heal with increasing temperature leaving 'scar' tissue, sometimes with melanisation of the area. The infection may become systemic and extensive petechial haemorrhage may develop in the ventral surface (Fig. 6.14) and internally in the peritoneum, adipose tissue, pyloric region and liver (Fig. 6.15).

Histopathological changes are variable depending on the period of ulcer development. Initial stages are characterized by oedema down to the compact layer of dermis and some inflammatory cell invasion. Ata later stage, lesions may reach the white muscle with inflammatory infiltrates between muscle bundles, haemorrhage and thrombosis of small vessels. Bacteria are typically found near the edges of the lesions. In the reparative phase, granulation tissue covers the ulcers starting from the edges and is gradually replaced by new epidermal and dermal layers, without scales. Diagnosis is

Fig. 6.13 Winter ulcers caused by *Moritella viscosa* in farmed Atlantic salmon

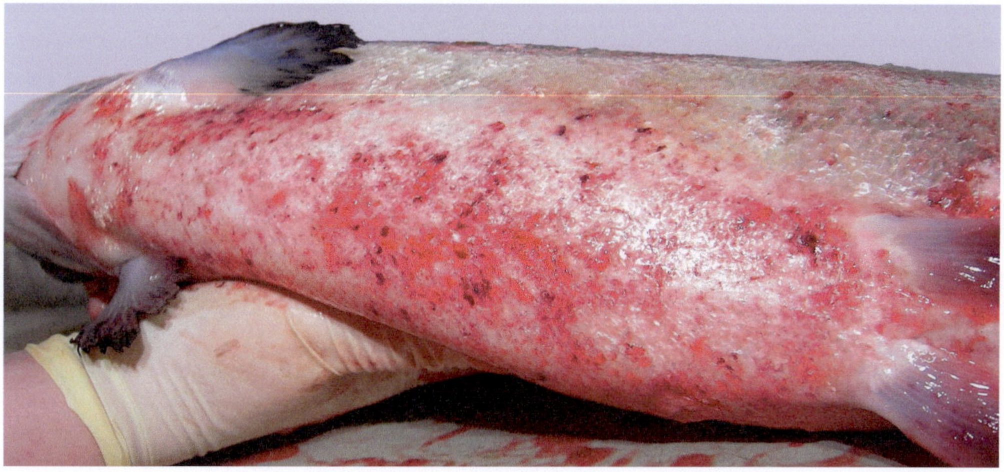

Fig. 6.14 Ventral haemorrhage in farmed Atlantic salmon with systemic *Moritella viscosa* infection

based upon clinical signs, and the isolation and identification of the bacteria. A differential diagnosis would include other *Vibrio* spp.

Moritella viscosa is a psychrophilic Gram negative, motile, flagellated curved rod. *M. viscosa* antigens are included in most of the multivalent injection vaccines used in the salmon industry, but the protection is apparently variable.

6.6 *Yersinia ruckeri*

Yersinia ruckeri is the causative agent of yersiniosis or enteric redmouth disease (ERM), an economically important condition in wild and farmed salmonids, both in fresh and sea water. *Y. ruckeri* has a wide host range and most salmonids are susceptible. Transmission occurs horizontally and many species of asymptomatic carriers as well as birds, being reservoirs of infection. Outbreaks are typically stress-mediated, e.g. poor water quality, increase in temperature, grading and handling. Clinical signs and pathology are similar to other Gram negative septicaemias and the disease may occur in peracute to acute or more chronic forms. In peracute cases, as in fry or fingerlings in freshwater, there may be high mortality with few or no external disease signs. In the more chronic cases, fish show pigment changes, disturbance of balance and lethargy. Other signs include ascites, exophthalmia, cutaneous petechiae and localized haemorrhage at the tip of the gill filaments. Hyperaemia of the oral cavity and jaw as a result of congestion of the submucosa, is not always evident, but has given the disease its popular name, enteric redmouth disease (Fig. 6.16). At necropsy, general congestion, intestinal haemorrhage, petechiae on serosa membranes, a swollen kidney and splenomegaly are common (Fig. 6.17). Histologically, haemorrhage, congestion, oedema and bacterial colonization of several tissues, including brain and gills is frequent. Necrosis associated with bacterial colonization is common in the kidney particularly the glomerulus, and the spleen (Fig. 6.18).

As gross and histopathological changes may be unspecific, the diagnosis requires confirmation by isolation of the causative agent on TSA or blood agar at 22 °C. *Y. ruckeri* is a Gram negative, motile rod-shaped bacterium. It is catalase positive and oxidase negative and several serotypes have been identified. On TSA *Y. ruckeri* colonies are rounded, translucent, glistening and buff-coloured colonies after 24 h incubation. Immunohistochemistry, fluorescent antibody and ELISA can be used for confirmation (Fig. 6.19). Both immersion and injection vaccines are available and provide good protection against clinical outbreaks.

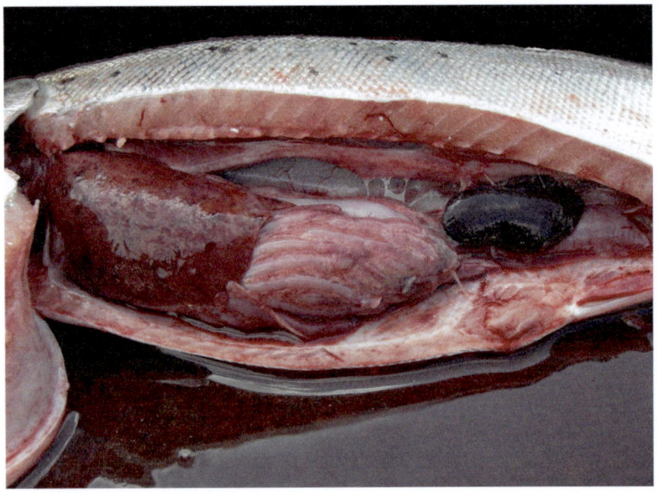

Fig. 6.15 Liver haemorrhage and swollen spleen in farmed Atlantic salmon with systemic *Moritella viscosa* infection

Fig. 6.16 Mandibular haemorrhage in rainbow trout fingerlings infected with *Yersinia ruckeri*

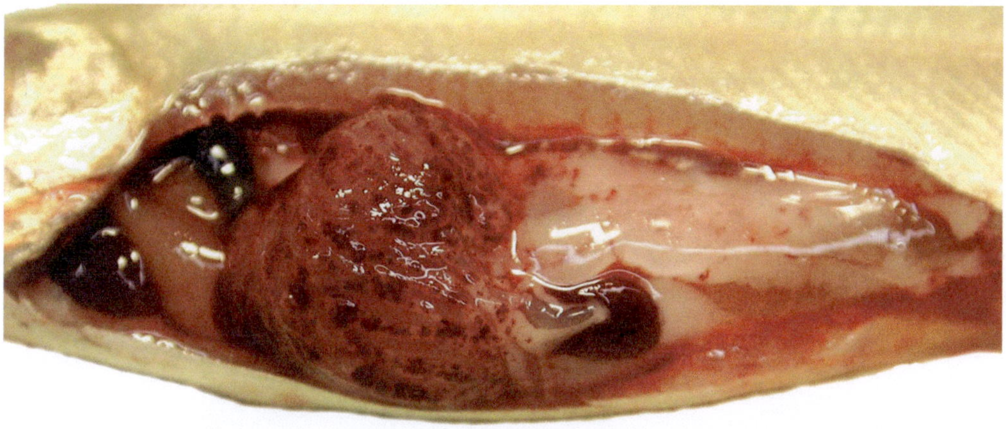

Fig. 6.17 Enteric redmouth in farmed rainbow trout; petecchia in peripancreatic tissues

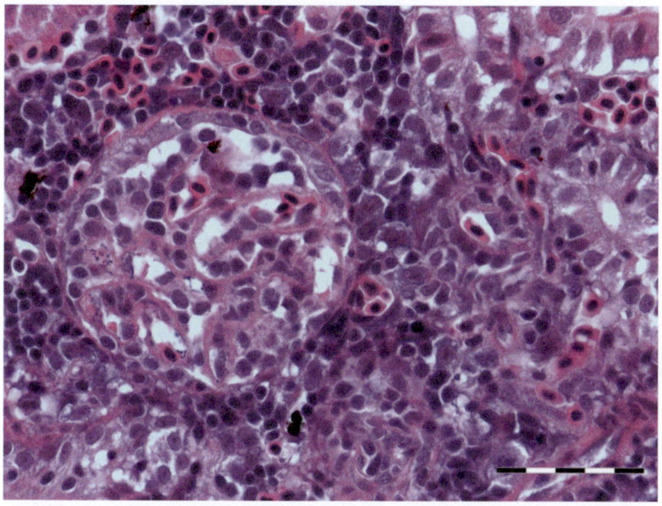

Fig. 6.18 Bacteria in glomerular vessels and thickened basal membrane in rainbow trout infected with *Yersinia ruckeri*. Bar = 50 μm

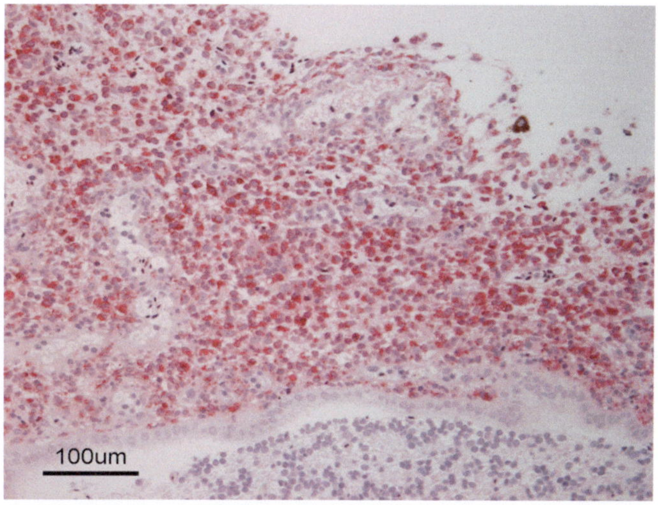

Fig. 6.19 Meningitis in farmed rainbow trout with enteric redmouth. Immunohistochemical stain

6.7 *Pseudomonas fluorescens*

The ubiquitous bacterium *Pseudomonas fluorescens* is generally considered as a non-pathogenic saprophyte, however, it can become an opportunistic pathogen causing disease in wild and farmed salmonids, particularly following stress and periods of poor water quality. Under farming conditions outbreaks can occur after vaccination or concurrent to other disease (e.g. infectious pancreatic necrosis). Smolts suffering from subclinical disease during smoltification and sea-water transfer may show clinical disease during the sea water phase. Affected fish may display many different manifestations of disease, from chronic, non-symptomatic, to acute haemorrhagic septicaemia, with high mortality (>15 %). External signs may include exophthalmia, frayed fins or fin rot, skin ulcerations with haemorrhagic edges, ventral petechiae and dark colouration (Fig. 6.20). At necropsy, changes are similar to other bacterial conditions, including ascites, petechiae on internal organs and pale liver, but a purulent epicarditis can also be observed (Fig. 6.21).

Histopathological lesions include septic thrombi with bacterial colonies in the gill lamellae, bacteria in spleen, kidney and heart plus epicarditis. Other *Pseudomonas* spp., including *P. anguilliseptica* can also cause disease in both wild and farmed salmonids with similar clinical and pathological manifestations.

Pseudomonas fluorescens is motile Gram negative rod-shaped bacterium found in soil, freshwater and on plant surfaces. It produces a water-soluble yellow-green pigment which fluoresces under UV-light. Diagnosis is based upon isolation and characterization of the causative bacterium on enrichment medium and immunohistochemistry.

Fig. 6.20 Farmed Atlantic salmon smolts with *Pseudomonas fluorescens* infection

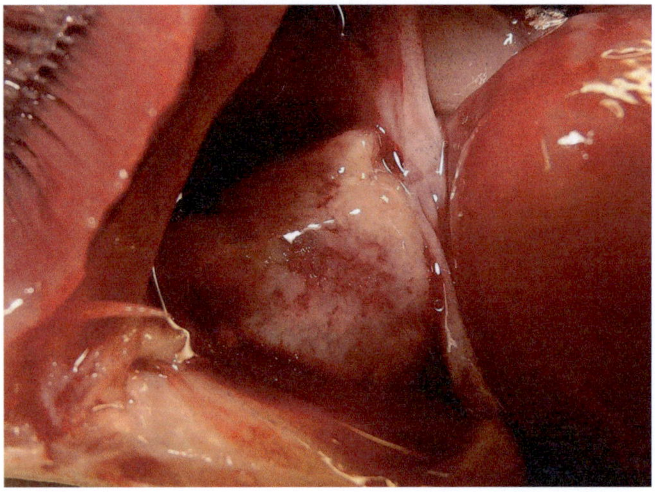

Fig. 6.21 Purulent epicarditis in farmed Atlantic salmon with *Pseudomonas fluorescens* infection

6.8 *Flavobacterium psychrophilum*

Flavobacterium psychrophilum is the causative agent of rainbow trout fry syndrome (RTFS) or bacterial cold water disease (BCW) is responsible for significant economic losses in salmonid aquaculture in fresh water at cooler temperatures. The following terms have also been historically used to describe the disease: fry mortality syndrome, peduncle disease and cold water disease. RTFS is mainly reported from juvenile fish and initially recognised by caudal or peduncle fin erosion with increased mucous (Fig. 6.22). Other clinical signs include increased lethargy and pigmentation, loss of balance, bilateral exophthalmia, abdominal swelling, pale gills, spinal deformities and a yellow skin discolouration (Fig. 6.23). Fish up to ~60 g may display subcutaneous lesions at one or more sites on the body surface (Fig. 6.24) and an underlying necrosis often involves the whole tail fin. Internal signs include ascites, intestinal inflammation, splenomegaly and liver discolouration.

Histologically a generalised necrosis occurs with filamentous bacteria interspersed throughout fins, gills and skin and in H&E sections *F. psychrophilum* shows weak staining (Fig. 6.25). Bacteria can also occur in the retina with subsequent inflammation associated with infiltration by polymorphic granulocytes. An initial infiltrative periportal response is also recorded in the liver, although recognised as a feature of other bacterial diseases. The spleen may show haemosiderosis, haemorrhage and necrosis with loss of definition of the splenic border and replacement by a loosely structured eosinophilic layer, fibrinous inflammation and intercellular oedema.

Infected fish shed bacteria into water where they survive for several months. Vertical transmission is reported and broodstock may serve as a reservoir for the bacterium. Similarly, dead fish release a large number of bacteria and infect other fish through tegument lesions. General septicaemia in trout fry is a differential diagnosis.

There are a number of diagnostic approaches and cultural methods overall, but are reliable but time-consuming. Diagnosis is therefore based upon clinical signs, and the isolation and identification of the bacteria. *Flavobacterium psychrophilum* is Gram negative, flexible, weakly retractile, slender bacteria and can be isolated on media such as Anacker and Ordal's medium. After 14 days at 15 °C, slightly raised, yellow-pigmented sticky colonies with thin spreading margins are recognized. A specific TaqMan polymerase

Fig. 6.22 Skin lesions and severe tail rot in farmed rainbow trout with *Flavobacterium psychrophilum* infection

Fig. 6.23 Dermal necrosis with yellow mats of bacteria in seawater farmed Atlantic salmon infected with *Flavobacterium psychrophilum*

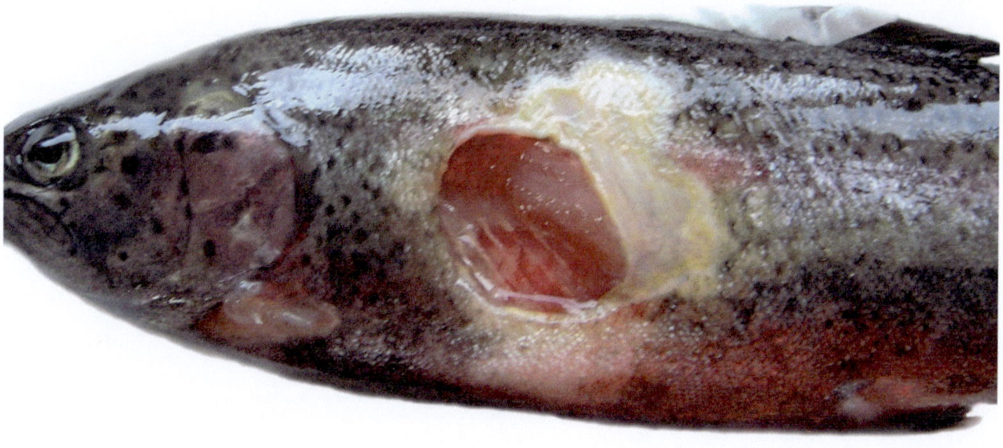

Fig. 6.24 Skin ulcer and yellow discolouration in rainbow trout infected with *Flavobacterium psychrophilum*

chain reaction (PCR) assay is available as a diagnostic tool and a fluorescent *in situ* hybridization (FISH) has the potential detect *F. psychrophilum* in infected tissues. *Flavobacterium psychrophilum* exists in at least three serotypes with recognised variation in their virulence.

6.9 *Flavobacterium columnare*

Flavobacterium columnare, formerly known as *Flexibacter columnaris*, is the causative agent of 'columnaris disease', a bacterial infection affecting freshwater fish including rainbow trout. Outbreaks are related to stress including high temperature, low levels of dissolved oxygen, increased ammonia and organic load. Infection is transmitted horizontally through direct contact with infected fish.

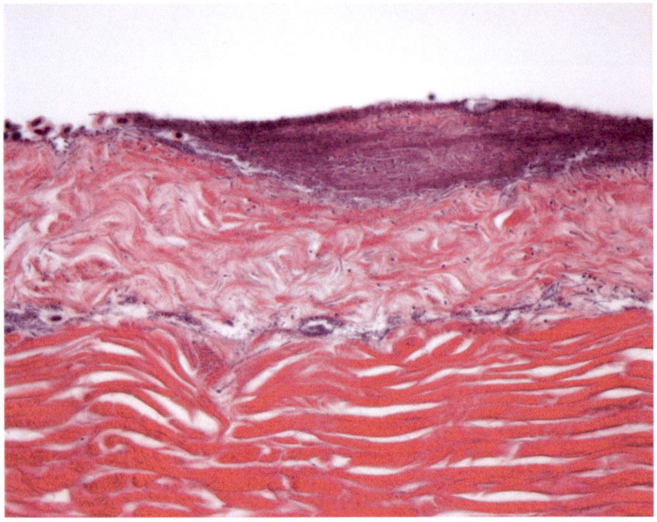

Fig. 6.25 Flexibacter-like bacteria covering skin lesion in farmed Atlantic salmon. Note absence of epidermis and scales. Medium power

Early clinical signs are nonspecific and include lethargy, inappetence, and swimming near the water surface with increased opercula movements. Additional and more characteristic signs include skin discoloration, dorsal fin damage (Fig. 6.26) and yellow necrotic gill lesions at the tips of the lamellae. As lesions progress to the mandible and maxillae the paucity of underlying tissue lesion leads to fatal osmotic changes. In addition, hypoxia caused by gill necrosis and biochemical disturbances resulting from skin ulceration, are a likely cause of death. Bacterial attachment to the gill tissue and entry into the fish is facilitated by physical injuries thus highlighting the importance of the gills in pathogenesis. Internally, ascites can be observed.

Histologically, lamellar shortening, epithelial and goblet cell hyperplasia with an associated moderate necrosis is reported. These lesions rapidly progress to a marked inflammation predominantly by neutrophils and severe gill necrosis. In acute infections, hypoxia and death may result from extensive damage to this tissue.

Diagnosis is based upon characteristic clinical signs and histopathology, and the isolation and identification of the bacteria. *Flavobacterium columnare* is a Gram negative thin rod-shaped bacterium which measures 3–5 μm in length. Isolation of the bacteria can be made from the gill, surface lesions as well as internal organs such as the kidney, liver and spleen. In wet mounts bacteria are uniquely arranged into columnar formations. A definitive diagnosis of *F.columnare* is dependent upon culture using a selective medium (e.g. *Cytophaga* or Shieh) with the presence of yellow pigmented rhizoid colonies. This is followed by biochemical testing using conventional methods.

Fig. 6.26 Farmed rainbow trout with characteristic saddleback lesions caused by *Flexibacter columnare*

6.10 *Tenacibaculum maritimum*

Tenacibaculum maritimum (formally *Flexibacter maritimus*) is the aetiological agent of an opportunistic 'marine flexibacteriosis' which is primarily a skin infection causing an ulcerative dermatitis in a range of fish species including, rainbow trout and Atlantic salmon. Infection is characterised by an eroded and haemorrhagic mouth, necrotic lesions on the body and head, frayed fins and fin rot, scale loss and oedema and to a lesser extent, involvement of the gills (Figs. 6.27 and 6.28). Filamentous mats of bacteria can occur on the liver and skin in chronic cases (Fig. 6.29). Histologically, early signs include fragmentation and degeneration of the epithelium with infiltration and occasionally intra-epithelial cellular inflammatory cells, plus congestion and haemorrhage of the superficial dermis. A necrotizing stomatitis may progress to cellulitis and eventually perforation of the jaw. Infected gill tissue shows a necrotizing branchitis and acute telangiectasia with a focal lamellar hyperplasia. In the skin a mild inflammation can occur in the scale pockets with some adherent bacteria before full epithelial erosion. Typically, there is a lack of inflammatory response and consequently bacteria may infect the connective tissue and occasionally the musculature.

Diagnosis is based upon clinical signs and the isolation and identification of the bacteria. *Tenacibaculum maritimum* is a Gram negative filamentous aerobic rod exhibiting gliding motility on wet surfaces. Bacterial colonies are flat, light

Fig. 6.27 Severe cranial erosion and panopthalmitis in farmed Atlantic salmon infected with *Tenacibaculum maritimum*

Fig. 6.28 Severe tail rot caused by *Tenacibaculum maritimum* in seawater farmed Atlantic salmon

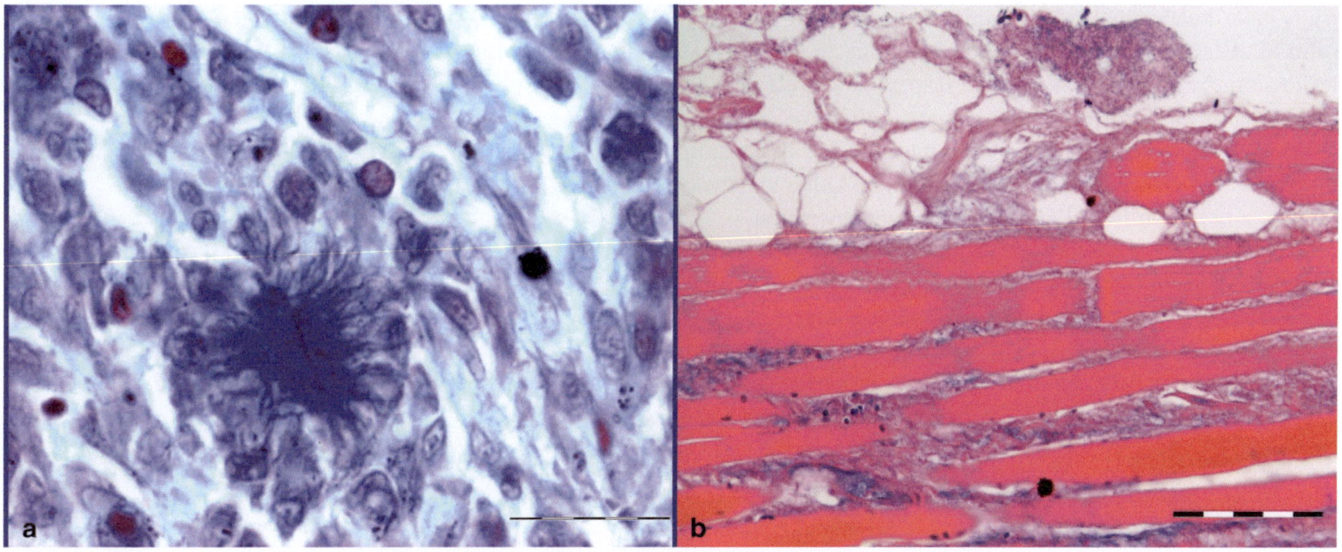

Fig. 6.29 (**a**) *Tenacibaculum* colony in kidney of farmed Atlantic salmon. High power. (**b**) Ulcerative dermatitis due to *Tenacibaculum maritimum* in Atlantic salmon. Bar = 100 μm

yellow-pigmented with uneven edges. Agent diagnosis requires the observation of thin filamentous rods in wet mounts, with culture on an appropriate medium such as Anacker and Ordal or Marine Agar, and supported by other methods such as nested PCR.

6.11 *Hafnia alvei*

Hafnia alvei is an opportunistic pathogen causing mortality in juvenile brown and rainbow trout in Europe and cherry salmon in Japan. Externally, diseased fish show a dark body surface, abnormal swimming and a swollen abdomen.

Histological changes in the kidney include necrosis, macrovacuolar degeneration in the liver and loss of the lymphatic tissue in the spleen, and overall typical signs of a generalized haemorrhagic septicaemia.

Diagnosis is based upon clinical signs and the isolation and identification of the bacteria. *H. alvei* is a Gram negative, facultative anaerobic, rod-shaped bacterium and taxonomically placed in the Enterobacteriaceae. This bacterium is similar morphologically and serologically to enteric red mouth disease (see *Yersinia ruckeri*) but can be differentiated biochemically.

6.12 *Chryseobacterium* spp.

The genus *Chryseobacterium* is widely distributed and can be recovered from a variety of environments. Although it has not been considered a relevant pathogen, there has been an increase in the frequency of clinical cases reported in Chile and Finland, in which several species or strains of the bacterium have been isolated from rainbow trout and Atlantic salmon. Diseased fish show skin and muscle ulcerative lesions on the flank and in the vent or peduncle area.

Histologically, fish exhibit degeneration of kidney tubules, with oedema in the renal interstitial tissues and proteinaceous casts within the tubular lumen. Other changes include a heterophilic cellulitis and myo-degeneration.

Isolates of the bacterium from internal organs are Gram negative, non-motile rods, catalase negative, and on the basis of 16S rRNA gene sequence analysis are classified into species of *Chryseobacterium*, with the proposed name of C. *piscicola* sp. nov., but followed by *C. viscerum* sp. nov. The differential diagnosis is *Flavobacterium psychrophilum*.

6.13 *Pasteurella skyensis*

A novel Gram negative bacterium, *Pasteurella skyensis* has been identified from farmed salmon in Scotland. However, relevant histopathology and isolation of the bacteria has been infrequent and currently not considered as a significant infection for farmed fish.

Histologically, multifocal granulomas are observed in the kidney, liver and spleen with mild fibrous encapsulation. Most granulomas contain multinucleate giant cells with an eosinophilic caseous necrotic core (Figs. 6.30 and 6.31). Some functional tissue surrounds the granulomas and some cells show karyohexis. The ventricle shows proliferative lesions between the compactum and spongiosum, and some granulomas at the junction of the two layers are observed with a fibrous pericarditis and loss of staining properties. There is also loss of structure in the bulbous arteriosus. The gill lamellae may show old aneurysms, lamellar lifting and basal hyperplasia, with overgrowth of new epithelial tissue. Giant cells and small granulomas are also observed. Thickening of the connective tissue layer around the primary lamellar rods and almost complete absence of the latter, is noted.

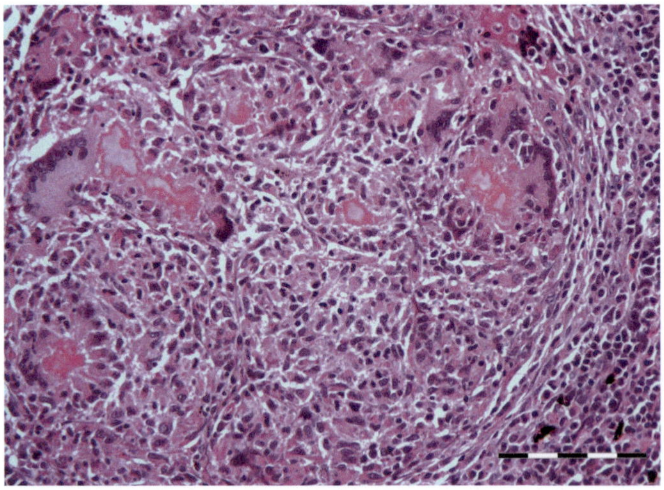

Fig. 6.30 Granulomatous inflammation with associated giant cells in farmed Atlantic salmon with *Pasteurella skyensis*. Bar = 100 μm

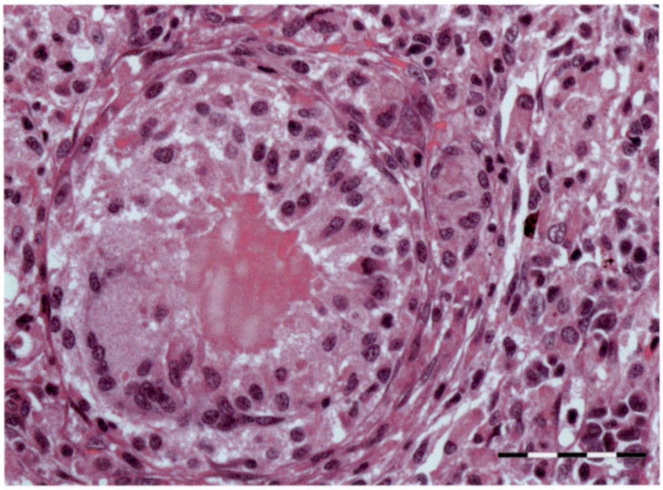

Fig. 6.31 Granulomatous inflammation with associated giant cells in farmed Atlantic salmon with *Pasteurella skyensis*. Bar = 50 μm

6.14 *Francisella noatunensis* subsp. *noatunensis*

Francisellosis was initially reported as a significant threat to the Norwegian cod farming industry. The disease develops as a systemic, chronic, granulomatous infection with high morbidity and resulting in varying degrees of mortality but cumulative losses ranging from 5 to 20 % have been reported. In 2006, francisellosis was also reported among Atlantic salmon parr held in freshwater cages in Lake Llanquihue, Chile but the disease develops in a similar fashion independent of host and regions.

Francisella noatunensis is a facultative intracellular Gram negative bacterium refractive to culture on standard laboratory media and recently reported that it requires cysteine enriched media. A PCR test has been developed and the 16S ribosomal RNA gene sequence show the Chilean isolate to be 98–100 % identical to *F. noatunensis* subsp. *noatunensis* from cod. Further studies are required to establish the risks to farmed salmonids.

6.15 *Renibacterium salmoninarum*

Renibacterium salmoninarum is the aetiological agent of bacterial kidney disease (BKD), a serious, usually chronic, condition of wild and farmed salmonids. The first record of BKD occurred in wild Atlantic salmon in Scotland during the 1930s but is now reported worldwide, namely from all major areas where these fish are farmed. Outbreaks occur in freshwater as well as in the marine environment. The disease is transmitted horizontally by cohabitation, and vertically via the eggs directly from ovarian tissue prior to ovulation, with the highest mortality recorded as temperatures reach 12 °C.

Clinical observations and external lesions are variable, however loss of balance, darkening and mottled appearance of the skin, distended abdomen, exophthalmia, petechiae and haemorrhaging around the base of the pectoral fins and the lateral line, are described. Superficial blisters with vesicle formation, ulceration and abscesses may develop in the tegument.

At necropsy, the gills and internal organs are pale giving an indication of anaemia. The most obvious internal lesion is a swollen kidney that may show greyish-white nodular lesions. Similar nodules also occur in heart, liver and spleen (Figs. 6.32, 6.33, 6.34 and 6.35). Petechial haemorrhage of the muscle, the peritoneum and ascites is reported. The pyloric caeca are similarly pale with a 'fat-like' appearance. A diffuse white membranous layer (pseudomembrane) covering internal organs is described in affected fish (see Fig. 6.34). A yellow, viscous fluid occurs in the intestine and may contain blood.

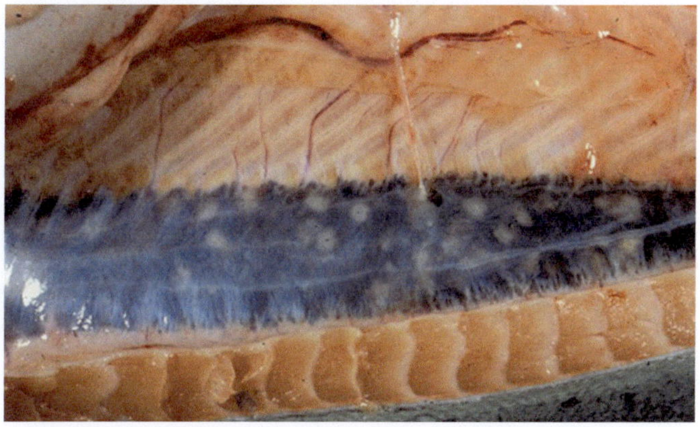

Fig. 6.32 Granulomas in kidney of Atlantic salmon with bacterial kidney disease

Histologically, a chronic proliferating granulomatous response affects the haematopoietic tissues. Multiple and often large necrotic areas and granulomas with a central caseous zone are bounded by epithelioid and other infiltrating lymphoid cells (Figs. 6.36, 6.37 and 6.38). During the early stages of infection *R. salmoninarum* may aggregate at the surface of the spleen and around pancreatic tissues, followed by widespread necrosis with pyknotic cells within the ellipsoids. A fibrous capsule may form around the spreading necrotic lesion that traps bacteria as well as phagocytes (Fig. 6.39). *R. salmoninarum* is recorded intra- and extracellularly and within the kidney, focal necrosis and glomerular oedema is observed, as well as a granulomatous inflammatory reaction and membranous glomerulopathy due to deposition of immune complexes. In cases where the bacteria are killed by host cells, such granulomas gradually resolve. *R. salmoninarum* can also be found in the gill filaments (Fig. 6.40).

A granulomatous peritonitis occurs in the pancreatic area with numerous bacteria accompanied by focal infiltration of leucocytes (Fig. 6.41). Small foci containing phagocytised *R. salmoninarum* can develop within the liver parenchyma which coalesce and become the centre of an inflammatory reaction. Deposits of fibrin and collagen accumulate around the swim bladder and intestine, with some hypertrophy and numerous phagocytic cells containing bacteria. A diptheric epicarditis comprising thin layers of fibrin, collagen and macrophages containing *R. salmoninarum* results in restrictive pericarditis.

In Canada, a seasonal 'spawning rash' has been reported in mature rainbow trout, where a pustulous dermatitis may cover large areas of the skin with many small blisters or raised haemorrhagic nodules within the epidermis. Granulomatous tissue invades adjacent scale pockets and extends longitudinally along the fibrous tissue layer of the dermis.

Where *R. salmoninarum* occurs within the central nervous system, haematogenous spread to the meninges has been

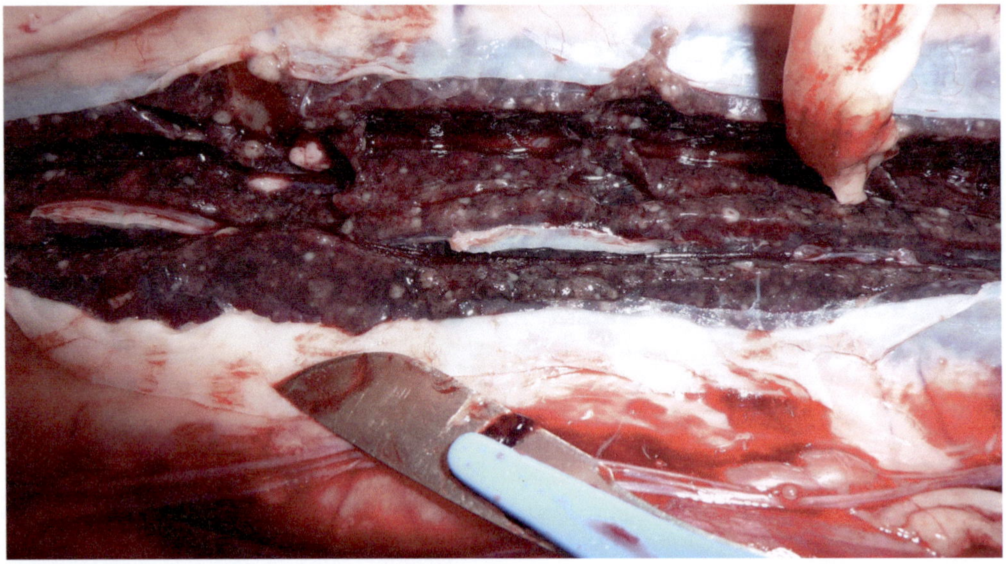

Fig. 6.33 Multiple granulomas in the kidney of Atlantic salmon with bacterial kidney disease

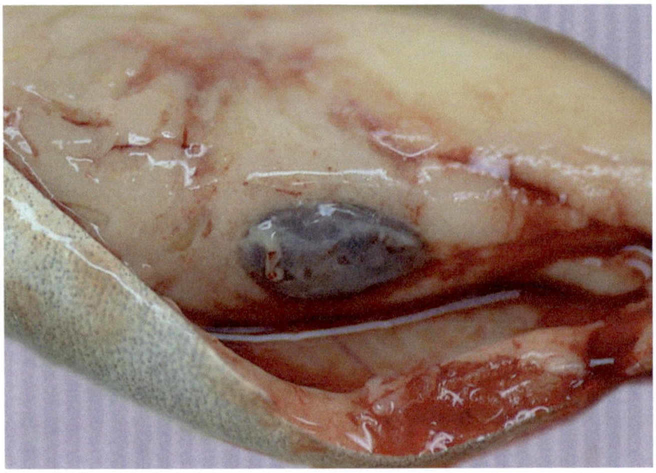

Fig. 6.34 Ascites and splenomegaly with a fibrinous coat in Atlantic salmon with bacterial kidney disease

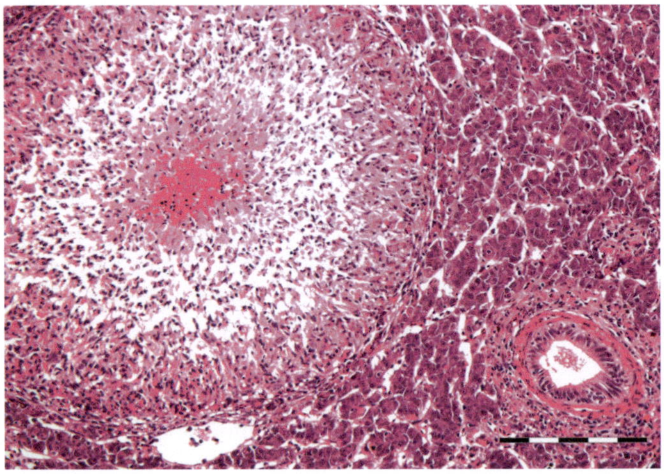

Fig. 6.36 Resolving granuloma in the liver of Atlantic salmon with bacterial kidney disease. Bar = 200 μm

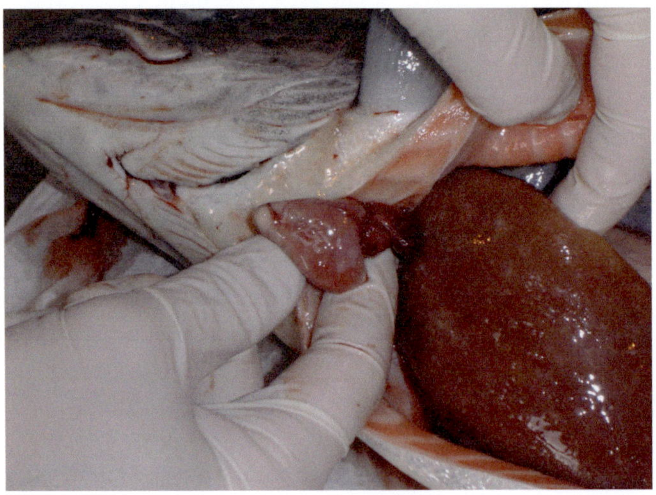

Fig. 6.35 Epicarditis and multiple liver granulomas in farmed Atlantic salmon infected with *Renibacterium salmoninarum*

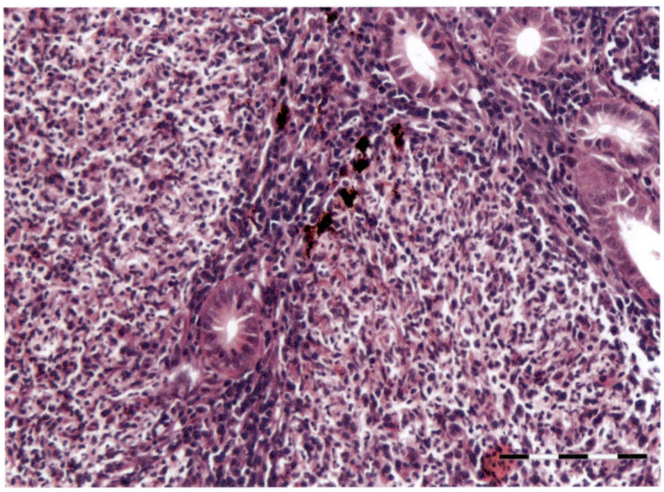

Fig. 6.37 Multiple granulomas in kidney with characteristic melanin associated with borders in Atlantic salmon infected with *Renibacterium salmoninarum*. Bar = 100 μm

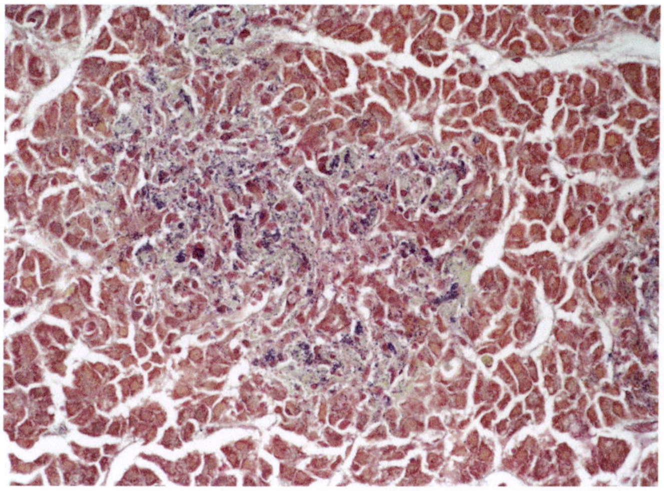

Fig. 6.38 Necrotic focus with aggregates of *Renibacterium salmoninarum* in liver of Atlantic salmon with bacterial kidney disease. Medium power

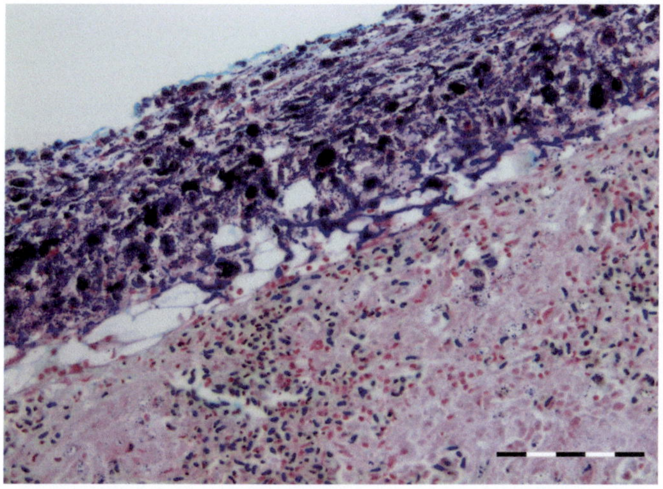

Fig. 6.39 Heavily melanised pseudomembrane associated with spleen capsule, note presence of large number of *Renibacterium salmoninarum*. Gram stain. Bar = 100 μm

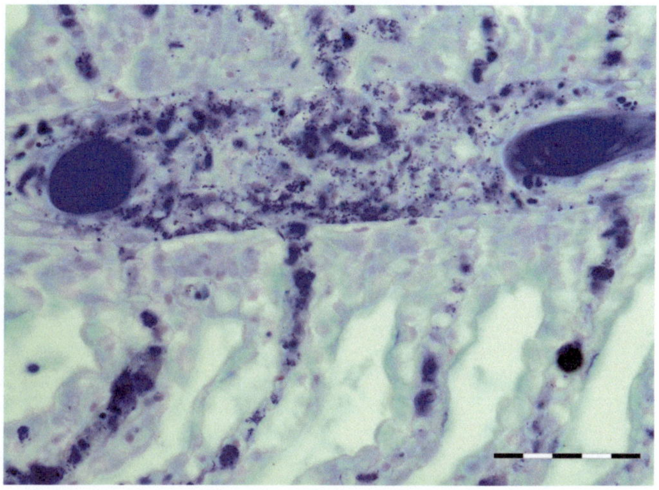

Fig. 6.40 *Renibacterium salmoninarum* in gill filament and lamellae of Atlantic salmon. Gram stain. Bar = 50 μm

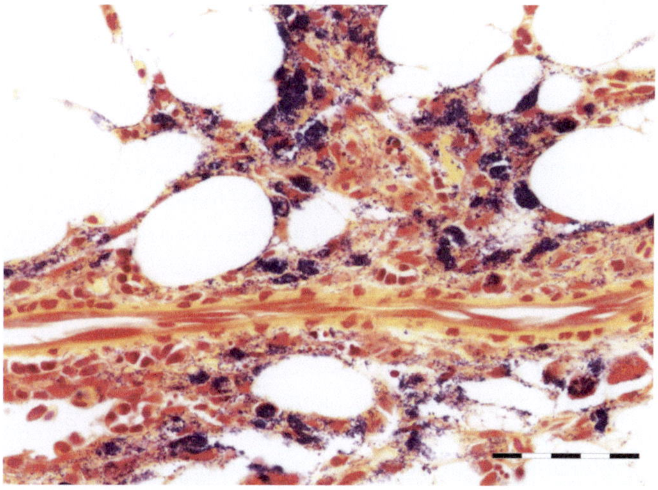

Fig. 6.41 *Renibacterium salmoninarum* colonies in pancreatic tissue of Chinook salmon. Gram stain. Bar = 50 μm

reported. Retrograde extension from the posterior uvea to the floor of the diencephalon along the epineurium and perineurium of the optic nerve, may also be a route of neural invasion.

Renibacterium salmoninarum occur within phagocytes and readily multiply within making it possible to avoid the immune mechanisms of the host and the effect of antibiotic therapy.

Diagnosis is based on the observation of typical clinical signs and the histological observation of Gram positive bacilli within tissues and phagocytes. *R. salmoninarum* can be detected by Gram and PAS staining, but not by H&E. *R. salmoninarum* is a small (0.5 × 1.0 μm), slow-growing Gram positive, non-acid fast, non-motile diplobacillus with optimal growth at 15 °C. The bacterium is proteolytic, produces catalase and has an absolute requirement for L-cysteine. Culture is achieved after several weeks on medium such as Mueller Hinton with added L-cysteine hydrochloride. An ELISA and a real time PCR are commonly used for diagnosis.

6.16 *Carnobacterium maltaromaticum*

Carnobacterium maltaromaticum (synonym *C. piscicola*) infections are responsible of the condition recognised as 'pseudo kidney disease'. The bacterium has been be recovered from rainbow trout, Chinook salmon and whitefish from North America and also, although less frequently, from fish within Europe, Australia and South America. The genus *Carnobacterium* incorporates *Lactobacillus piscicola* and related lactic acid

Fig. 6.42 *Carnobacterium maltaromaticum* infection in farmed rainbow trout; dark coloured fish with splenomegaly and petecchia in liver

species. *C. piscicola* can be isolated from seemingly healthy fish and probably part of the normal microbiota of the gastrointestinal tract. Isolation has been reported from salmonids which are over a year old, and up to and including broodstock held in fresh water. Handling, post-spawning and other forms of stress appear to be predisposing factors to what may become a chronic condition and although infrequently, there are reports of significant outbreaks. The bacterium has also been reported as causing autolytic changes in cold-smoked salmon.

Clinical signs are somehow inconsistent but overall indicative of a septicaemia, including general darkening, sub dermal blisters, abdominal distension, pronounced bilateral exophthalmia with periocular haemorrhaging (Fig. 6.42). At necropsy, peritonitis, splenomegaly and diffuse haemorrhaging involving the musculature, liver and swim bladder, may be observed. Histologically, focal necrosis and vacuolation of the kidney tubular endothelium, liver sinusoid congestion, hyaline droplet degeneration and pancreatic acinar cell necrosis are reported.

Diagnosis is based on clinical signs, tissue sections and isolation of the aetiological agent. *C. maltaromaticum* is Gram positive, non-motile bacilli which grows well on TSA at 22 °C occurring singly and in short chains. A differential diagnosis would include *Renibacterium*.

6.17 *Streptococcus phocae*

Streptococcus phocae is reported as an emerging pathogen for farmed Atlantic salmon smolts and adult fish in Chile located in estuaries and marine waters. Outbreaks occur during the summer when temperatures are above 15 °C, reaching in some occasions a cumulative mortality up to 25 % of the affected population. Infected fish show exophthalmia with accumulation of purulent and haemorrhagic fluid around eyes, and ventral petechial haemorrhage. At necropsy, haemorrhage in the abdominal fat, pericarditis and enlarged liver, spleen and kidney are expected pathological changes.

The bacteria are Gram positive and beta-haemolytic. Additional studies will be necessary to determine the clinical significance of this species for the salmon industry.

6.18 *Mycobacterium* spp.

Mycobacterium spp. in teleosts is largely represented by three species, *M. chelonae*, *M. fortuitum* and *M. marinum*. They can infect and cause disease in a wide range of fish species including salmonids, developing as a typically chronic condition which may take several years to progress into clinical disease.

Common clinical signs are emaciation, ocular lesions with exophthalmia, change in skin colour, fin rot and ulceration of the skin. At the external examination, a subcutaneous erythema and granulomatosis may occur on the ventral aspect of the cranial abdomen. At necropsy, ascites and greyish-white visceral nodules are seen in many organs but particularly, the heart, kidney, liver and spleen, becoming progressively swollen and fused by white membranes (Fig. 6.43). Histologically, chronic or proliferative forms of focal granulomas composed of various types of immune

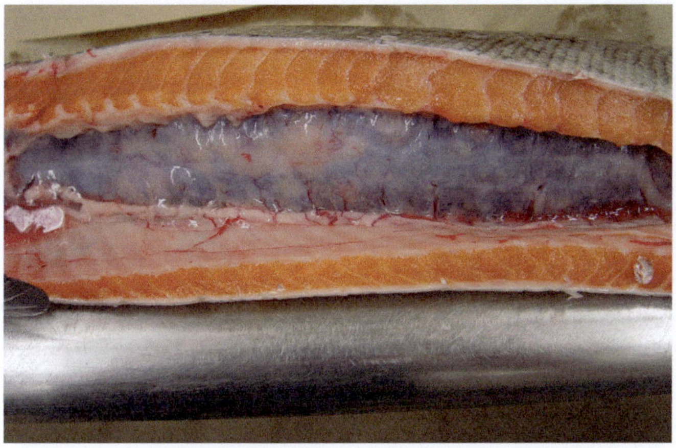

Fig. 6.43 Granulomas and necrotic areas in the kidney of Atlantic salmon infected with *Mycobacterium* sp.

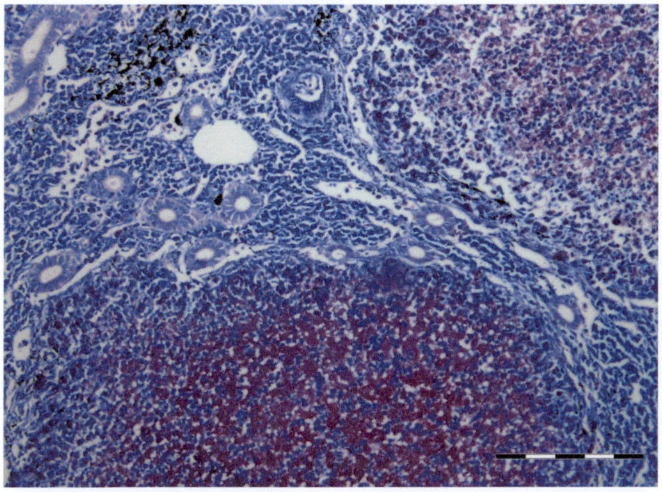

Fig. 6.44 Mycobacterial infection in farmed Atlantic salmon. Acid-fast bacteria are present in large numbers within a kidney granuloma. Ziehl-Nielsen stain. Bar = 200 μm

cells that include fibrocytes, granulocytes, centrally located epithelioid cells and macrophages are described. Acid-fast bacteria may be demonstrated within these lesions and in phagocytic cells (Fig. 6.44). Both melanisation and vacuolation are reported around these granulomas.

Diagnosis is based on the characteristic lesions and the demonstration of acid-fast bacteria in histological sections, further supported by isolation. *Mycobacterium* spp. are Gram-positive, aerobic, straight to slightly curved, non-motile rods. Many isolates of *Mycobacterium* are difficult to establish in culture indicating the fastidious nature of these pathogens. A PCR is useful for detecting and speciation of *Mycobacterium* in infected fish. The differential diagnosis would be *R. salmoninarum* and oomycete nephritis.

6.19 *Nocardia* sp.

Nocardiosis is caused by a Gram-positive, partially acid-fast, aerobic, filamentous bacterium causing nodular lesions in gills, spleen, kidney and liver with or without multiple skin ulcers. *Nocardia* are rarely attributed to infection in salmonids and therefore not covered in detail in this book. However it represents an important differential diagnosis for *Mycobacterium*, but reliable tests are available to differentiate these genera.

6.20 *Piscirickettsia salmonis*

Piscirickettsia salmonis is the causative agent of salmonid rickettsial septicaemia (SRS) or piscirickettsiosis, and recognised as a serious pathogen primarily in farmed stock reared in sea water, with occasional outbreaks among rainbow trout reported in fresh water from Chile. In other countries outbreaks have had less significance. Clinical signs of SRS generally include lethargy, swimming near the surface with erratic movements and dark skin colouration. Raised scales with associated haemorrhagic skin lesions can be observed. Some fish may nevertheless appear normal. At necropsy, gills show anaemia and internally, ascites, splenomegaly, cream-coloured, focal sub-capsular nodules in the liver, fibrinous epicarditis and a swollen, grey kidney are reported (Figs. 6.45, 6.46 and 6.47). Histopathological changes occur in most organs including brain, heart, kidney, liver, ovary and spleen. Gills can show epithelial hyperplasia with occasional necrosis. Within the kidney, extensive necrosis of the haematopoietic tissue with oedema and increase in inflammatory cells, glomerulonephritis and enlargement of the Bowman's space occur. Normal haematopoietic and lymphoid tissues can be replaced by inflammatory cells. Liver lesions include a focal to diffuse necrotizing hepatitis sometimes with granuloma formation (Fig. 6.48). Similar focal granulomas are reported for the spleen (Fig. 6.49). Meningitis, endocarditis, peritonitis, pancreatitis, and branchitis may be seen with accompanying chronic inflammatory vascular changes, similar to those in the liver. Cardiac changes include a mild endocarditis with variable degrees of epicarditis or pericarditis. Petechial haemorrhaging is frequently observed on the swim bladder and intestinal tract, with necrosis and inflammation of the lamina propria. Mild inflammatory and thrombotic lesions are noted in the brain, pancreas and adipose tissue. An apparent neutrophilia is associated with severely anaemic fish.

Diagnosis is based upon characteristic clinical signs, histopathology and the isolation and identification of the bacteria. These bacteria can be observed by light microscopy

Fig. 6.45 Fibrinous epicarditis and panopthalmitis in farmed Atlantic salmon infected with *Piscrickettsia salmonis*

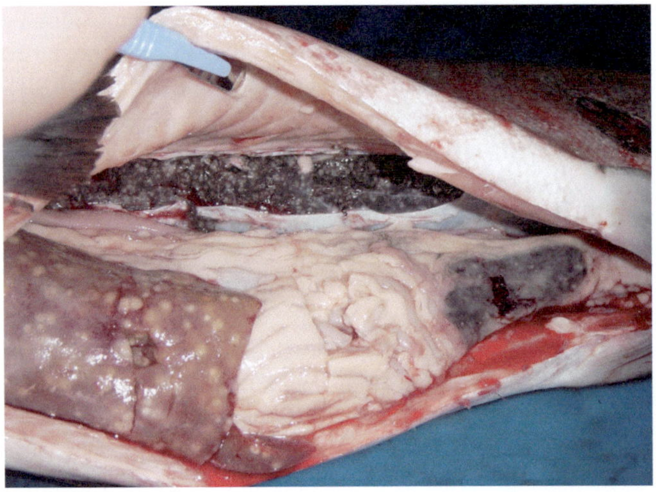

Fig. 6.46 Multiple granulomas in liver, kidney and spleen in Atlantic salmon with *Piscrickettsia salmonis*

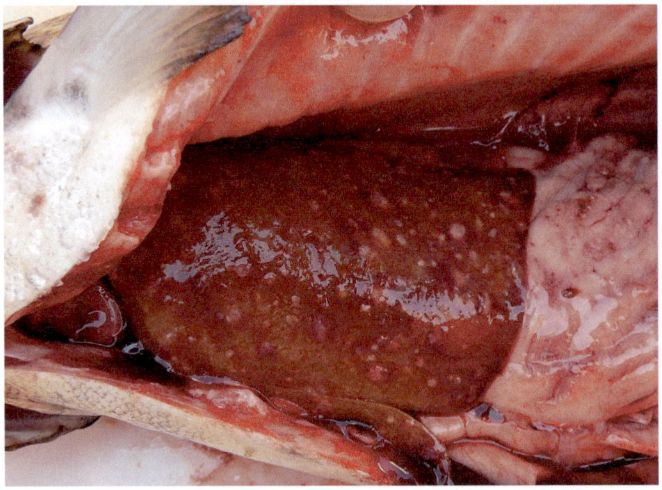

Fig. 6.47 Multiple granulomas in liver from Atlantic salmon with *Piscrickettsia salmonis*

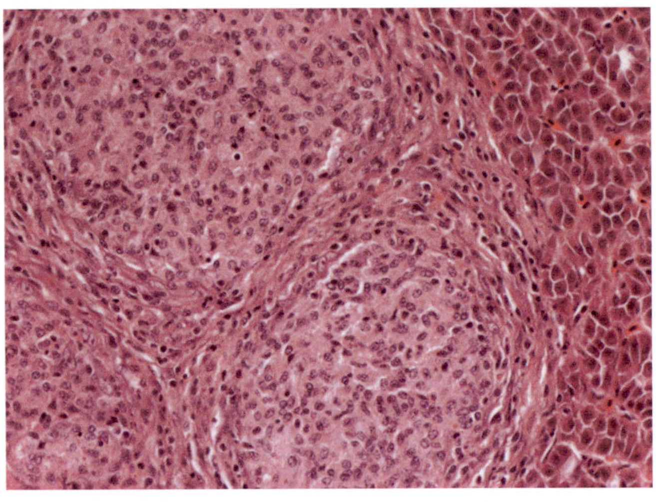

Fig. 6.48 Granulomatous lesion in liver of Atlantic salmon with *Piscrickettsia salmonis*. Medium power

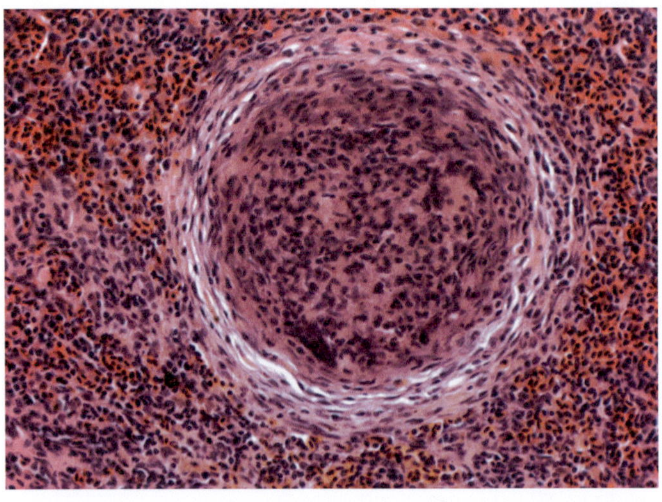

Fig. 6.49 Focal granuloma with peripheral fibrosis in spleen of Atlantic salmon with *Piscrickettsia salmonis*. Medium power

within membrane-bound cytoplasmic vacuoles using H&E, methylene blue or Giemsa stained sections, and provide a presumptive diagnosis. Similarly, macrophages containing *P. salmonis* can be detected in stained peripheral blood imprints. *P. salmonis* is a Gram negative, acid-fast, non-motile, predominantly coccoid, non-capsulated (although often pleomorphic) organism. Several cell lines including CHSE-214 and RTG-2 and antibiotic free media, have been used successfully for the primary isolation of *P. salmonis* from kidney. Confirmation can be achieved through the use of an ELISA, presence in cell culture, a PCR assay or *in situ* hybridization. Culture media including a marine based broth supplemented with L-cysteine appears to allow the successful culture independently of the more costly and time consuming isolation on cell lines. Additionally, it avoids the difficulty of eliminating the contamination with host cell debris. A differential diagnosis would include viral haemorrhagic septicaemia.

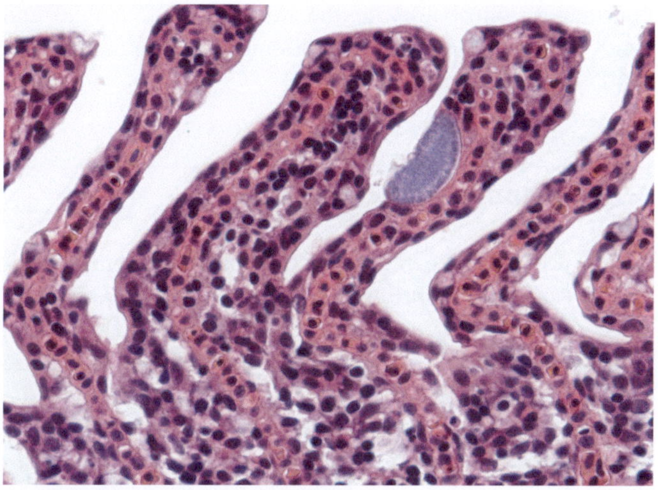

Fig. 6.50 Epitheliocyst on gill lamella of farmed Atlantic salmon. Medium power

6.21 Epitheliocystis

At least three bacterial species may be involved in the condition known as epitheliocystis, two of which, *Candidatus* Piscichlamydia salmonis which is mainly from sea waterfish in Ireland and Norway, and *Candidatus* Clavochlamydia salmonicola from freshwater fish, belong to the Phylum Chlamydiae. This group is a cosmopolitan class of intracellular granular, basophilic Gram-negative bacteria, considered mostly as opportunistic rather than primary pathogens. Both species have been reported from wild freshwater brown trout in Switzerland, and *Candidatus* Piscichlamydia salmonis has been diagnosed in freshwater-reared Arctic char in North America and in-farmed Atlantic salmon in Ireland and Norway. Clavochlamydia salmonicola related epitheliocysts have been reported to disappear 6 weeks after transfer to sea water. As gill diseases often have a complex aetiology, the exact role of the different pathogens and environmental factors involved may be difficult to ascertain.

Clinical signs in affected fish include lethargy induced by the severe hyperplastic gill inflammation, leading to hyperventilation, flared opercula and increased mucus production. Mortality levels in sea-farmed salmon may be highly variable but up to 80 % has been recorded, although this also depends on environmental conditions and the concurrence of other pathogens. A seasonal occurrence typically peaking in the autumn months indicate water temperature as an important risk factor.

Histologically, the affected bacteria-containing cells can be seen as round structures circumscribed by an eosinophilic hyaline capsule (Figs. 6.50 and 6.51). Pathological changes may vary but it is considered that the following would be expected in a diagnosis: circulatory disturbances, epithelial hyperplasia, inflammation in sub-epithelial and epithelial tissue with mild to severe hypertrophy and associated hyperplasia, increased mucus cells with fusion of lamellae, telangiectasia and infiltration of macrophages.

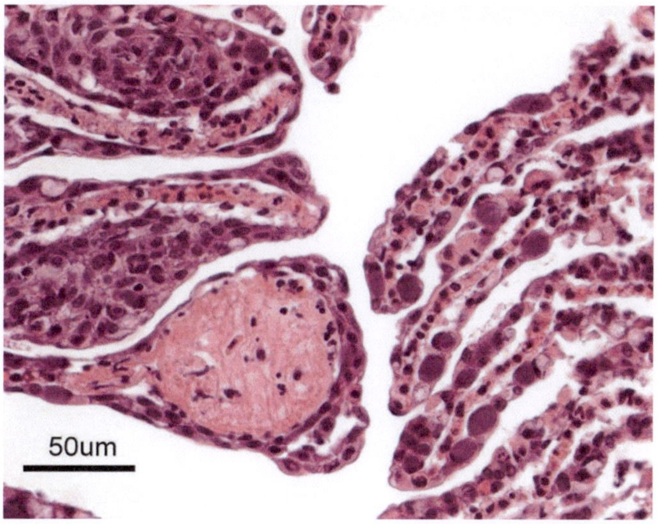

Fig. 6.51 Gill of farmed Atlantic salmon with epitheliocystis. Old, organized aneurism (*left*) and numerous characteristic epitheliocysts (*right*)

Diagnosis is based upon clinical signs, histopathology with demonstrating the characteristic lesions and epitheliocysts, plus real time (RT)-PCR assay.

Candidatus Branchiomonas cysticola, a non-chlamydial bacterium has also been associated with epitheliocyst formation in sea water-farmed Atlantic salmon in Ireland and Norway. These organisms target the epithelial cell of the gill lamellae of several fish species in both fresh and sea water. The hypertrophied epithelial cells filled with bacteria may range in size from 10 to 400 μm. Epitheliocysts are considered contributory to the multifactorial, maybe end stage condition known as proliferative gill inflammation (PGI) in Atlantic salmon.

6.22 *Candidatus arthromitus*

Rainbow trout gastroenteritis (RTGE) is an emerging condition in trout culture in Europe and linked to the presence of large numbers of a segmented filamentous bacterium *Candidatus arthromitus* in the distal intestine and/or pyloric caeca of the digestive tract, suggesting this is preferred site for these bacteria. Affected fish show lethargy, reduced appetite and accumulation of mucoid faeces particularly in the summer months, with diffuse haemorrhage (Fig. 6.52). Histopathological changes include enterocyte detachment and congestion of the lamina propria and adventitial layers (Fig. 6.53). The bacteria are not always adjacent to the areas

Fig. 6.52 Rainbow trout gastroenteritis; dilated and hyperaemic intestine

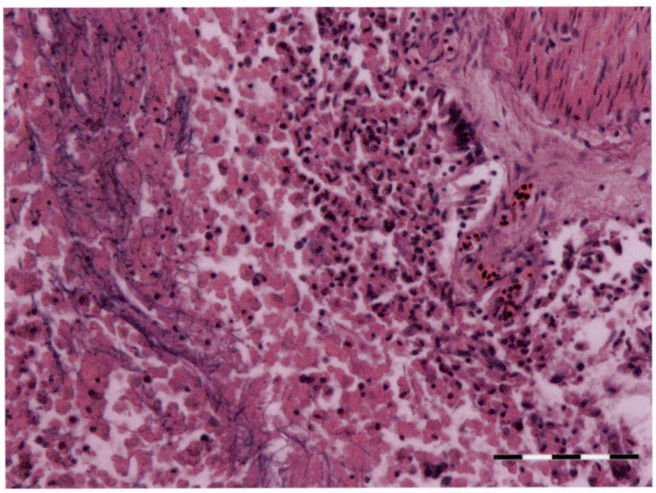

Fig. 6.53 Rainbow trout gastroenteritis; haemorrhage and characteristic bacteria in intestinal wall. Bar = 100 μm

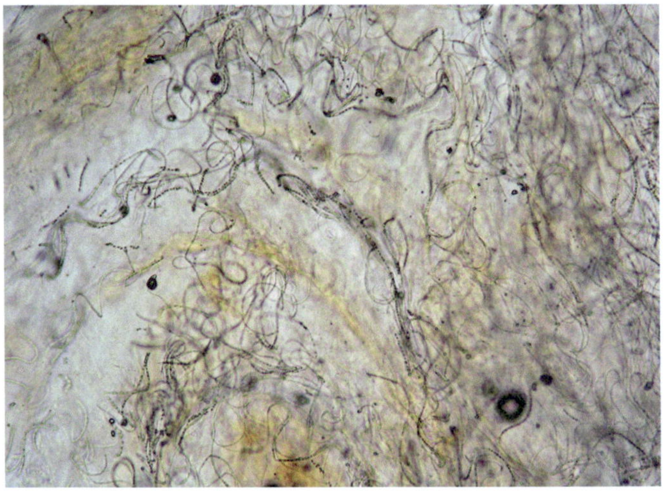

Fig. 6.54 Rainbow trout gastroenteritis; fresh mount from intestinal content showing characteristic bacteria. High power phase image

with pathological changes, suggesting that if these organisms play a role in the pathogenesis, extracellular products, T cells or apoptosis may also be involved in the development and pathogenesis of RTGE. Ultrastructural changes included loss of microvillar structure, membrane blebbing, hydropic mitochondrial damage and basal hydropic degeneration of enterocytes. The exposure of large areas of the lamina propria probably results in a compromised osmotic balance and facilitates the entry of other pathogens. The examination of fresh material (Fig. 6.54) and molecular methods have helped to detect a low number of bacteria, but further studies will be required to establish the link between *C. arthromitus* and RTGE.

6.23 Red Mark Syndrome = Cold Water Strawberry Disease

Red mark syndrome (RMS) is a skin condition affecting farmed rainbow trout and was first recorded in Scotland in 2003. Distinct lesions with single to multifocal reddish coloured patches, particularly over the flanks and generally below the lateral line are reported. Lesions vary in size from a few millimetres to ~4 cm across and are frequently devoid of scales in the centre. The absence of scales likely contributes towards the susceptibility to mechanical damage and occasionally, lesions become ulcerated after normal farming practices. No behavioural changes have been associated to the condition. RMS currently affects trout farmed in the UK, USA and some regions of continental Europe, with reports from Switzerland, Austria, Germany and France.

RMS generally occurs in fish larger than ~100 g and typically observed at temperatures below 15 °C, with clinical disease regressing at higher temperatures. Morbidity can be high but RMS is usually a non-lethal condition where appetite and growth are unaffected. Severe lesions may, however, adversely affect the welfare of the fish.

At necropsy, early observations reveal discrete, swollen, and circular to oval and well-demarcated areas of slightly raised skin, which are opaque or light creamy coloured (Figs. 6.55, 6.56 and 6.57). As lesions develop, they show a marked hyperaemia giving the condition its popular name.

Fig. 6.55 Skin lesions in farmed rainbow trout with early red mark syndrome

Histologically, a severe dermatitis characterised by a marked inflammation and thickening of all dermal layers including the hypodermis is observed, occasionally with foci of infiltration into the superficial muscular layer. The diffusely infiltrated dermis shows distended scale pockets due to moderate to severe oedema, and infiltration by macrophages, lymphocytes and granular cells. Scales are seen as degraded or are completely absent (resorption) in association with the presence of osteoclasts (Fig. 6.58). Congestion, haemorrhage and small amounts of pigment cells can be present. Advanced lesions show increased number of inflammatory cells in the region immediately below the basal membrane and between the stratum compactum and the underlying adipose tissue, with degeneration and necrosis of the connective tissue. Less frequently, epidermal involvement with lymphocytic infiltration and multifocal erosion has been reported. Evidence of scale regeneration has been observed in healing lesions. Changes in internal organs such as renal degeneration, focal liver necrosis or inflammatory lesions including heart, intestinal smooth muscle and connective tissues have been reported from affected fish. However, field cases could represent mixed conditions where the changes in the internal organs and association with RMS remain to be proven.

RMS is reported to be responsive to antibiotic treatment although scientific proof for the treatments effectiveness has not been published to date and affected stocks also show

Fig. 6.56 Farmed rainbow trout with red mark syndrome; characteristic vertical banded skin lesions

Fig. 6.57 Close up of red mark lesion in farmed rainbow trout; characteristic vertical bright haemorrhagic lesion

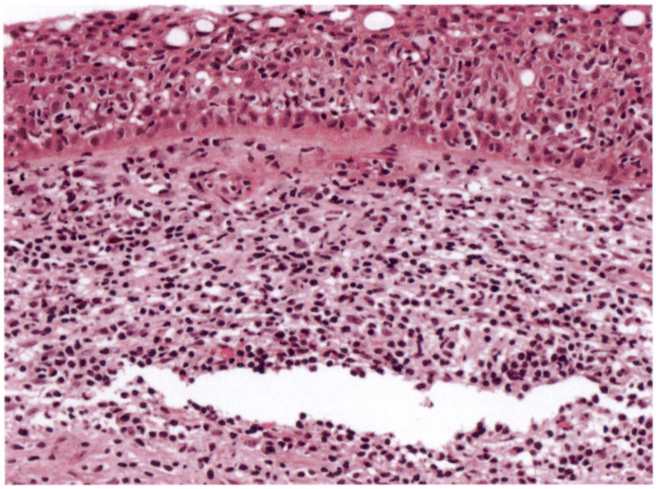

Fig. 6.58 Typical dermatitis with digested scale from farmed rainbow trout with red mark syndrome. Medium power

spontaneous recuperation without intervention. The condition has nevertheless a significant economic impact mainly due to increased labour costs and product rejection or downgrading at slaughter, lowering its market value.

RMS has shown to be transmissible, both experimentally and by observed spreading in field cases through live fish movement, supporting the condition has an infectious aetiology. A rickettsia-like organism (RLO) has been reported associated with RMS, however, conclusive evidence of the agent being the responsible agent has not been provided.

RMS resembles many features of 'strawberry disease', a condition described in the UK for rainbow trout at warmer waters. Moreover, a disease with similar clinical signs to RMS has was been reported concurrently in the USA and named strawberry disease (SD-USA). Recently, a case definition decided that SD-USA assimilates to RMS and should be referred to as 'cold water strawberry disease', while the former SD-UK is recognised as 'warm water strawberry disease', thus differentiating the conditions by the temperature window range and avoiding further confusion. Currently RMS is diagnosed through light microscopy.

Further Reading

Apablaza P, Løland AD, Brevik ØJ, Iiardi P, Battaglia J, Nylund A (2013) Genetic variation among *Flavobacterium psychrophilum* isolates from wild and farmed salmonids in Norway and Chile. J Appl Microbiol. doi:10.1111/jam.12121

Barnes ME, Brown ML (2011) A review of *Flavobacterium psychrophilum* biology, clinical signs, and bacterial cold water disease prevention and treatment. Open Fish Sci J 4:40–48

Bernardet JF, Kerouault B (1989) Phenotypic and genomic studies of *Cytophaga psychrophila* isolated from diseased rainbow trout (*Oncorhynchus mykiss*) in France. Appl Environ Microbiol 55:1796–1800

Birbeck TH, Bordevik M, Frøystad MK, Baklien Å (2007) Identification of *Francisella* sp. from Atlantic salmon, *Salmo salar* L., in Chile. J Fish Dis 30:505–507

Birkbeck TH, Laidler LA, Grant AN, Cox DI (2002) *Pasteurella skyensis* sp. nov., isolated from Atlantic salmon (*Salmo salar* L.). Int J Syst Evol Microbiol 52:699–704

Birrell J, Mitchell S, Bruno DW (2003) *Piscirickettsia salmonis* in farmed Atlantic salmon *Salmo salar* in Scotland. Bull Eur Assoc Fish Pathol 23:213–217

Bohle H, Tapia E, Martínez A, Rozas M, Figueroa A, Bustos P (2009) *Francisella philomiragia*, a bacteria associated with high mortalities in Atlantic salmon (*Salmo salar*) cage-farmed in Llanquihue lake. Arch de Med Vet 41:237–244

Bradley TM, Newcomer CE, Maxwell KO (1988) Epitheliocystis associated with massive mortalities of cultured lake trout *Salvelinus namaycush*. Dis Aquat Org 4:9–17

Bricknell IR, Bruno DW, Stone J (1996) *Aeromonas salmonicida* infectivity studies in goldsinny wrasse, *Ctenolabrus rupestris* (L). J Fish Dis 19:469–474

Bruno DW (1986a) Scottish experience with bacterial kidney disease in farmed salmonids between 1976 and 1985. Aquac Fish Manag 17:185–190

Bruno DW (1986b) Histopathology of bacterial kidney disease in laboratory infected rainbow trout, *Salmo gairdneri*, Richardson, and Atlantic salmon, *Salmo salar* L. J Fish Dis 9:523–537

Bruno DW (1988) The relationship between auto-agglutination, cell surface hydrophobicity and virulence of the fish pathogen *Renibacterium salmoninarum*. FEMS Microbiol Lett 51:135–140

Bruno DW (1990) Presence of a saline extractable protein associated with virulent strains of the fish pathogen, *Renibacterium salmoninarum*. Bull Eur Assoc Fish Pathol 10:8–10

Bruno DW (2004) Prevalence and diagnosis of bacterial kidney disease (BKD) in Scotland between 1990 and 2002. Dis Aquat Org 59:125–130

Bruno DW, Brown LL (1999) The occurrence of *Renibacterium salmoninarum* within vaccine adhesion components from Atlantic salmon, *Salmo salar* L. and Coho salmon, *Oncorhynchus kisutch* Walbaum. Aquaculture 170:1–5

Bruno DW, Munro ALS (1986) Observations on *Renibacterium salmoninarum* and the salmonid egg. Dis Aquat Org 1:83–87

Bruno DW, Munro ALS (1989) Immunity in Atlantic salmon, *Salmo salar* L., fry following vaccination against *Yersinia ruckeri*, and the influence of body weight and infectious pancreatic necrosis virus (IPNV) on the detection of carriers. Aquaculture 89:205–211

Bruno DW, Hastings TS, Ellis AE (1986) Histopathology, bacteriology and experimental transmission of a cold water vibriosis in Atlantic salmon *Salmo salar*. Dis Aquat Org 1:163–168

Bruno DW, Griffiths J, Mitchell CG, Wood BP, Fletcher ZJ, Drobniewski FA, Hastings TS (1998a) Pathology attributed to *Mycobacterium chelonae* infection among farmed and laboratory-infected Atlantic salmon *Salmo salar*. Dis Aquat Org 33:101–109

Bruno DW, Griffiths J, Petrie J, Hastings TS (1998b) *Vibrio viscosus* in farmed Atlantic salmon *Salmo salar* in Scotland, field and experimental observations. Dis Aquat Org 34:161–166

Del-Pozo J, Crumlish M, Ferguson HW, Turnbull JF (2009) A retrospective cross-sectional study on "*Candidatus arthromitus*" associated rainbow trout gastroenteritis (RTGE) in the UK. Aquaculture 290:22–27

Del-Pozo J, Turnbull JF, Crumlish M, Ferguson HW (2010a) A study of gross, histological and blood biochemical changes in rainbow trout, *Oncorhynchus mykiss* (Walbaum), with rainbow trout gastroenteritis (RTGE). J Fish Dis 33:301–310

Del-Pozo J, Turnbull JF, Ferguson HW, Crumlish M (2010b) A comparative molecular study of the presence of "*Candidatus arthromitus*" in the digestive system of rainbow trout, *Oncorhynchus mykiss* (Walbaum), healthy and affected with rainbow trout gastroenteritis. J Fish Dis 33:241–250

Draghi A, Bebak J, Daniels S, Tulman ER, Geary J, West AB, Popov V, Frasca S Jr (2010) Identification of *Candidatus* Piscichlamydia salmonis in Arctic charr *Salvelinus alpinus* during a survey of charr production facilities in North America. Dis Aquat Org 89:39–49

Egidius E, Andersen K, Clausen E, Raa J (1981) Cold-water vibriosis or "Hitra disease" in Norwegian salmonid farming. J Fish Dis 4:353–354

Egidius E, Wiik R, Andersen KA, Hjeltnes B (1986) *Vibrio salmonicida* sp. nov., a new fish pathogen. Int J Syst Bacteriol 36:518–520

Faisal M, Loch TP, Fujimoto M, Woodiga SA, Eissa AE, Honeyfield DC, Wolgamood M, Walker ED, Marsh TL (2011) Characterization of novel *Flavobacterium* spp. Involved in the mortality of Coho salmon (*Oncorhynchus kisutch*) in their early life stages. Aquac Res Dev S2:005. doi:10.4172/2155-9546.S2-005

Fryer JL, Hedrick RP (2003) *Piscirickettsia salmonis*: a gram-negative intracellular bacterial pathogen of fish. J Fish Dis 26:251–262

Holten-Andersen L, Dalsgaard I, Buchmann K (2012) Baltic salmon, *Salmo salar*, from Swedish river Lule Älv is more resistant to furunculosis compared to rainbow trout. PLoS One 7(1):e29571. doi:10.1371/journal.pone.0029571

Keeling SE, Johnston C, Wallis R, Brosnahan CL, Gudkovs N, McDonald WL (2012) Development and validation of real-time PCR for the detection of *Yersinia ruckeri*. J Fish Dis 35:119–125

Loch TP, Kumar R, Xu W, Faisal M (2011) *Carnobacterium maltaromaticum* infections in feral *Oncorhynchus* spp. (Family Salmonidae) in Michigan. J Microbiol 49:703–713

Lunder T (1992) "Winter ulcer" in Atlantic salmon: a study of pathological changes, transmissibility and bacterial isolates. Dr. Scient. thesis. Norwegian School of Veterinary Science

Mauel MJ, Giovannoni SJ, Fryer JL (1996) Development of polymerase chain reaction assays for detection, identification, and differentiation of *Piscirickettsia salmonis*. Dis Aquat Org 26:189–195

Mitchell SO, Rodger HD (2011) A review of infectious gill diseases in marine salmonid fish. J Fish Dis 34:411–432

Mitchell SO, Steinum T, Holland C, Rodger H, Colquhoun DJ (2010) Epitheliocystis in Atlantic salmon, *Salmo salar* L., farmed in fresh water in Ireland is associated with "*Candidatus* Clavochlamydia salmonicola" infection. J Fish Dis 33:665–673

Nese L, Enger Ø (1993) Isolation of *Aeromonas salmonicida* from salmon lice *Lepeophtheirus salmonis* and marine plankton. Dis Aquat Org 16:79–81

Nilsen H, Johansen R, Colquhoun DJ, Kaada I, Bottolfsen K, Vågnes Ø, Olsen AB (2011) *Flavobacterium psychrophilum* associated with septicaemia and necrotic myositis in Atlantic salmon *Salmo salar*: a case report. Dis Aquat Org 97:37–46

Nowak BF, La Patra S (2006) Epitheliocystis in fish. J Fish Dis 29:573–588

Oidtmann B, Verner-Jeffreys D, Pond M, Peeler EJ, Noguera PA, Bruno DW, LaPatra SE, St-Hilaire S, Schubiger CB, Snekvik K, Crumlish M, Green DM, Metselaar M, Rodger H, Schmidt-Posthaus H, Galeotti M, Feist SW (2013) Differential characterisation of emerging skin diseases of rainbow trout – a standardised approach to capturing disease characteristics and development of case definitions. J Fish Dis. doi:10.1111/jfd.12086

Olsen AB, Nilsen H, Sandlund N, Mikkelsen H, Sørum H, Colquhoun DJ (2011) *Tenacibaculum* sp. associated with winter ulcers in sea-reared Atlantic salmon *Salmo salar*. Dis Aquat Org 94:189–199

Ostland VE, Byrne PJ, Hoover G, Ferguson HW (2000) Necrotic myositis of rainbow trout, *Oncorhynchus mykiss* (Walbaum): proteolytic characteristics of a crude extracellular preparation from *Flavobacterium psychrophilum*. J Fish Dis 23:329–336

Pacha RE, Porter S (1968) Characteristics of myxobacteria isolated from the surface of freshwater fish. Appl Microbiol 16:1901–1906

Poppe T, Håstein T, Salte R (1985) 'Hitra disease' (haemorrhagic syndrome) in Norwegian salmon farming: present status. In: Ellis AE (ed) Fish shellfish pathology. Academic Press, London, pp 223–229

Powel P, Carson J, van Gelderen R (2004) Experimental induction of gill disease in Atlantic salmon *Salmo salar* smolts with *Tenacibaculum maritimum*. Dis Aquat Org 61:179–185

Rhodes LD, Rice CA, Greene CM, Teel DJ, Nance SL, Moran P, Durkin CA, Gezhegne SB (2011) Nearshore ecosystem predictors of a bacterial infection in juvenile Chinook salmon. Mar Ecol Prog Ser 432:161–172

Rodríguez L, Gallardo C, Acosta F, Nietop P, Real F (1998) *Hafnia alvei* as an opportunistic pathogen causing mortality in brown trout, *Salmo trutta*. J Fish Dis 21:365–370

Schmidt-Posthaus H, Polkinghorne A, Nufer L, Schifferli A, Zimmerman D, Segner H, Steiner P, Vaughan L (2011) A natural freshwater origin for two chlamydial species, *Candidatus* Piscichlamydia salmonis and *Candidatus* Clavochlamydia salmonicola, causing mixed infections in wild brown trout (*Salmo trutta*). Environ Microbiol. doi:10.1111/j.1462-2920.2011.02670.x

Sørum H, Poppe TT, Olsvik Ø (1988) Plasmids in *Vibrio salmonicida* isolated from salmonids with haemorrhagic syndrome (Hitra disease). J Clin Microbiol 26:1679–1683

Starliper CE (2010) Bacterial coldwater disease of fishes caused by *Flavobacterium psychrophilum*. J Adv Res 2:97–108

Steinum T, Kvellestad A, Colquhoun D, Heum M, Mohammad S, Grøndtvedt RN, Falk K (2010) Microbial studies of proliferative gill inflammation in Norwegian salt water reared Atlantic salmon. Dis Aquat Org 91:201–211

Toenshoff E, Kvellestad A, Mitchell SO, Steinum TM, Falk K, Colquhoun DJ, Horn M (2012) A novel Betaproteobacterial agent of gill epitheliocystis in salt water farmed Atlantic salmon (*Salmo salar*). PLoS One 7(3):e32696. doi:10.1371/journal.pone 0032696

Verner-Jeffreys DW, Algoet M, Feist SW, Bateman K, Peeler EJ, Branson EJ (2006) Studies on red mark syndrome. Finfish News 1:19–22

Verner-Jeffreys DW, Pond MJ, Peeler EJ, Rimmer GSE, Oidtmann B, Way K, Mewett J, Jeffrey K, Batemann K, Reese RA, Feist SW

(2008) Emergence of cold water strawberry disease of rainbow trout *Oncorynchus mykiss* in England and Wales: outbreak investigations and transmission studies. Dis Aquat Org 79:207–218

Wiklund T, Madsen L, Bruun MS, Dalsgaard I (2000) Detection of *Flavobacterium psychrophilum* from fish tissue and water samples by PCR amplification. J Appl Microbiol 88:299–307

Yañez AJ, Valenzuela K, Silva H, Retamales J, Romero A, Enriquez R, Figueroa J, Claude A, Gonzalez J, Avendaño-Herrera R, Carcamo JG (2012) Broth medium for the successful culture of the fish pathogen *Piscirickettsia salmonis*. Dis Aquat Org 97:197–205

Zamora L, Vela AI, Palacios MA, Domínguez L, Fernández-Garayzábal JF (2012) First isolation and characterization of *Chryseobacterium shigense* from rainbow trout. BMC Vet Res 8:77–81

Zerihun MA, Nilsen H, Hodeland S, Colquhoun DJ (2011) *Mycobacterium salmoniphilum* infection in farmed Atlantic salmon, *Salmo salar* L. J Fish Dis 34:769–781

Fungal and Related Oomycete Infections

7

Abstract

The oomycetes (fungal-like) occur as primary and secondary agents of infection among wild and farmed salmonids in fresh and sea water. These infections can prove fatal, at least for farmed fish, if preventative measures are not undertaken. Molecular sequencing has improved our understanding of the phylogenetic relationships within this taxonomically diverse group with current evidence indicating these organisms evolved from simple holocarpic marine parasites. This chapter covers the commonly encountered oomycetes infections in salmonids.

Keywords

Oomycetes • Fungus • Salmon • Trout

The oomycetes (fungal-like) are common and occur as primary and secondary agents of infection among wild and farmed salmonids in fresh and sea water (Fig. 7.1). These infections can prove fatal, at least for farmed fish, if preventative measures are not undertaken. Molecular sequencing has improved our understanding of the phylogenetic relationships within this taxonomically diverse group with current evidence indicating these organisms evolved from simple holocarpic marine parasites. Examples of key infections are summarised in Table 7.1.

7.1 *Saprolegnia* spp.

Oomycete infections (family Saprolegniaceae) may be commensal or act directly as primary infectious agents and this group, in general terms, is represented by two species, *S. parasitica* and *S. diclina* affecting wild and farmed fish at all developmental stages in fresh water (Figs. 7.2, 7.3, 7.4 and 7.5). The presence of superficial cotton-wool-like tufts, particularly on the integument and gills of host fish or eggs, is likely to be the result of *Saprolegnia* infection. Thin, white or grey threads with circular or crescent-shaped colonies grow by radial extension until adjacent lesions merge (Figs. 7.6 and 7.7). Diseased fish become increasingly lethargic with a loss of equilibrium shortly before death, following lethal haemodilution. Respiratory difficulties may also feature when infection is associated with the gills.

Microscopic examination shows the characteristic filamentous, coenocytic mycelium of non-septate hyphae with many zoosporangia. An early infection comprises rapid degenerative changes in the muscle resulting in a diffuse oedema. As the infection radiates from the focus of infection, more of the epidermis is destroyed, and consequently hyphae can penetrate the basement membrane, with growth continuing into the dermis, hypodermis and musculature (Fig. 7.8). Thrombi are frequently observed in blood-vessels as a result of the penetrating hyphae. There is loss of integrity of the integument and oedema of the hypodermis accompanied by marked myofibrillar degenerative changes. Severe infection causes swelling in the inter-myotomal connective tissue, loss of nuclei and a minor host reaction. More aggressive lesions show deeper myofibrillar and focal cellular necrosis, spongiosis or intracellular oedema and ultimate sloughing of the epidermis.

A marked lymphocytopenia and significant impairment of haematopoietic tissue is reported. Lymphoid cell degeneration, cell depletion, vascular alterations within blood-vessels

D.W. Bruno et al., *A Colour Atlas of Salmonid Diseases*,
DOI 10.1007/978-94-007-2010-7_7, © Springer Science+Business Media B.V. 2013

Ernest Hemingway 1927: Big two-hearted river In: The Nick Adams stories

'He had wet his hand before he touched the trout, so he would not disturb the delicate mucus that covered him. If a trout was touched with a dry hand, a white fungus attacked the unprotected spot. Years before when he had fished crowded streams, with fly fishermen ahead of him and behind him, Nick had again and again come on dead trout, furry with white fungus, drifted against a rock, or floating belly up in some pool. Nick did not like to fish with other men on the river. Unless they were of your party, they spoiled it.'

Fig. 7.1 Brook trout with severe *Saprolegnia* infection following handling

Table 7.1 Principal fungal and related oomycete infections of salmonids

Pathogen	Family	Principal salmonid host	Environment
Saprolegnia spp.	Saprolegniaceae	All	FW
Exophiala salmonis, *E. psycrophila*	Herpotrichiellaceae	Rainbow trout, Atlantic salmon	SW
Phialophora sp.	Herpotrichiellaceae	Rainbow trout, Atlantic salmon	FW
Isaria farinosa (*Paecilomyces farinosus*)	Cordycipitaceae	Atlantic salmon parr	FW
Phoma herbarum	Didymellaceae	Chinook salmon fingerlings	FW

FW freshwater, SW sea water

Fig. 7.2 Adult Atlantic salmon returning to spawn in freshwater and infected with *Saprolegnia*

Fig. 7.3 *Saprolegnia* infection on head and at the base of pectoral fins of a wild adult Atlantic salmon

and hypertrophy of sinusoidal endothelial cells also take place. Considerable changes occur in the structure of the parenchyma, with large areas showing a marked decrease in cellular density.

Saprolegnia spp. can be isolated on a variety of media and the identification of non-septate hyphae from culture or suspected lesions is used to support a diagnosis. PCR methods are also available.

7.2 *Isaria farinosa* (formally *Paecilomyces farinosus*)

Isaria farinosa (*Paecilomyces farinosus*) is a soil associated fungus that has been recorded as the cause of sporadic mortality among Atlantic salmon parr. Affected fish are

Fig. 7.4 Extensive *Saprolegnia* infection comprising head, ventral body and tail in wild adult Atlantic salmon

Fig. 7.5 Spawning whitefish with patchy *Saprolegnia* infection on head, body flank and fins

Fig. 7.6 Adult Atlantic salmon with extensive *Saprolegnia* infection on dorsal fin and adjacent skin

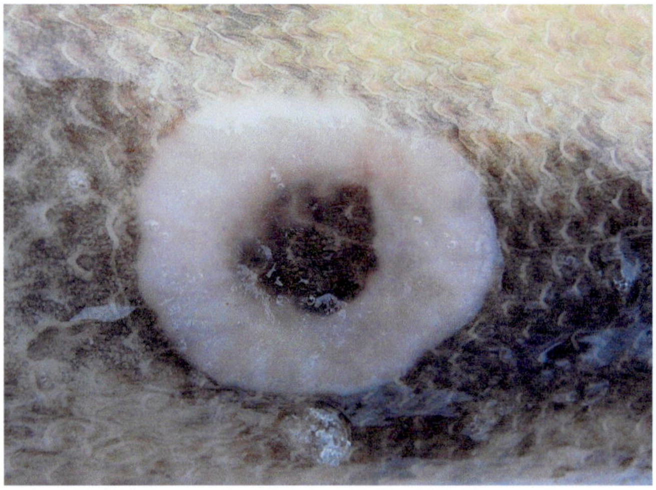

Fig. 7.7 Atypical circular infection on body flank of spawning wild Atlantic salmon

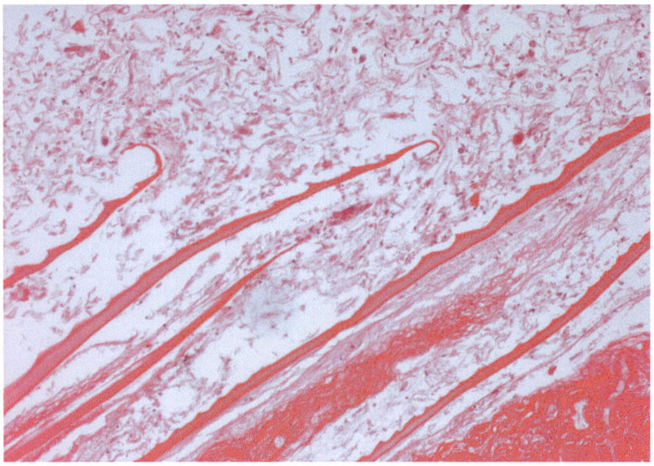

Fig. 7.8 *Saprolegnia* mycelium covering the scales of rainbow trout. Low power

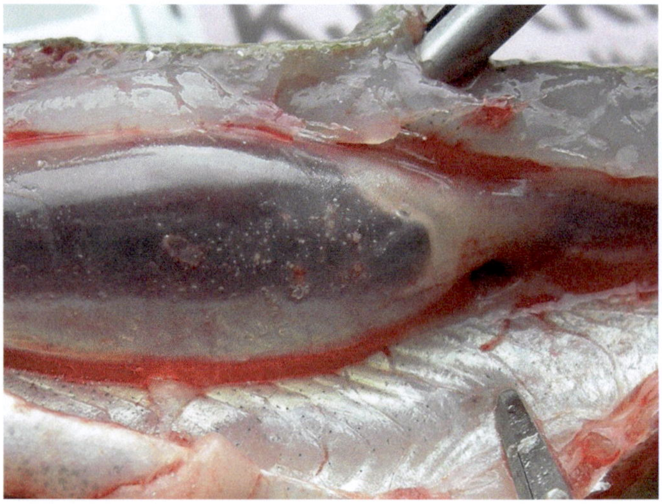

Fig. 7.9 Swim bladder mycosis in the posterior region of the organ in Atlantic salmon parr

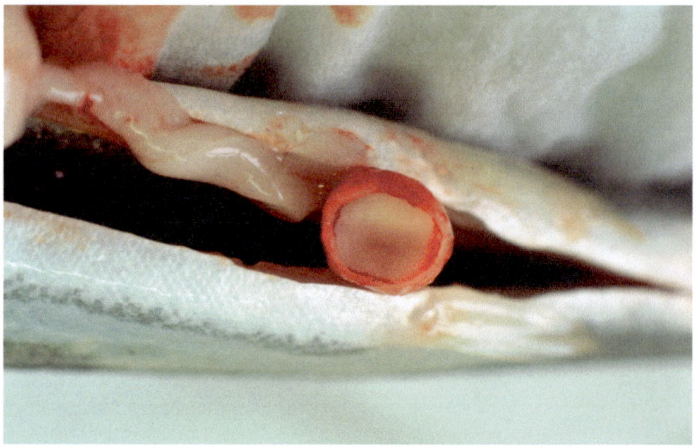

Fig. 7.10 *Isaria farinosa* in Atlantic salmon parr. The swim bladder has been cut transversely and the myceium has occluded the lumen

generally darker, show some loss of balance and an enlarged abdomen with swelling and reddening of the vent. The fungus initially develops in the swim bladder which becomes thickened and filled with a white coloured mass of hyphae (Figs. 7.9 and 7.10). In severe cases it can also affect the gut, peritoneum and skeletal muscle.

In stained sections the hyphae are seen to penetrate from the swim bladder lumen to the outer fibrous coat completely destroying the tissue (Figs. 7.11 and 7.12). Hyphae have not been observed in other organs.

I. farinosa grows moderately well on a variety of media forming a dense basal felt from which conidophores arise, the colonies become slightly granular and tufted with the development of the conidia. *Isaria* is a branching, septate fungus and a ubiquitous insect pathogen. A provisional diagnosis may be possible at fish autopsy with identification

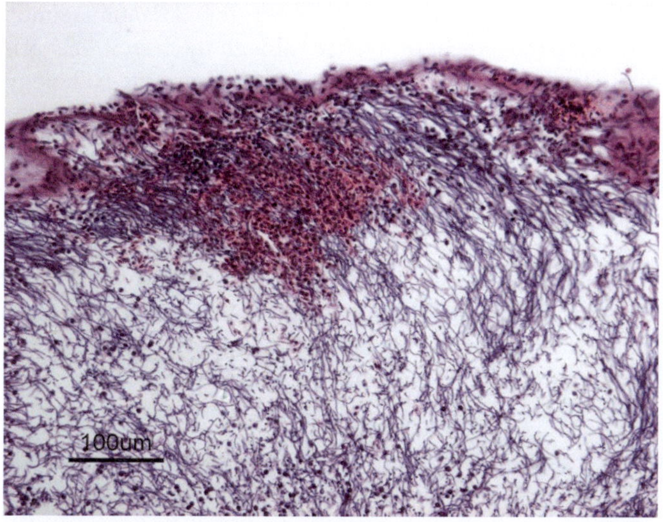

Fig. 7.11 Swimbladder mycosis in Atlantic salmon parr. There is haemorrhage in the wall of the mycelium-filled organ. Medium power

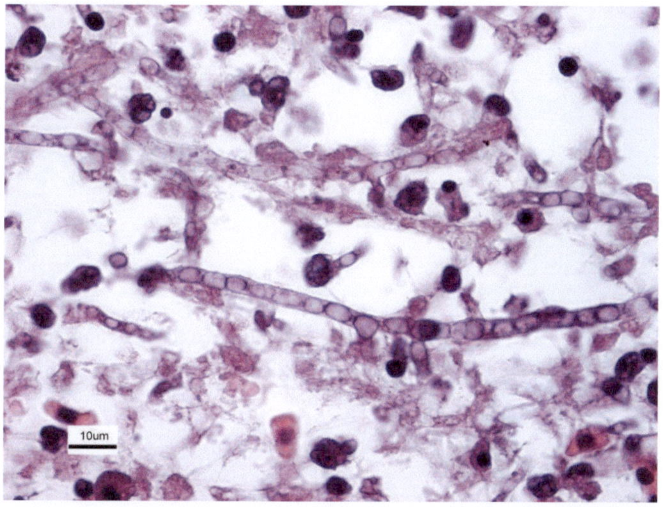

Fig. 7.12 Swimbladder mycosis in Atlantic salmon parr. High power

of the fungi from morphological characteristics and rDNA ITS sequence data.

7.3 *Phialophora* sp.

Phialophora is both parasitic and saprophytic and described as a systemic although infrequent infection of rainbow trout and Atlantic salmon at low temperatures. In rainbow trout, infection has been linked with cerebral mycetoma. Salmon show fin haemorrhage and petechiae along the ventral surface. The internal organs are pale, the swim bladder filled with a whitish, mucoid material and adherent to the body wall and other internal organs, including the posterior kidney and intestine. Apart from the haemorrhage in the adipose tissue and mesentery, limited tissue response is recorded to the hyphae. Microscopic examination reveals a dense mycelium with septate and branching hyphae. The route of infection is believed to be via the pneumatic duct when fry fill the swim bladder with air from the surface. The source of infection is unknown, but the genus *Phialophora* has a wide distribution in the environment. In culture this fungus is slow growing, with a thin-walled branching mycelium.

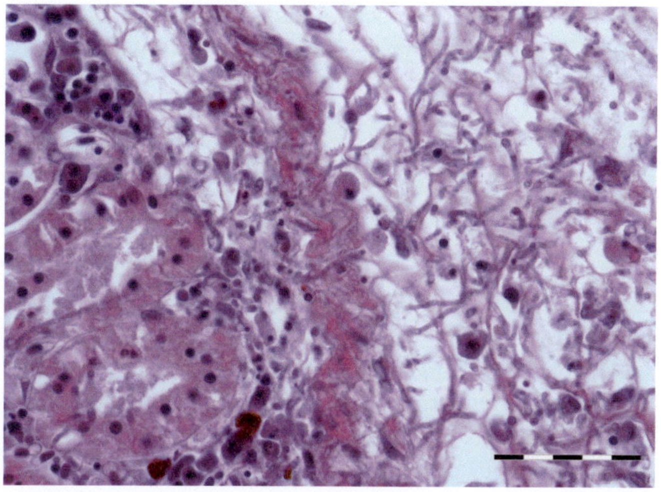

Fig. 7.13 Kidney of Chinook salmon infiltrated with *Phoma herbarum*. Note widespread necrosis. Medium power

7.4 *Phoma herbarum*

Systemic infections of *Phoma* occur sporadically but particularly in hatchery-reared Chinook salmon fingerlings from the Pacific Northwest coast of America. Affected fish show abnormal swimming behaviour, loss of equilibrium, exophthalmia and sometimes haemorrhagic protrusion of the vent. Internally, the most characteristic lesions comprise a whitish creamy viscous mass in the swim bladder and thickening of the wall, ascites and adhesions between visceral organs. Microscopically infection is associated with necrosis and haemorrhage, with mild to moderate lymphocytic and histiocytic infiltration. Infection can spread to other organs resulting in nephritis and a severe, systemic, granulomatous reaction which includes the wall and lumen of the dorsal aorta (Fig. 7.13).

Phoma spp., are saprophytic on a wide range of organic material, including soil, plants and sewage. Infection is believed to be via the pneumatic duct, however the absence of pycnidia indicates that fish are not a natural site for development. On Sabouraud medium with added dextrose growth occurs as brownish colonies and microscopically broad septate-branched hyphae that are hyaline, and 5–12 μm in diameter are reported. Diagnosis is based on morphological characterization, the histological demonstration of PAS-positive hyphae in affected organs which can be supported through molecular tools.

7.5 *Exophiala* spp.

Exophiala salmonis is an anamorphic black yeast, a fungus from the family Herpotrichiellaceae. This species causes a low prevalence internal systemic mycosis of marine-reared salmonids e.g. Atlantic salmon. Infected fish may continue to feed normally, but display erratic swimming movements which may be followed by whirling behaviour. Exophthalmia and cranial cutaneous ulcers are common, although these clinical signs are not considered pathognomonic. Considerable distension of the abdomen is reported. Internally, an opaque capsule and enlargement of the kidney is typical with large raised, greyish-white nodules containing variable quantities of hyphae (Figs. 7.14 and 7.15). The host attempts to limit vascular invasion with the development of a marked systemic granulomatous response involving macrophages and multinucleate giant cells (Figs. 7.16, 7.17, 7.18 and 7.19). Fibrosis and atrophy develop as the hyphae penetrate the kidney tubules and blood vessels, as well as other organs such as the heart, spleen and liver, where an acute multi-focal hepatitis can be observed. An eosinophilic gastritis and enteritis occurs within the gut. In severe infections the musculature may be discoloured. A concurrent infection with polycystic liver has been reported.

In fresh water fish the species, *E. psycrophila* has been described from rainbow trout but also from Atlantic salmon in sea water from Norway. A cranial location for *E. psycrophila* has been reported for Atlantic salmon following movement of hyphae through the lateral line system. Healing lesions are fibrous in nature and the pathology associated with *E. psycrophila* is similar to that described for *E. salmonis*.

A presumptive diagnosis may be made from gross lesions and the presence of pigmented septate hyphae readily observed in H&E sections. Similarly, staining infected tissue with periodic acid-Schiff's is a useful diagnostic tool. Cultures of *E. salmonis* on Sabouraud's agar appear grey, with a darker reverse, abundant spores and a colony growth of 5–8 mm at 25 °C after approximately 14 days. Growth is not recorded at 37 °C.

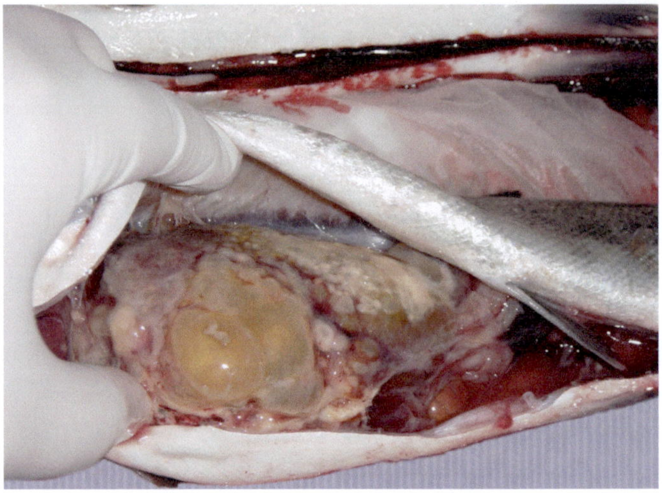

Fig. 7.15 Extensive granulomatous response due to *Exophiala* infection and polycystic liver in Atlantic salmon post smolt

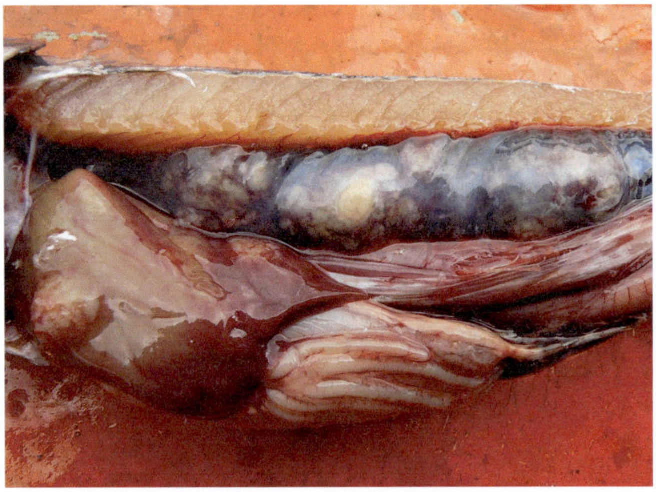

Fig. 7.14 Systemic infection with *Exophiala* sp. in farmed Atlantic salmon post smolt. Extensive granulomatous response in kidney and liver

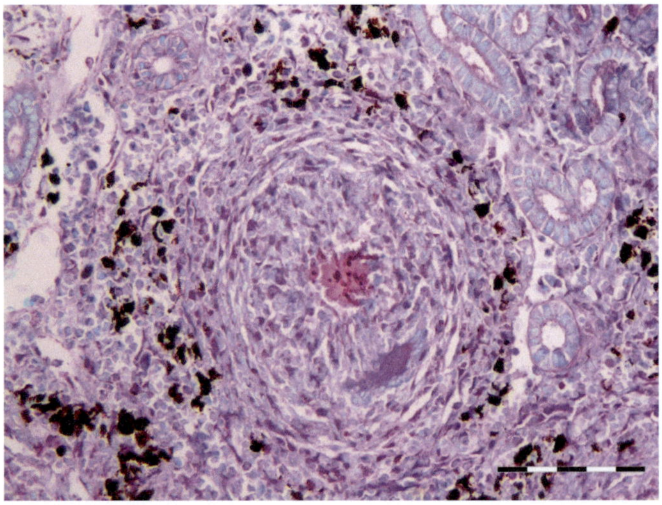

Fig. 7.16 *Exophiala* infection in farmed Atlantic salmon. Granuloma with central hyphae in kidney. Bar = 100 μm

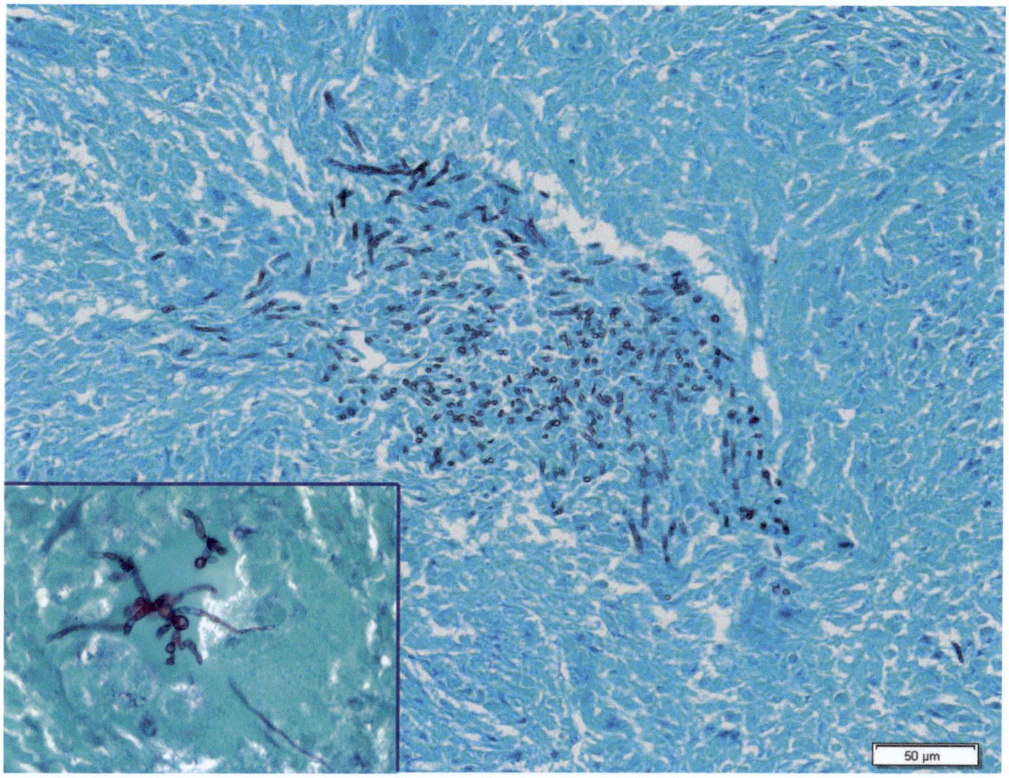

Fig. 7.17 *Exophiala* in kidney of Atlantic salmon. *Insert* shows detail of hyphae. High power

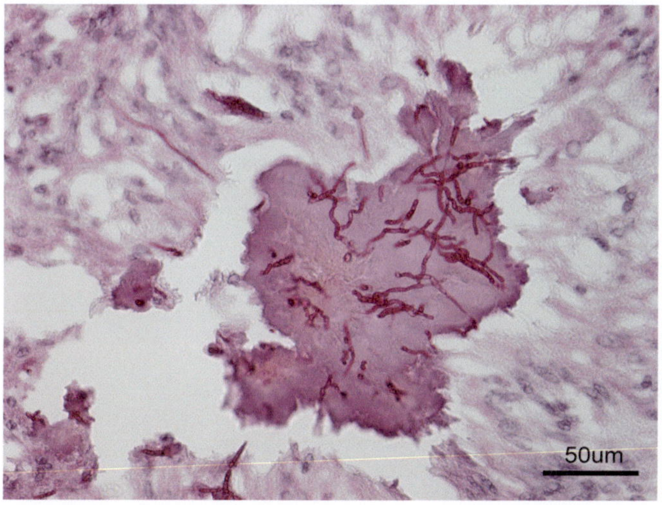

Fig. 7.18 *Exophiala* infection with giant cell formation in farmed Atlantic salmon. Medium power

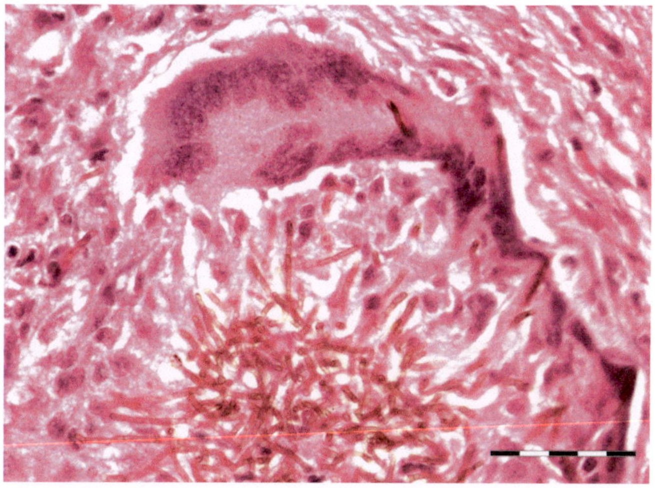

Fig. 7.19 *Exophiala* infection in the liver of a farmed Atlantic salmon. Well-defined necrosis and granulomatous response with centrally located multinucleated giant cell. Medium power

Further Reading

Blazer VS, Wolke RE (1979) An *Exophiala*-like fungus as the cause of a systemic mycosis of marine fish. J Fish Dis 2:145–152

Bruno DW (1989) Observations on a swim bladder fungus of farmed Atlantic salmon, *Salmo salar* L. Bull Eur Assoc Fish Pathol 9:7–8

Bruno DW, Stamps DJ (1987) Saprolegniasis of Atlantic salmon *Salmo salar* L. fry. J Fish Dis 10:513–517

Carmichael JW (1966) Cerebral mycetoma of trout due to a *Phialophora*-like fungus. Sabouraudia 6:120–123

Ellis AE, Waddell IF, Minter DW (1983) A systemic fungal disease in Atlantic salmon parr, *Salmo salar* L., caused by a species of *Phialophora*. J Fish Dis 6:511–523

Faisal M, Elsayed E, Fitzgerald SD, Silva V, Mendoza L (2007) Outbreaks of phaeohyphomycosis in the chinook salmon (*Oncorhynchus tshawytscha*) caused by *Phoma herbarum*. Mycopathologia 163:41–48

Khoshkho Z, Matin RH (2013) Efficacy of medication therapy to control of Saprolegniasis on rainbow trout (*Oncorhynchus mykiss*) eggs. Global Vet 10:80–83

Pedersen O, Langvad F (1988) *Exophiala psychrophila* sp. nov., a pathogenic species of the black yeasts isolated from farmed Atlantic salmon. Mycol Res 92:53–156

Richards RH, Holliman A, Helgason S (1978) *Exophiala salmonis* infection in Atlantic salmon *Salmo salar* L. J Fish Dis 1:357–368

Ross AJ, Yasutake WT, Leek S (1975) *Phoma herbarum*, a fungal plant saprophyte, as a fish pathogen. J Fish Res Board Can 32:1648–1652

Thoen E, Evensen Ø, Skaar I (2012) Pathogenicity of *Saprolegnia* spp. to Atlantic salmon, *Salmo salar* L., eggs. J Fish Dis 34: 601–608

Protists

8

Abstract

The Protists are a large group of eukaryotic microorganisms with a taxonomy that is under constant revision. Molecular studies show the group includes diverse and sometimes distantly related phyla that share relatively simple levels of organization that can be unicellular or multicellular, but without specialized tissues. In fish, Protists range from true parasites that may cause significant mortality, to those that show commensalism. Conditions such as overcrowding or poor water quality, as well as other changes in environmental conditions may allow parasites to rapidly increase in number and as a result fish become vulnerable to infection with an increased chance of invasion by secondary pathogens. This chapter presents a selection of the most common Protists reported from salmonid species.

Keywords

Protist • Salmon • Trout

The Protist are a group of eukaryotic microorganisms with a changing taxonomy. Current molecular information indicates that the group includes diverse and sometimes distantly related phyla, that share relatively simple levels of organization and can be unicellular or multicellular but without specialized tissues. In general they are not closely related through evolution and have different life cycles, trophic levels, modes of locomotion and cellular structures. In fish, Protists range from true parasites that may cause significant mortality, to those that show commensalism. Conditions such as overcrowding or poor water quality and other changes in environmental conditions may allow parasites to rapidly increase in number and as a result, the host may lose weight, suffer from osmotic distress and become susceptible to predation with an increased chance of invasion by secondary pathogens. Chosen examples from this group affecting salmonids are presented with emphasis on the host effects, while details of the parasite ecology and life cycle are only included where relevant to the understanding of the infection or pathogenesis. Common Protists, the Phyla or class, location on the fish and environment are summarised in Table 8.1.

8.1 Amoebozoa

8.1.1 *Paramoeba perurans*

Amoebic gill disease (AGD) is attributed to colonization by *Paramoeba perurans* and of global significance to the aquaculture industry. *Neoparamoeba* has recently been reclassified to *Paramoeba*. Infestation is associated with a severe gill proliferative response and consequently, significant mortalities can occur in Atlantic salmon reared in sea water. AGD outbreaks normally occur from late summer to early winter with 10–12 °C believed to be the threshold for clinical signs. However, outbreaks have occurred at lower temperatures and recent studies show high salinity as a more relevant risk factor. The amoeba is usually confined to the gill surface but can penetrate the epithelium and occur internally. Heavily infected fish are lethargic and congregate at the water surface. The opercula are flared with excessive mucous from the gills and the presence of focal, whitish patches (Fig. 8.1). Fresh preparations of amoebae removed from infected gills show large distinct pseudopodia, a well-delineated hyaloplasm and measure approximately 15–40 µm. High salinity (low rainfall),

D.W. Bruno et al., *A Colour Atlas of Salmonid Diseases*,
DOI 10.1007/978-94-007-2010-7_8,

Table 8.1 Principal Protists described for salmonids

Name	Class or group	Common location	Environment
Paramoeba perurans	Amoebozoa	Gills	SW
Capriniana piscium	Ciliophora	Gills	FW
Chilodonella piscicola	Ciliophora	Gills, external surface	FW
Epistylis spp.	Ciliophora	Gills, external surface	FW
Ichthyophthirius multifiliis	Ciliophora	Oral cavity, gills, external surface	FW
Trichodina truttae	Ciliophora	Gills, external surface	FW, SW
Ichthyophonus hoferi	Mesomycetozoea	Musculature	FW, SW
Sphaerothecum destruens	Mesomycetozoea	Macrophages particularly in the spleen and kidney	FW
Dermocystidium salmonis	Mesomycetozoea	Oral cavity, external surface	FW, SW
Kabatana takedai	Microsporidia	Musculature, gills	FW, SW
Loma salmonae	Microsporidia	Gills	FW
Paranucleospora theridion (*Desmozoon lepeophtherii*)	Microsporidia	Many organs	SW
Cryptobia salmositica	Sarcomastigophora	Blood	FW
Ichthyobodo spp.	Sarcomastigophora	Gills, external surface	FW, SW
Spironucleus salmonicida	Sarcomastigophora	Systemic	FW, SW
Spironucleus salmonis	Sarcomastigophora	Intestine	FW
Spironuclueus barkhanus	Sarcomastigophora	Gall bladder, intestine	FW

FW freshwater, SW sea water

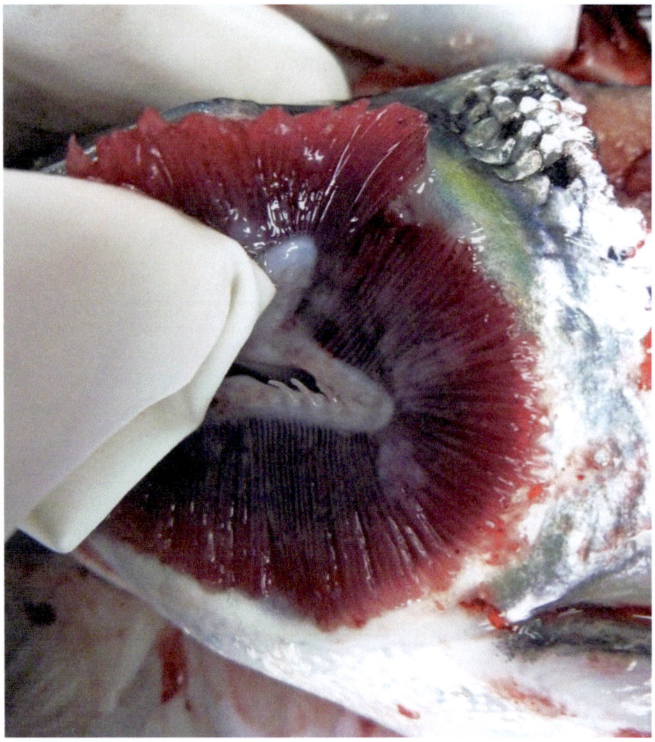

Fig. 8.1 Amoebic gill disease in seawater-farmed Atlantic salmon. Note pale gills and nodular, slimy patches

warmer temperatures, suspended organic matter, cage fouling, high stocking densities and previous gill damage, have also been reported as risk factors.

Primary attachment by amoeba is associated with branchial irritation and localized host cellular alterations including squamation-stratification of the epithelium (Fig. 8.2). This is followed by epithelial hypertrophy and stratification of epithelia at lesion surfaces, with recruitment of mucous cells to affected areas, reduced chloride cells and the formation of large interlamellar lacunae or vesicles. Small to medium sized lacunae may contain amoeba, whereas larger lacunae are generally clear of any cellular debris (Fig. 8.3). Very similar gill lesions associated with amoeba infections on brook trout reared in freshwater, have also been reported (Fig. 8.4). Proactive treatment by freshwater bathing is carried out when gross gill assessment (i.e. a gill score) indicates a moderate level of infestation in a population. However, re-infections have been reported at 2 weeks post bathing. During the post-bath period, non-AGD lesions including haemorrhage, necrosis and regenerative hyperplasia have been observed.

Paramoeba can be detected from wet preparations of gill tissue, although subsequent histological examination is recommended to improve diagnostic accuracy. Histologically, hyperplasia, lamellar fusion, vesicles or lacunae, flattened epithelial cells ('pavement cells') and the presence of amoeba form the basis of a diagnosis. Species identification requires molecular testing using a quantitative duplex real-time PCR. Differential diagnosis includes other Protists, water borne irritants and bacteria.

8.2 Ciliophora

8.2.1 *Capriniana piscium*

Capriniana (previously known as *Trichophrya*), is an ectocommensal organism that commonly occurs on the gills of several fresh water fish species. The shape of the

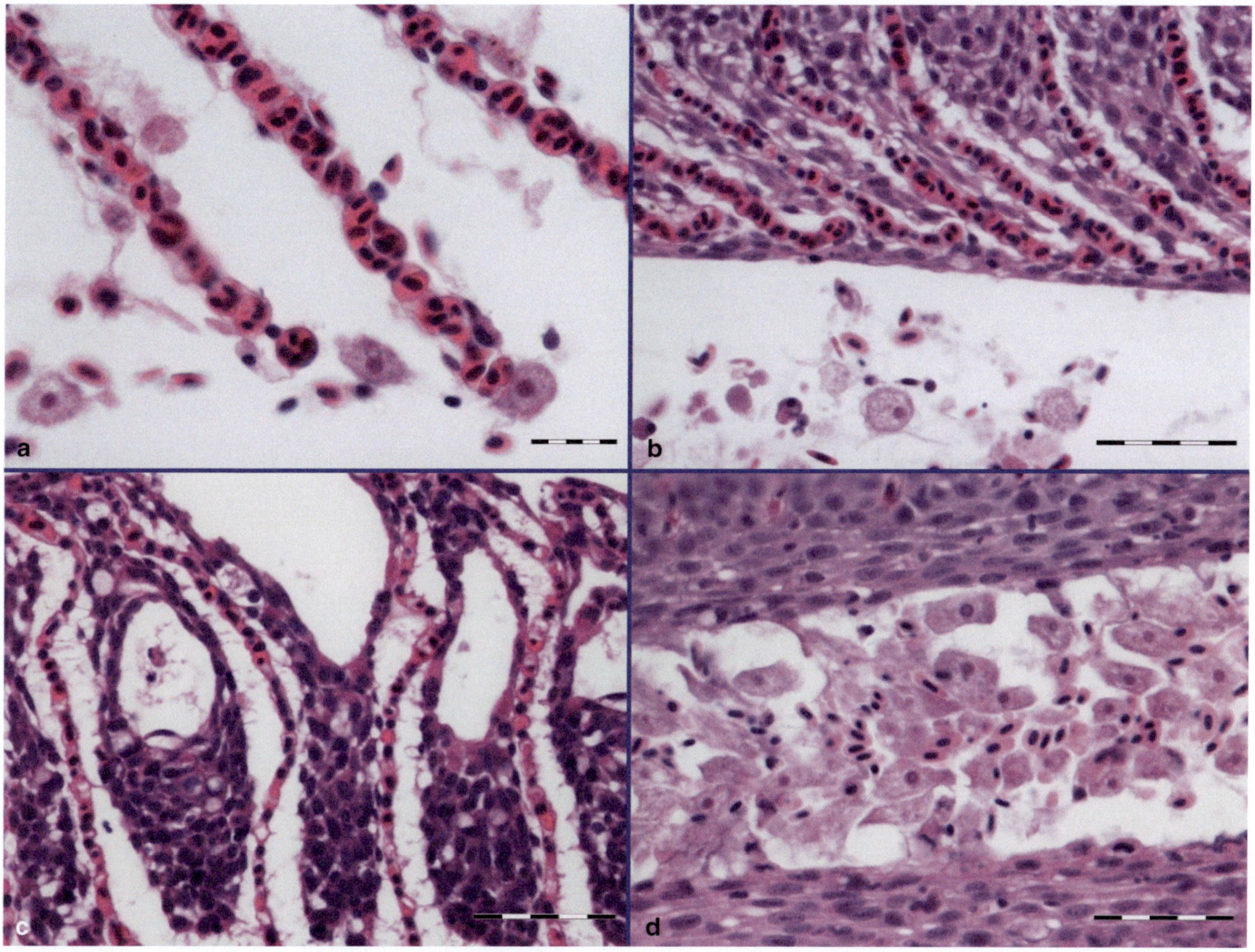

Fig. 8.2 Amoebic gill disease in farmed Atlantic salmon. (**a**) Lamella congestion with associated amoeba. (**b**) Hyperplasia of respiratory epithelium and fusion of lamellae. (**c**) Interlamellar lacunae or vesicles. (**d**) Heavy infestation of amoeba adjacent to the epithelial surface. (a) Bar = 20 μm. (b, c, d) Bar = 50 μm

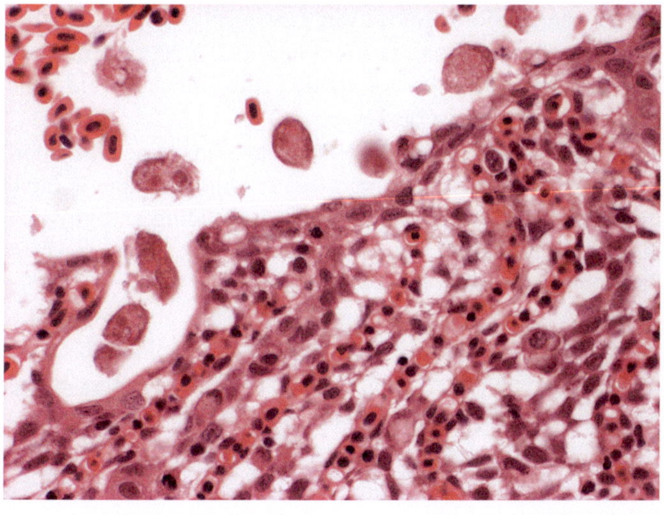

Fig. 8.3 *Paramoeba perurans* on hyperplastic gill epithelium of farmed Atlantic salmon with amoebic gill disease. Medium power

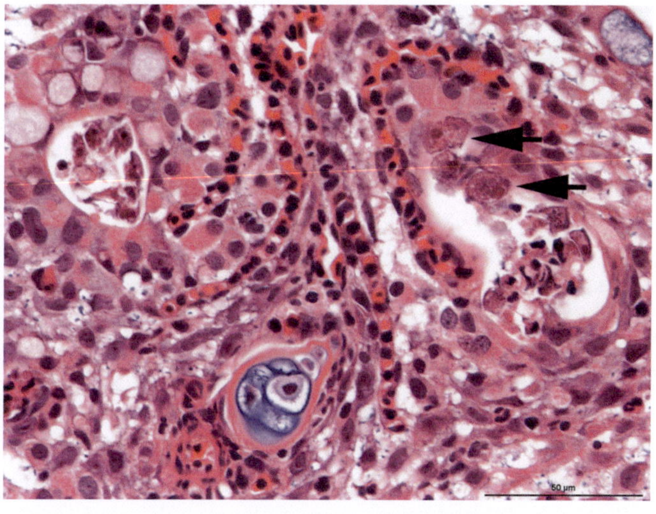

Fig. 8.4 Amoebae causing gill hyerplasia in freshwater-farmed brook trout. Amoebae are visible in cavernae in the hyperplastic tissue (*arrows*). Medium power

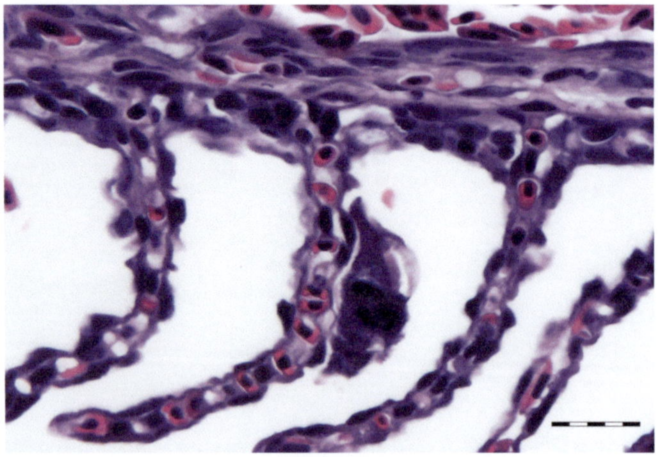

Fig. 8.5 *Capriniana* sp. attached to gill lamella of coho salmon. Bar = 20 μm

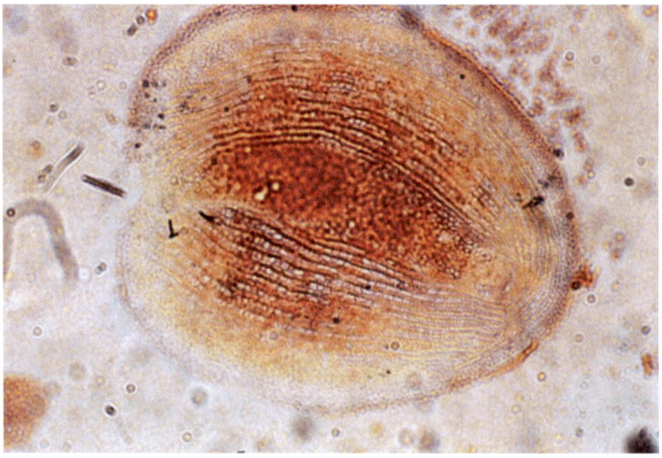

Fig. 8.6 *Chilodonella piscicola*. Silver impregnated specimen showing characteristic *oval shape* and notch at the posterior end. High power

parasite is variable but generally, it is sac-like (50–100 μm), elongated structure with 10–35 tentacles projecting from the cytoplasm opposite to the site of attachment (Fig. 8.5). The body adheres to the lamellae through a flattened attachment surface, termed the scopuloid. *C. piscium* feeds on free living cells and is not considered pathogenic to fish. However, where this ciliate occurs in large numbers, they can impede water flow and oxygen uptake by the gills and cause some irritation to the mucous membranes. Diagnosis is made by microscopic examination of smears from the skin and gills of live fish although they can also be observed during histological examination.

8.2.2 *Chilodonella piscicola*

This cosmopolitan, holotrich, ectoparasite is a serious pathogen on the gills and skin of several fish species in fresh and brackish water in North and South America, Europe and Japan. This parasite is a problem among wild fish and hatchery-reared salmonids, although mortality is particularly evident in cases where farm husbandry is poor.

This flattened ovoid to pear shaped parasite is up to 70 μm in length with an indentation at the posterior margin, and its surface is fully covered by rows of cilia. Live *C. piscicola* can be seen moving in a gliding manner over the epithelial surface on which it feeds, resulting in significant damage. Under certain conditions the parasite may encyst and remain viable for long periods. Clinical signs include increased mucus, hypoxia and reduced growth. Typical gill lesions include hyperplasia, necrosis and impaired gill function, followed by infiltration of eosinophilic granulocytes. Respiratory failure due to diffuse hyperplasia and inflammation is considered to be the primary cause of fish mortality. Diagnosis is based on the identification of the characteristic ciliate from fresh skin and gill smears (Fig. 8.6) and can be completed with histological examination.

8.2.3 *Ichthyophthirius multifiliis*

Ichthyophthirius multifiliis, 'white spot disease' or 'ich', is one of the most frequently encountered ciliates, principally of farmed fish. The geographical distribution of the parasite is widespread and almost all fish in fresh water can be considered susceptible with observations indicating that parasitism affects the growth of the host.

Numerous small white spots, less than 0.5 mm in diameter, are visible on the body surface of affected fish including the gill and the lining of the oral cavity (Fig. 8.7). Clinically, fish show frayed fins, dark colouration, increased mucous and rapid respiration. Internally, splenomegaly and pale mottled liver may be observed.

Mature *I. multifiliis* are round to oval in shape with short cilia covering the entire surface. *I. multifiliis* has a rapid developing life cycle that includes the feeding and growing attached phase observed in the fish, the 'throphonts', which occur under the skin or gill epithelium and visible as 'white spots'. As they reach maturity e.g. in the gills (Fig. 8.8), or under the host skin just beneath the epidermis, they eventually emerge breaking through the cysts as a free-living form, a 'pro tromont', and usually sink to the bottom or substrate, but a sticky capsule also allows them to attach to surfaces as plants or nets. The cycle continues with an encapsulated dividing stage or cyst formation while at the bottom of the tank or natural substrate. The 'tomont'

Fig. 8.7 Rainbow trout heavily infested with *Ichthyophthirius multifilis*. The fish is dark in colour

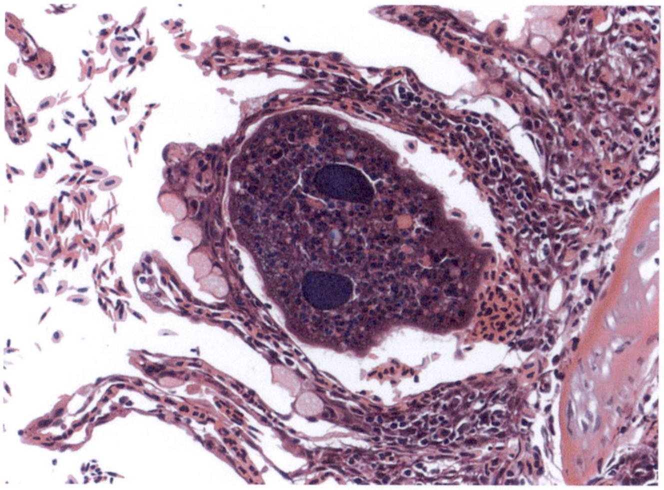

Fig. 8.8 *Ichthyophthirius multifilis* lodged between two fused lamellae in rainbow trout. Medium power

multiplication is temperature mediated, giving raise to thousands of new individuals, the 'tomites'. These break the cyst and become free swimming pear-shaped infecting forms, 'theronts', that have to reach a new host within a limited time period of ~2–4 days. Theronts penetrate the host skin becoming 'throphozoites', which will grow again into throphonts. This can be seen constantly turning and moving under the skin, feeding on the dead cells and fluids produced by their own action.

Histologically, *I. multifiliis* has a macronucleus and cilia that are readily recognised. The parasites can be observed within an interstitial space just adjacent to the basement membrane. The trophont growth will gradually lift and displace the epithelial cell layers causing these cells to become hydropic and vacuolated, with resulting necrosis. As the parasite emerges through the skin the epidermis shows erosion with subsequent dermatitis, desquamation and hyperplasia, with the likelihood of a secondary infection. In the gill, lamellar hyperplasia with reduction in the interlamellar spaces is observed, which become enclosed by the proliferation of epithelium and surrounded by congested tissue, diffuse lymphocytic infiltration and oedema, associated with the peripheral layer.

Diagnosis can be carried out by examining wet mounts from gill, tail, fins or the body surface, and the demonstration of a large (200–800 μm) ciliated trophont with a characteristic horseshoe-shaped macronucleus which is considered a pathognomonic feature.

Infection induces a protective immune response in rainbow trout survivors and a similar protection can be conferred by intra peritoneal injection of live theronts, but overall, there has only been limited progress with a vaccine.

8.2.4 *Trichodina truttae*

Trichodinids are widespread, peritrichous ciliates typically found on the gills, skin and fins of fish. Sometimes they can also be found in the lateral line canal and within the urogenital system. Most species within the genus have a direct life cycle and are ectozoic commensals where the fish acts as substrate for attachment, but some species are primary pathogens on the gills and body surface of fresh water and marine fish. *T. truttae* is considered to be specific for salmonid fish and reported from juvenile coho salmon in British Columbia and chum salmon fry in Japan.

Trichodinids usually feed on suspended bacteria and when loads are high, it provide abundant food availability and the parasite proliferates. Therefore when trichodinids become a problem under farming conditions, this often indicates poor water quality and an eutrophication issue.

Fish with severe infections show signs of listlessness, erratic swimming and inappetence, a greenish sheen to the body, loosened scales and osmoregulatory difficulty. The

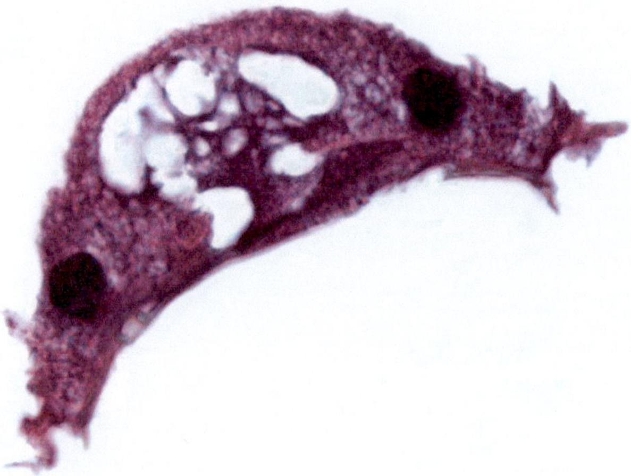

Fig. 8.9 Cross section showing *Trichodina* sp. from gill of Atlantic salmon. High power

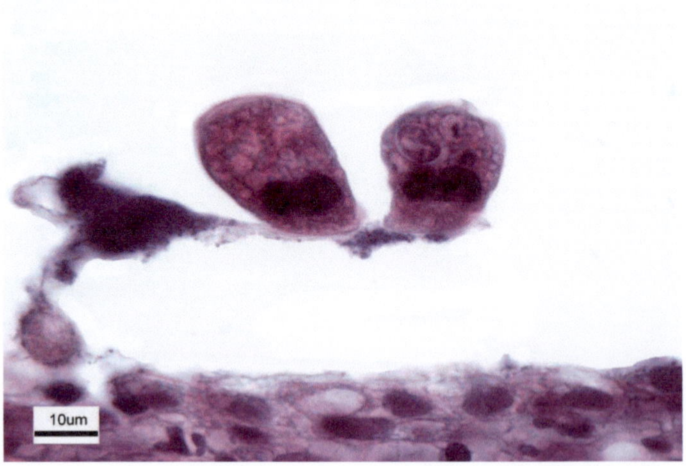

Fig. 8.11 *Riboscyphidia* sp. attached to lamellae of Atlantic salmon parr

Fig. 8.10 Aboral view of *Trichodina* sp. Normarski interference contrast. High power

parasite is mobile and does not attach permanently, but irritation of the gill lamellae can cause respiratory distress associated with excessive mucous. Histologically, mild hyperplasia and epithelial sloughing is observed. The differential diagnosis would be problems with water quality

Trichodinids are round and disc shaped parasites (Figs. 8.9 and 8.10). *T. truttae* is distinguished from other fresh water species by its large body diameter of 114–179 µm, as well as the presence of radial ridges on the oral surface and two markedly different lengths of cilia. The oral-aboral axis is shortened with a prominent basal disc (usually at the aboral pole).

Diagnosis is achieved by examining wet mounts from gill or the body surface, with demonstration of a cup or dome shaped organisms that move in a characteristic circular manner with a spiral of oral cilia at the anterior pole. The diagnosis of trichodinids can be completed with histological sections and morphology of the denticles in the adhesive discs.

8.2.5 *Scyphidia* (*Riboscyphidia, Ambiphyra*)

Scyphidia, *Riboscyphidia* and *Ambiphyra* are considered as synonyms, although there is little consensus about their 'correct' taxonomical status. These ectocommensal organisms are sessile peritrichs and their occurrence on freshwater fish can result in small wounds through the release of 'protein enzymes', opening areas for bacterial infection. Mortality has been associated with chronic infections of the gills by *Ambiphyra*, inducing mechanical blockage of the respiratory epithelium. Diagnosis is dependent upon identification from skin or gill scrapings or histopathology (Fig. 8.11).

8.3 Mesomycetozoea

8.3.1 *Ichthyophonus hoferi*

Ichthyophonus hoferi causes a granulomatous systemic disease primarily in marine fish, with several well reported epizootics occurring in wildfish including rainbow trout and Chinook salmon. Farmed salmonids are susceptible to infection resulting in poor growth rates. Clinical signs and pathology vary, but largely dependent upon the organs affected and the degree of infection. Behavioural anomalies, including lethargy and uncoordinated swimming movements, have been reported in salmon, particularly, where infection is located in the central nervous system.

A reservoir is believed to be present in marine fish and the common route of infection for farmed rainbow trout is probably through the ingestion of infected material. Ingested multinucleate spherical bodies will germinate and penetrate the gastric mucosa, entering the bloodstream and are thereby, spreading to several organs where secondary cysts are formed.

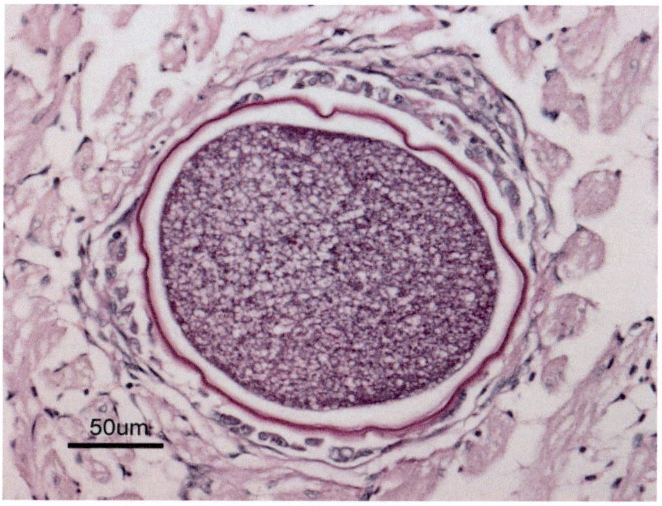

Fig. 8.12 *Ichthyophonus hoferi* surrounded by moderate granulomatous response in myocardium of wild Chinook salmon. PAS stain

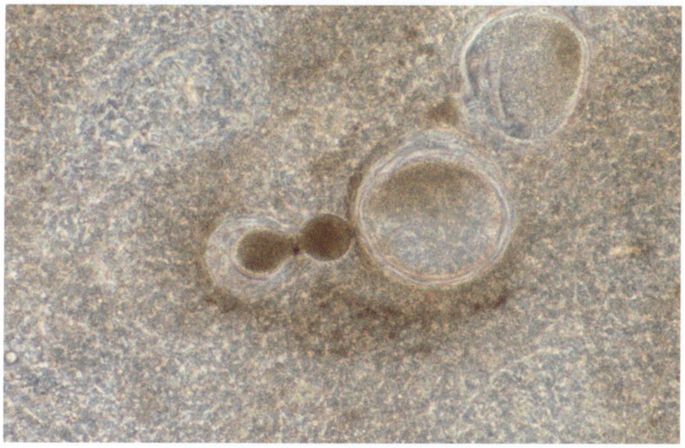

Fig. 8.13 *Ichthyophonus hoferi*. Squash mount showing characteristic budding from spore. Phase contrast. High power

At necropsy, whitish nodules may be seen in many organs, primarily the heart, muscle, kidney, liver and spleen. Microscopically, a severe granulomatous response is characteristic, often with large numbers of macrophages and multinucleated giant cells. Several developmental stages may be observed, but a spore or resting stage in several organs is common. This stage is spherical, with a double wall that is strongly PAS positive and measures 10–250 μm (Fig. 8.12). Germinating spores of *I. hoferi* may also be seen histologically when samples are taken during post-mortem examination. These spores are characterised by a budding cytoplasm protruding through the thick outer wall of the resting spore (Fig. 8.13).

I. hoferi belongs to a group of microorganisms with a diffuse taxonomical status. Historically, *I. hoferi* has been located incorrectly into the Haplosporidia and named *Ichthyosporidium gastrophilum*. Currently, and based upon 18S small-subunit ribosomal DNA, this pathogen is placed among members of the Protoctistan Mesomycetozoa clade.

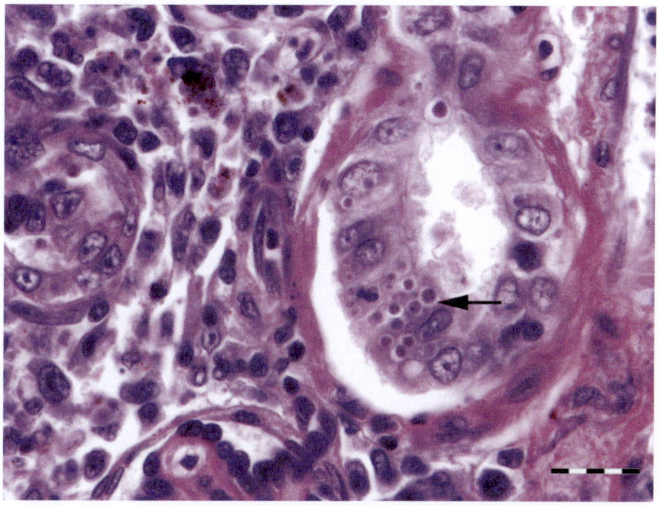

Fig. 8.14 *Sphaerotecum destruens* in kidney tubule of Chinook salmon (*arrow*). Bar =20 μm

Diagnosis is based on pathological and histological findings. Wet mounts of fresh kidney can also be examined microscopically. The presence of hyphae protruding through the outer spore wall is a definitive characteristic. A differential diagnosis should include bacterial diseases accompanied by granulomatous response.

8.3.2 *Sphaerothecum destruens*

Sphaerothecum destruens is an obligate unicellular eukaryotic parasite that was formally described as the 'rosette agent'. Several salmonids species can host *S. destruens* but the histopathology of the infection can differ between hosts. Infection results in morbidity and high mortality in Chinook salmon in aquaculture facilities in the USA, with chronic mortalities that are higher in the summer and autumn. In the UK, experimental challenges to determine their potential threat to wild Atlantic salmon concluded that spores could replicate and were associated with increased mortality (up to 90 %), when injected intraperitoneally.

External gross signs are generally not evident, however splenomegaly and nephromegaly are reported. In addition, infected salmon may be anaemic and slightly emaciated in advanced infections.

S. destruens primarily infects the spleen and kidney (Fig. 8.14), but in heavy infections these may occur in other organs with intracellular development of spore stages eliciting a host granulomatous response. *In vitro*, the spore stage (2–6 μm) replicates in a salmonid cell line by sequential asexual division, giving rise to daughter cells.

The genus *Sphaerothecum* is closely related to *Dermocystidium* and classified as a member of the Class Mesomycetozoea (formerly Ichthyosporea) based on phylogenetic analyses of the small subunit ribosomal DNA. Detection of the parasite can be recorded histologically and in tissues of naturally exposed adult fish using a nested PCR test.

8.3.3 *Dermocystidium salmonis*

Dermocystidium spp. affects many fish in fresh and sea water. *D. salmonis* occurs in all life stages and affects Atlantic salmon but also reported in Chinook, coho, and sockeye salmon. Although generally not fatal, *Dermocystidium* has been associated with mortality in young or juvenile fish, particularly when water temperatures are low.

The disease usually affects the gill lamellae, oral cavity and skin, and occasionally, systemic infections have been reported. When large cyst form in the gills, they may prevent closure of the opercula resulting in anoxia and mortality. In fresh mounts of infected gills or skin, numerous spherical to oval spores (**7**–12 μm across) can be seen. The morphological distinctive kinds of spores probably representing different developmental stages of the parasite but the most characteristic type are the so called 'signet ring cell' or hypnospore, with a large refringent vacuole and a narrow rim of cytoplasm. Other spores are irregularly vacuolated and equipped with one or multiple prominent nuclei.

Internally, the swollen abdomen harbours small, round (~1 mm in diameter) white cysts visible within the cavity, and large numbers of cysts may protrude through the body wall (Fig. 8.15). Splenic and hepatic lesions are characterised by well-circumscribed granulomas and an increase in leucocytes (Fig. 8.16). Each cyst contains a number of uninucleate spores which evoke a granulomatous dermatitis, with marked inflammatory response, haemorrhage, hyperplasia and hydropic degeneration.

Dermocystidium are currently classified under the Class Mesomycetozoea (formerly Ichthyosporea), a rather diverse group situated near the evolutionary division between animals and fungi, within the recently assigned assemblage *Opisthokonta*. *Dermocystidium* stain PAS positive and can be diagnosed microscopically (Fig. 8.17).

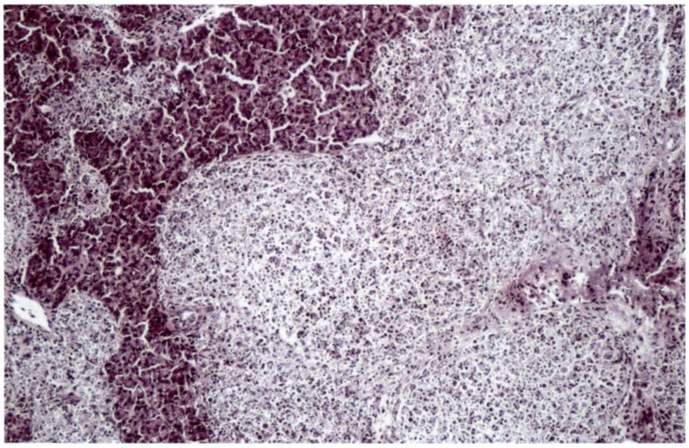

Fig. 8.16 *Dermocystidium* in liver of rainbow trout. Low power

Fig. 8.15 *Dermocystidium* in rainbow trout. The abdominal wall has been opened to reveal cysts and marked inflammatory response

8.4 Microsporidia

8.4.1 *Kabatana takedai*

Kabatana takedai (formally *Microsporidium takedi*) affects the muscle and heart of various species including chum, masou, pink and sockeye salmon, rainbow and brown trout, and Japanese char. A seasonal prevalence of this parasite is recognised, with the initial outbreak during the summer at around 15 °C.

The parasite invades the striated skeletal and heart muscle (Fig. 8.18) as well as smooth muscle of the host. Acute cases result in a high mortality and are characterised by a massive occurrence of whitish, spindle to ovoid-shaped proliferating microsporidian, which measure 2.5–4.0 μm. The affected tissue becomes granulomatous and spores are phagocytised by macrophages, followed by degeneration of the myofibrils and proliferation of the connective tissue (Fig. 8.19). Fibrinoid degeneration occurs in the marginal area of the foci. In chronic cases the heart shows an extreme hypertrophy and deformation of the tissue, with inflammatory oedema.

The route of transmission is unclear. Provisional identification is based on dissection and gross examination of the musculature and confirmed by PCR or by microscopical examination of the cysts.

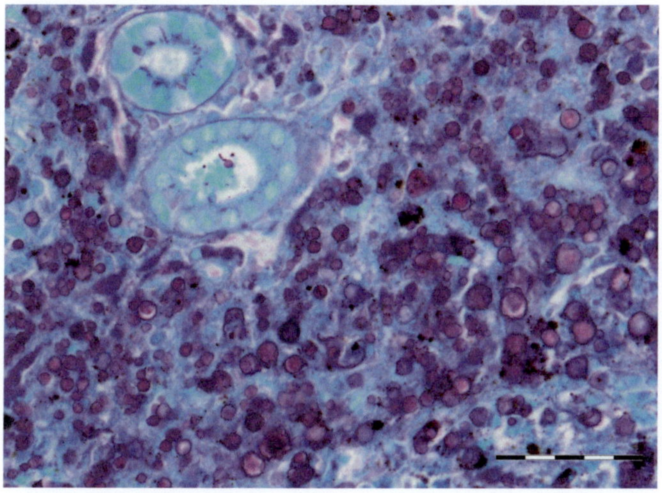

Fig. 8.17 *Dermocystidium* in kidney of rainbow trout. PAS stain. Bar = 50 μm

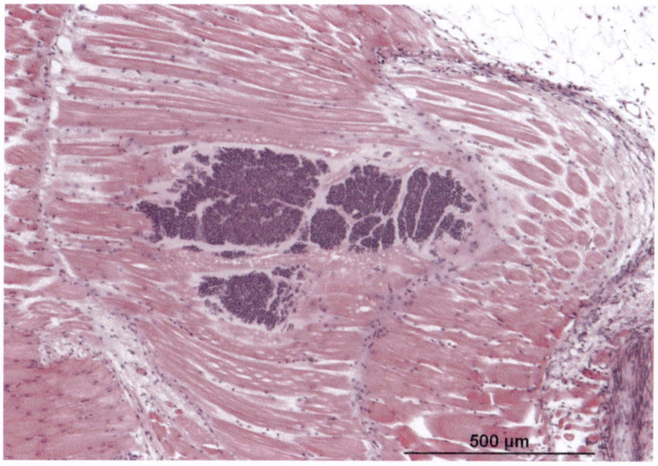

Fig. 8.19 *Kabatana takedai* xenoma in muscle of rainbow trout. Note muscle degeneration

Fig. 8.18 *Kabatana takedai*. Multiple cysts in the trunk muscle of rainbow trout

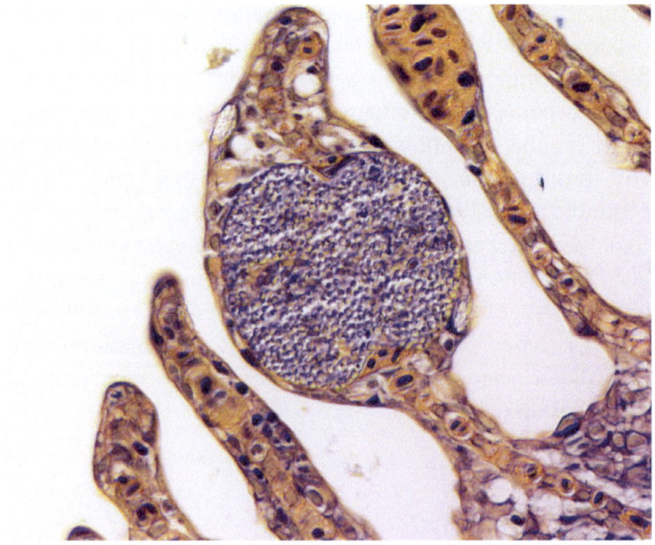

Fig. 8.20 *Loma* sp. in lamellae of brown trout. HE with Normarski interference contrast. Medium power

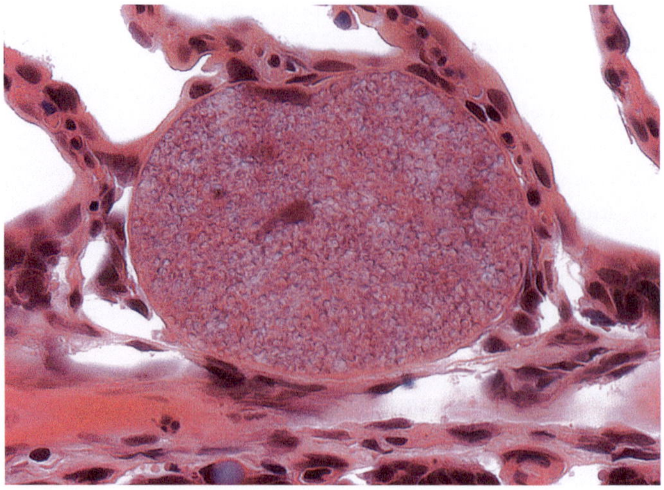

Fig. 8.21 *Loma* sp. xenoma in the gills of rainbow trout. High power

8.4.2 *Loma salmonae*

Loma salmonae is an economical important obligate gill pathogen. It has been detected from farmed rainbow trout, Chinook, coho and sockeye salmon reared in fresh water, and among some species reared in sea water cages. Affected fish may show respiratory distress and impaired swimming, with reduced growth rates. Gross signs include exophthalmia, ascites and petechiae on the opercula. Small, round, white cyst-like formations (xenoma) up to 0.5 mm in diameter can be observed within various tissues, but principally in the gill lamellae and generally in close association with the pillar cell system, during sporogony. The target cells for *L. salmonae* include pillar and endothelial cells, or leucocytes that migrate through the basement membrane of a blood vessel. Significant pathological changes occur as the infected cell undergoes hypertrophy and marked hyperplasia (Figs. 8.20 and 8.21). These cells can rupture with obliteration of the capillary lumen, multifocal areas of granulation or fibrous tissue, culminating in a persistent inflammatory response. The latter includes neutrophil infiltration and a vascular thrombosis. Parasites and associated lesions have also been reported in kidney, spleen and pseudobranch. A few presporogonic stages of the parasite can be found in the heart endothelium prior to xenoma formation in the gills, this is followed by pericarditis and hyperplasia involving the muscle, with inflammation around the coronary arteries. Recovering fish show multiple focal areas of chronic perivasculitis in the gill lamellae.

The characteristic xenoma wall appears as a chromophilic layer up to 1.5 μm thick and when the spores mature, the xenoma wall ruptures with the release of spores into the surrounding environment. Fish are therefore infected directly by ingesting spores. The vegetative stages of *L. salmonae* are unicellular and measure approximately 3 × 8 μm in fresh preparations. *Loma fontinalis* is described from gill lamellae of brook trout and is similar to *L. salmonae*.

A presumptive diagnosis of *Loma* involves the detection of the spores by light microscopy or from wet mounts. The spores stain PAS positive. A specific polymerase chain reaction assay is also available as a diagnostic test.

8.4.3 *Candidatus* Paranucleospora theridon (=*Desmozoon lepeophtheiri*)

Candidatus Paranucleospora theridon is an intracellular microsporidean found in Atlantic salmon, rainbow and brown trout, and in the lice *Lepeophtheirus salmonis* and *Caligus elongatus*. In salmon, two developmental cycles has been described, one producing spores in the cytoplasm of phagocytes or epidermal cells, and the other in the nuclei of epidermal cells. The former spores are small and thin walled with a short polar tube and believed to be auto infective forms, while the larger oval intranuclear spores have a thick endospore and a longer polar tube, and are thought to be responsible for transmission from salmon to *Lepeophtheirus salmonis*.

Candidatus Paranucleospora theridon can be found in most organs including gill (Fig. 8.22), heart, kidney, pancreas and spleen. It has been demonstrated in fish suffering from diseases such as salmonid pancreas disease and heart and skeletal muscle inflammation, but the exact role of the parasite is unknown. In addition, the parasite has also been associated with severe cases of post-vaccination peritonitis. Infected cells and free spores provoke a strong inflammatory response dominated by macrophages where necrotic foci of Malpighian and goblet cells may be present. In the gills, melanisation near the base of the primary lamellae and epithelial hyperplasia with inflammatory infiltrates is common. Farmed salmon are probably infected in the summer and autumn months when

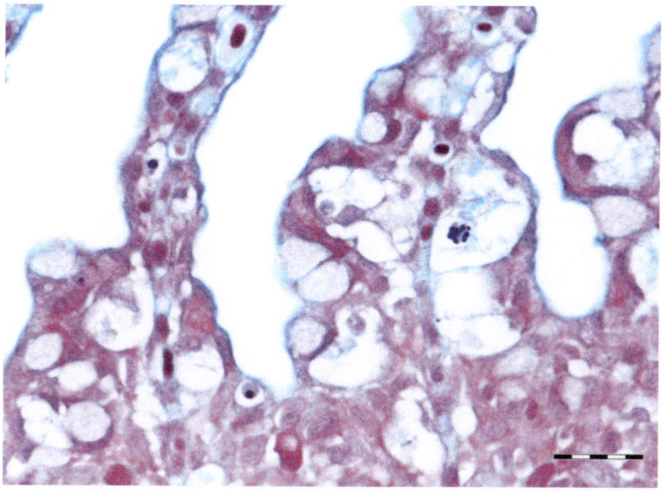

Fig. 8.22 *Candidatus* Paranucleospora theridon (=*Desmozoon lepeophtheiri*) in gill epithelium from Atlantic salmon. Gram stain. Bar = 20 μm

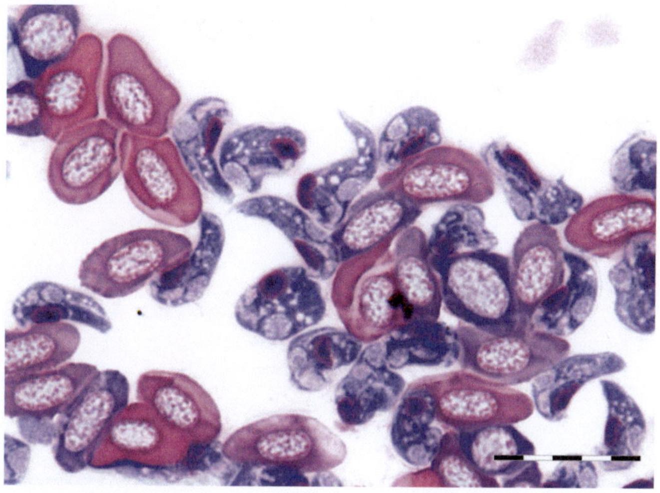

Fig. 8.23 *Cryptobia salmositica* in blood smear from Chinook salmon. Erythrocytes are *red*, parasites are *blue*. Diff Quick stain. Bar = 20 μm

temperatures are >15 °C, while clinical disease and peak mortality typically occurs between September and February. The diagnosis is based upon demonstration of spores in histological sections and supported by PCR.

8.5 Sarcomastigophora

8.5.1 *Cryptobia salmositica*

Cryptobia salmositica is a haemoflagellate that causes disease in freshwater fish. The infected fish are typical cold-water species occurring in streams with gravel beds and moderate to swift flowing currents. Among salmonids, it has been described from the gills, body surface and digestive system of coho and pink salmon in fresh water. Brook trout can also become infected but they do not appear to develop cryptobiasis, and therefore, may act as a reservoir host. *C. salmositica* is normally transmitted by the leech, *Piscicola salmositica*, although direct transmission has also been demonstrated.

The first clinical sign is anaemia, followed by exophthalmia, oedema, splenomegaly, hepatomegaly and abdominal distension resulting from ascites. Microcytic and hypochromic anaemia is correlated with increasing parasitaemia and extravascular localization of the parasite. Histologically, the initial lesions occur in the liver, gills and spleen, and comprise focal haemorrhage with congestion of blood vessels and oedema in the glomeruli. In addition, there is dilatation and oedematous swelling of the glomeruli followed by endovasculitis and mononuclear infiltration. Necrosis in the liver and kidney with depletion of haematopoietic tissue in the acute phase causes the mortality. A mucosal and submucosal granulomatous gastritis has also been reported.

C. salmositica measures 6–25 × 2–4 μm, and possesses a prominent kinetoplast either anterior to, or beside the round nucleus, with both an anterior and posterior flagella. Diagnosis can be achieved using stained blood films where (Fig. 8.23). Additionally, a monoclonal antibody against the 47- kDa antigen has been used in an antigen-capture enzyme-linked immunosorbant assay.

8.5.2 *Ichthyobodo* spp.

Ichthyobodo necator ('*Costia*') is an important, obligate parasite, capable of infecting a broad range of wild and farmed fresh water fish. The parasite is associated with skin epithelium, gills and in cavities, within hyperplastic interlamellar tissue. Affected fish become emaciated, lethargic with flared opercula and they rub against the walls or bottom of the tank. The fish may appear greyish due to excessive skin mucous, and show focal skin haemorrhage. Infestation are often associated with poor husbandry and mortality may be high if the fish are left untreated.

A free-living and an attached form can be observed, the first measures 10–15 μm in length and is usually oval or kidney shaped, while the attached stage is cuneiform or pear-like. *I. necator* has two pairs of flagella and swims in a jerky spiral manner.

Histologically, there is a reduction of mucous cells early in the infection, while lamellar hyperplasia occurs in the recovery phase (Figs. 8.24 and 8.25). Other lesions include erosive and ulcerative dermatitis. Gill lesions include exhaustion of the goblet cells, diffuse hyperplasia, sometimes with characteristic cavitation, and degeneration of epithelial and mucous cells, fusion of adjacent lamellae

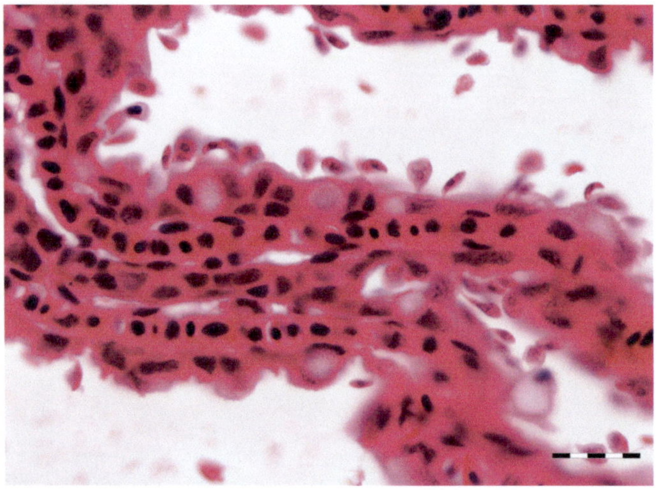

Fig. 8.24 *Ichthyobodo salmonis* attached to the gill epithelium in Atlantic salmon. Bar = 20 μm

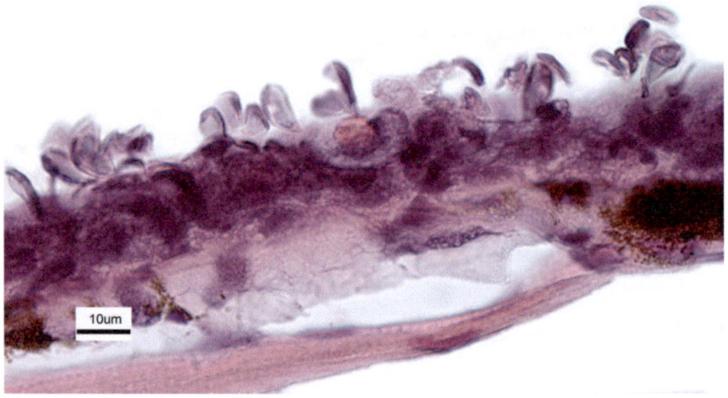

Fig. 8.25 *Ichthyobodo necator* attached to the skin of farmed brown trout fry

and cell sloughing. Damage to sub-surface cells show dramatic degeneration of the cytoplasm, although the nucleus usually remains intact. In the recovery phase, large numbers of eosinophilic granular cells may be seen within the lamellae.

Morphometric data has proven that *Ichthyobodo* from freshwater and sea water display a different cell shape and data from electron microscopy show that marine forms of *Ichthyobodo* possess ridge-like projections along the cytostome process, which are smooth in parasites from Atlantic salmon in fresh water. Phylogenetic analyses of SSU rDNA sequences were able to prove the existence of two *Ichthyobodo* species able to infect Atlantic salmon based on differences in the attachment region and the presence of spine-like surface projections, and the name *Ichthyobodo salmonis* sp. n. has been proposed.

The diagnosis of *Ichthyobodo* is based on the microscopic examination of fresh tissues and the identification of characteristic motile flagellates in mucous from the gills or skin. Stained sections also demonstrate the attached parasites.

8.5.3 *Spironucleus* spp.

Diplomonad flagellates are reported from several fish spices worldwide and most of them are commensals feeding on bacteria and on food digested by the host. However, some of them are pathogenic and among salmonid species, they may occur as enteric commensals or parasites (*S. salmonis* and *S. barkhanus*), or they may cause severe systemic disease (*S. salmonicida*). Intestinal diplomonads including infection of the gallbladder occur commonly as opportunistic parasites. In moderate numbers they seldom cause any harm, but heavily infected fry and fingerlings, especially of fresh water brook, brown, lake and rainbow trout, may show nonspecific locomotive disorders, emaciation, catarrhal enteritis, abdominal distension and exophthalmia. Gut contents may be yellowish and fish produce a pseudo faeces.

Spironucleus salmonis, previously known as *Hexamita salmonis*, had been known as a health issue during the early life stages of the fresh water reared rainbow trout for a long time, with high levels of morbidity and associated mortality. Infections of the intestine of rainbow trout cause weakness, anorexia and emaciation. Internally, enteritis, intestinal haemorrhage, yellow mucus and necrosis of hepatocytes may be observed.

S. barkhanus has been described from grayling and Arctic char. In Northern Norway, systemic infection with *S. salmonicida* has caused losses in several Atlantic salmon sea farms. A large proportion of the large fish in the population may be affected leading to rejects and downgrading. External lesions may include ascites and exophthalmia, while internally, haemorrhagic boil-like lesions in the muscle and necrotic patches in the kidney, spleen and liver, has been reported for *S. salmonicida* and S. *barkhanus* (Figs. 8.26 and 8.27). Affected fish often have an unpleasant, putrid odour at necropsy. Diffuse epicarditis or whitish cysts containing vast numbers of parasites may be found on the ventricular wall (Fig. 8.28). Microscopy also reveals widespread liquefactive muscle necrosis with haemorrhage. Parasites may be present in large numbers typically in the gut (Fig. 8.29) and characterized by their pear-shape and paired anterior nuclei ('eyes'). Multifocal necrosis may also be found in kidney, liver and spleen. The inflammatory response is variable depending on temperature and age of lesions. Parasite aggregates may be found in blood vessels and in the spongy myocardium. Purulent pericarditis with vast numbers of parasites and inflammatory cells may also occur (Fig. 8.30). *S. salmonicida* has also caused systemic disease in sea-farmed Arctic char in northern Norway, and in farmed Chinook salmon in BC, Canada. In these cases, the parasites have been found in large numbers

Fig. 8.26 Liver of farmed Atlantic salmon with *Spironucleus salmonicida*. Necrosis and granulomatous response throughout the organ

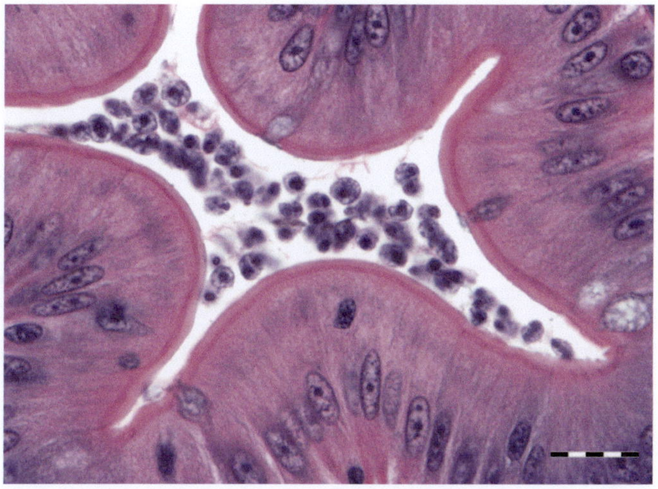

Fig. 8.29 *Spironucleus* sp. in the gut of farmed Chinook salmon. Bar = 20 μm

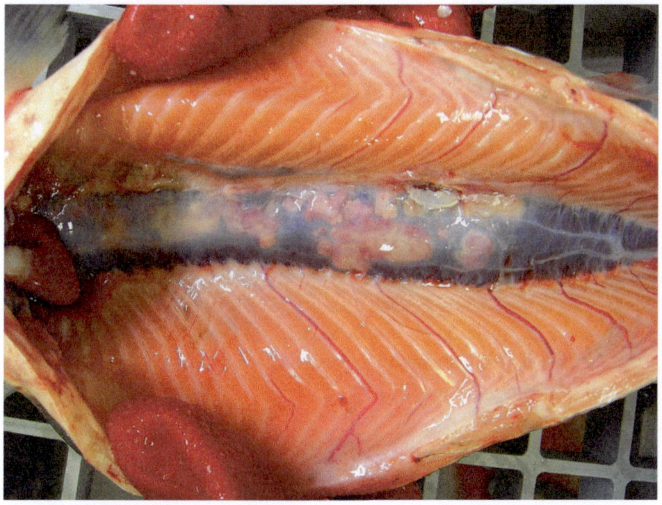

Fig. 8.27 *Spironucleus salmonicida* in farmed Atlantic salmon; necrosis and granulomatous inflammation throughout the kidney

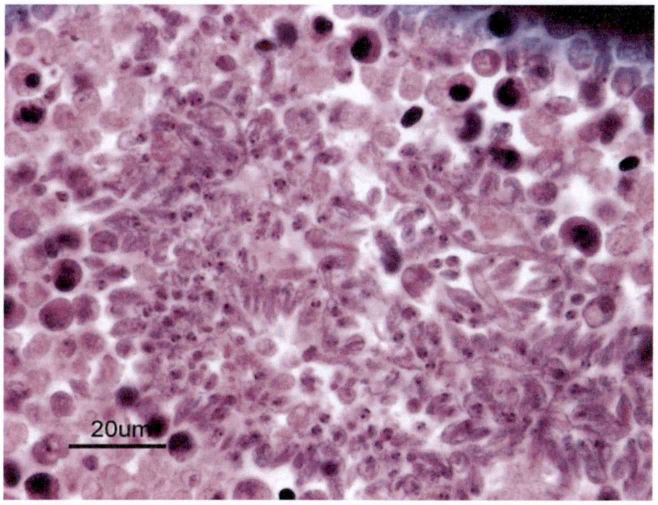

Fig. 8.30 Liver abscess of farmed Atlantic salmon filled with *Spironucleus salmonicida*

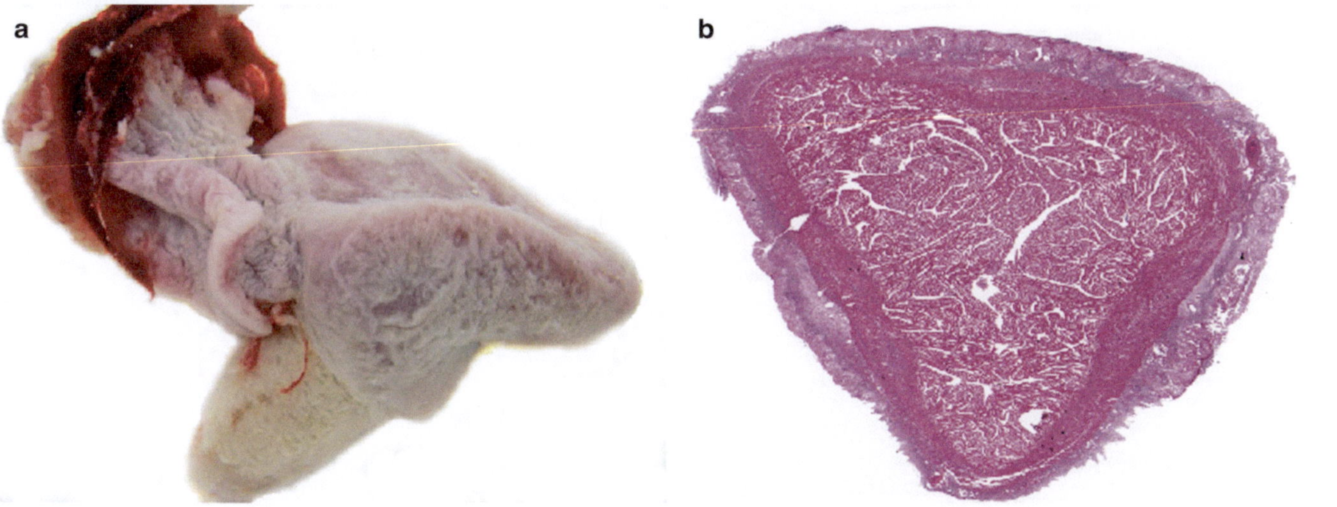

Fig. 8.28 (**a**) Purulent epicarditis caused by *Spironucleus salmonicida* in farmed Atlantic salmon. (**b**) Histological section of ventricle of the same heart. Low power

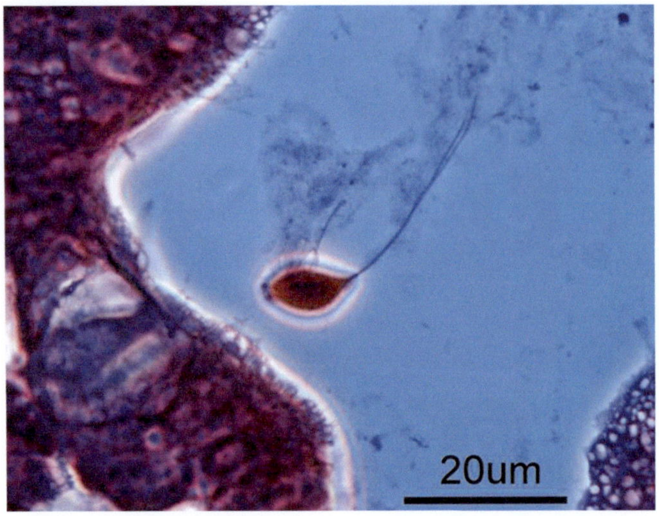

Fig. 8.31 *Spironucleus* sp. in the gut of farmed rainbow trout. Notice flagella. Phase contrast

in blood vessels, but with relatively few and mild lesions in organs. The inflammatory response is variable, depending on the duration of the disease and temperature. A mucosal and sub-mucosal granulomatous gastritis may found in more chronic cases. In the compact/spongy interface of the ventricular myocardium, aggregates of the parasite may be found. Individual parasitic cells, alone or in clusters may also be recorded in blood vessels virtually anywhere in the body, but frequently occur in vessels of the choroid and in coronary branches. In the brain, vast numbers of the parasite may be found infiltrating the meninges. Typically, the inflammatory response is minimal.

The parasite has an alternating life cycle between trophozoite and cyst. They multiply by longitudinal binary fission. The trophozoites are motile by 6 anterior flagella arranged in two groups, and two posterior flagella (Fig. 8.31), with size ranging from 6 to 35 μm. Diagnosis is assisted by light microscopy to identify characteristic flagellates and lesions, while accurate identification of species must be based upon molecular analysis.

Further Reading

Adams MB, Crosbie PBB, Nowak BF (2012) Preliminary success using hydrogen peroxide to treat Atlantic salmon, *Salmo salar* L., affected with experimentally induced amoebic gill disease (AGD). J Fish Dis 35:839–848

Allen RL, Meekin TK, Pauley GB, Fujihara MP (1968) Mortality among chinook salmon associated with the fungus *Dermocystidium*. J Fish Res Board Can 25:2467–2475

Andreou D, Arkush KD, Guégan J-F, Goxlan RE (2012) Introduced pathogens and native freshwater biodiversity: a case study of *Sphaerothecum destruens*. PLoS One 7(5):e36998. doi:10.1371/journal.pone.0036998

Ardelli BF, Forward GM, Woo PTK (1994) Brook charr, *Salvelinus fontinalis* (Mitchill), and cryptobiosis: a potential salmonid reservoir host for *Cryptobia salmositica* Katz, 1951. J Fish Dis 17:567–577

Awakura A (1974) Studies on the microsporidian infection in salmonid fishes. Sci Rep Hokkaido Fish Hatch 29:1–95

Bower SM, Evelyn TPT (1980) Acquired and innate resistance to the haemoflagellate *Cryptobia salmositica* in sockeye salmon (*Oncorhynchus nerka*). Dev Comp Immunol 12:749–760

Bruno DW (1992) *Ichthyobodo* sp., on farmed Atlantic salmon, *Salmo salar* L., reared in the marine environment. J Fish Dis 15:349–351

Bruno DW (2001) *Dermocystidium* sp. in Scottish Atlantic salmon, *Salmo salar*: evidence for impact on fish in marine fish farms. Bull Eur Assoc Fish Pathol 21:209–213

Bruno DW, Collins R, Morrison CM (1995) The occurrence of *Loma salmonae* sp., (Protozoa: Microspora) in farmed rainbow trout, *Oncorhynchus mykiss* Walbaum in Scotland. Aquaculture 133:341–344

Bruno DW, Nowak B, Elliott DG (2006) A guide to the identification of fish protozoan and metazoan parasites in stained tissue sections. Dis Aquat Org 70:1–36

Bustos PA, Young ND, Rozas MA, Bohle HM, Ildefonso RS, Morrison RN, Nowak BF (2011) Amoebic gill disease (AGD) in Atlantic salmon (*Salmo salar*) farmed in Chile. Aquaculture 310:281–288

Crosbie PBB, Bridle AR, Cadoret K, Nowak BF (2012) In vitro cultured *Neoparamoeba perurans* causes amoebic gill disease in Atlantic salmon and fulfils Koch's postulates. Int J Parasitol 42:511–545

Draghi A, Bebak J, Popov VL, Noble AC, Geary S, West AB, Byrne P, Frasca S (2007) Characterisation of a Neochlamydia-like bacterium associated with epitheliocystis in cultured Arctic charr *Salvelinus alpinus*. Dis Aquat Org 76:27–38

Ellis AE, Wootten R (1978) Costiasis of Atlantic salmon, *Salmo salar* L. smolts in salt water. J Fish Dis 1:389–393

Ferguson JA, St-Hilaire S, Peterson TS, Rodnick KJ, Kent ML (2011) Survey of parasites in threatened stocks of coho salmon (*Oncorhynchus kisutch*) in Oregon by examination of wet tissues and histology. J Parasitol 97:1085–1098

Isaksen TE, Karlsbakk E, Sundnes GA, Nylund A (2010) Patterns of *Ichthyobodo necator* sensu stricto infections on hatchery-reared Atlantic salmon *Salmo salar* in Norway. Dis Aquat Org 88:07–214

Jørgensen A, Sterud E (2006) The marine pathogenic genotype of *Spironucleus barkhanus* from farmed salmonids redescribed as *Spironucleus salmonicida* n. sp. J Eukaryot Microbiol 53:531–541

Jørgensen A, Torp K, Bjørland MA, Poppe TT (2011) Wild Arctic char *Salvelinus alpinus* and trout *Salmo trutta*: hosts and reservoir of the salmonid pathogen *Spironucleus salmonicida* (Diplomonadida; Hexamitidae). Dis Aquat Org 97:57–63

Kent ML, Speare DJ (2005) Review of the sequential development of *Loma salmonae* (Microsporidia) based on experimental infections of rainbow trout (*Oncorhynchus mykiss*) and Chinook salmon (*O. tshawytscha*). Folia Parasitologia 52:63–68

Kent ML, Sawyer TK, Hedrick RP (1988) *Paramoeba pemaquidensis* (Sarcomastigophora: Paramoebidae) infestation of the gills of coho salmon *Oncorhynchus kisutch* reared in salt water. Dis Aquat Org 5:163–169

Kent ML, Elliot DG, Groff JM, Hedrick RP (1989) *Loma salmonae* (Protozoa: Microspora) infections in salt water reared coho salmon *Oncorhynchus kisutch*. Aquaculture 80:211–222

Kent ML, Ellis J, Fournie JW, Dawe SC, Bagshaw JW, Whitaker DJ (1992) Systemic hexamitid (Protozoa: Diplomonadida) infection in salt water pen-reared chinook salmon *Oncorhynchus tshawytscha*. Dis Aquat Org 14:81–89

Kocan RP, Hershberger P, Winton J (2004) Ichthyophoniasis: an emerging disease of Chinook salmon in the Yukon river. J Aquat Animal Health 16:58–72

Kocan R, LaPatra S, Gregg J, Winton J, Hershberger P (2006) *Ichthyophonus*-induced cardiac damage: a mechanism for reduced swimming stamina in salmonids. J Fish Dis 29:521–527

Kocan R, Hershberger P, Sanders G, Winton J (2009) Effects of temperature on disease progression and swimming stamina in *Ichthyophonus*-infected rainbow trout, *Oncorhynchus mykiss* (Walbaum). J Fish Dis 32:835–843

Kube PD, Taylor RS, Elliott NG (2012) Genetic variation in parasite resistance of Atlantic salmon to amoebic gill disease over multiple infections. Aquaculture 364–365:165–172

Lamas J, Bruno DW (1992) Observations on the ultrastructure of the attachment plate of *Ichthyobodo* sp., from Atlantic salmon, *Salmo salar* L., reared in the marine environment. Bull Eur Assoc Fish Pathol 12:171–173

Leibovitz L (1980) Ichthyophthiriasis. J Am Vet Med Assoc 176:30–31

Lom J, Nilsen F, Urawa S (2001) Redescription of *Microsporidium takedai* (Awakura, 1974) as *Kabatana takedai* (Awakura, 1974) comb. n. Dis Aquat Org 44:223–230

Markey PT, Blazer VS, Ewing MS, Kocan KM (1994) *Loma* sp. in salmonids from the Eastern United States: associated lesions in rainbow trout. J Aquat Animal Health 6:318–328

McVicar AH (1977) *Ichthyophonus* as a pathogen in farmed and wild fish. Bull Office Int Epizoot 87:517–519

McVicar AH, Wootten R (1980) Disease in farmed juvenile Atlantic salmon caused by *Dermocystidium* sp. In: Ahne W (ed) Fish diseases. Third COPRAQ-Session. Springer, Berlin, pp 165–173

Miyajima S, Urawa S, Yokoyama H, Ogawa K (2007) Comparison of susceptibility to *Kabatana takedai* (Microspora) among salmonid fishes. Fish Pathol 42:149–157

Mo TA, Poppe TT, Iversen L (1990) Systemic hexamitosis in salt-water reared Atlantic salmon. Bull Eur Assoc Fish Pathol 10:69–70

Morrison CM, Sprague V (1983) *Loma salmonae* (Putz, Hoffman and Dunbar, 1965) in the rainbow trout, *Salmo gairdneri* Richardson, and *L. fontinalis* sp. nov. (Microsporida) in the brook trout, *Salvelinus fontinalis* (Mitchill). J Fish Dis 6:345–353

Morrison RN, Koppang EO, Hordvik I, Nowak BF (2006) MHC class II^+ cells in the gills of Atlantic salmon (*Salmo salar* L.) affected by amoebic gill disease. Vet Immuno Immunopathol 109:297–303

Munday BL, Zilberg D, Findlay V (2001) Gill disease of marine fish caused by infection with *Neoparamoeba pemaquidensis*. J Fish Dis 24:497–507

Nylund S, Nylund A, Watanabe K, Arnesen CE, Karlsbakk E (2010) *Paranucleospora theridon* n. gen., n. sp. (Microsporidia, Enterocytozoonidae) with a life cycle in the salmon louse (*Lepeophtheirus salmonis*, Copepoda) and Atlantic salmon (*Salmo salar*). J Eukaryot Microbiol 57:95–114

Olsen RE, Dungagan CF, Holt RA (1991) Water-borne transmission of *Dermocystidium salmonis* in the laboratory. Dis Aquat Org 12:41–48

Olsen MM, Kania PW, Heinecke RD, Skoedt K, Rasmussen KJ, Buchmann K (2011) Cellular and humoral factors involved in the response of rainbow trout gills to *Ichthyophthirius multifiliis* infections: molecular and immunohistochemical studies. Fish Shell Immunol 30:859–869

Poppe TT, Mo TA (1993) Systemic, granulomatous hexamitosis of farmed Atlantic salmon: interaction with wild fish. Fish Res 17:147–152

Poppe TT, Mo TA, Iversen L (1992) Disseminated hexamitosis in sea-caged Atlantic salmon, *Salmo salar*. Dis Aquat Org 14:91–97

Poynton SL, Fard RS, Jenkins J, Ferguson HW (2004) Ultrastructure of pathogenic diplomonad flagellates from fish: characterization of *Spironucleus salmonis* n. comb. from Northern Irish rainbow trout *Oncorhynchus mykiss*, and a diagnostic guide for recognition of species. Dis Aquat Org 60:49–64

Ramsay JM, Speare DJ, Dawe SC, Kent ML (2002) Xenoma formation during microsporidial gill disease of salmonids caused by *Loma salmonae* is affected by host species (*Oncorhynchus tshawytscha, O. kisutch, O. mykiss*) but not by salinity. Dis Aquat Org 48:125–131

Roubal FR, Bullock AM, Robertson DA, Roberts RJ (1987) Ultrastructural aspects of infestation by *Ichthyobodo necator* (Henneguy, 1883) on the skin and gills of the salmonids *Salmo salar* L. and *Salmo gairdneri* Richardson. J Fish Dis 10:181–192

Roubal FR, Lester RJG, Foster CK (1989) Studies on cultured and gill-attached *Paramoeba* sp. (Gymnamoebae: Paramoebidae) and the cytopathology of paramoebic gill disease in Atlantic salmon, *Salmo salar* L., from Tasmania. J Fish Dis 12:481–492

Schmidt-Posthaus H, Polkinghorne A, Nufer L, Schifferli A, Zimmermann DR, Segner H, Steiner P, Vaughan L (2012) A natural freshwater origin for two chlamydial species, *Candidatus* Piscichlamydia salmonis and *Candidatus* Clavochlamydia salmonicola, causing mixed infections in wild brown trout (*Salmo trutta*). Environ Microbiol 14:2048–2057

Steinum T, Kvellestad A, Rønneberg LB, Nilsen H, Asheim A, Nygård SM, Olsen AB, Dale OB (2008) First cases of amoebic gill disease (AGD) in Norwegian salt water farmed Atlantic salmon, *Salmo salar* L., and phylogeny of the causative amoeba using 18S cDNA sequences. J Fish Dis 31:205–214

Sterud E, Mo TA, Poppe TT (1977) Ultrastucture of *Spironucleus barkhanus* n. sp. (Diplomonadida: Hexamitidae) from grayling *Thymallus thymallus* (L.) (Salmonidae) and Atlantic salmon *Salmo salar* L. (Salmonidae). J Eukaryot Microbiol 44:399–407

Sterud E, Mo TA, Poppe TT (1988) Systemic spironucleosis in sea-farmed Atlantic salmon *Salmo salar* L, caused by *Spironucleus barkhanus* transmitted from feral Arctic char *Salvelinus alpinus*? Dis Aquat Org 33:63–66

Sterud E, Poppe TT, Bornø G (2003) Intracellular infection with *Spironucleus barkhanus* (Diplomonadida: Hexamitidae) in farmed Arctic char *Salvelinus alpinus*. Dis Aquat Org 56:155–161

Taylor RS, Kube PD, Muller WJ, Elliott NG (2009) Genetic variation of gross gill pathology and survival of Atlantic salmon (*Salmo salar* L.) during natural amoebic gill disease challenge. Aquaculture 294:172–179

Urawa S (1989) Seasonal occurrence of *Microsporidium takedai* (Microsporodia) infection in masou salmon, *Oncorhynchus masou*, from the Chitose river. Physiol Ecol Jpn Spec 1:587–598

Urawa S (2006) Microsporidian infection. In: Hatai K Ogawa K (eds) New atlas of fish diseases. Midori Shobo, p 38 (In Japanese)

Young ND, Dyková I, Snekvik K, Novak BF, Morrison RN (2008) *Neoparamoeba perurans* is a cosmopolitan aetiological agent of amoebic gill disease. Dis Aquat Org 78:217–223

9 Metazoa

Abstract

Metazoa parasites are multicellular organisms where cells have differentiated into organised tissues and organs. These parasites may be found in all fish organ systems both in wild and in farmed salmonids, in fresh and sea water. Recently, parasites that would have been included as Protists are now considered metazoans, as the case of Myxozoa. The use of molecular tools are able to identify many Metazoan parasites and 12S ribosomal DNA primers have significant potential for metagenetic analysis, but should not detract or replace traditional histology to asses their impact on the host. In this book we have used the term infestation to represent ectoparasitic conditions and infection, as referring to those that are endoparasitic. This chapter will cover a selection of the many species of Metazoa that can be found in salmonids.

Keywords

Metazoa • Salmon • Trout

Metazoa parasites are multicellular organisms where cells have differentiated into organised tissues and organs. These parasites occur in all organ systems of wild and farmed salmonids worldwide, in both fresh and sea water. Molecular methods are available for many Metazoaparasites and 12S ribosomal DNA primers have significant potential for metagenetic analysis. However, this should not detract or replace traditional histology. Recently, parasites that would have been included in Protist are now considered as Myxozoa and therefore they are covered in this chapter. We have used infestation to represent external ectoparasitic infestations and infection as referring to internal endoparasitic conditions. Where appropriate, Metazoa life cycles are summarised to improve understanding and the examples covered in this chapter are summarized in Table 9.1.

9.1 Myxozoa

9.1.1 *Ceratomyxa shasta*

Ceratomyxa shasta is a significant parasite of hatchery and wild juvenile anadromous salmonids on the Pacific coast of North America. Differences in susceptibility are reported between species of salmon as well as the range of responses to infection. Significant losses can occur in hatcheries, but *C. shasta* is also implicated as a primary cause of mortality among wild stocks. For example, the survival of Chinook salmon is reduced in locations where parasite densities are highest. Infection results in exophthalmia, lethargy and a dark body. Abdominal distension and haemorrhaging are common around the vent region with severe inflammation and necrosis of the intestine. Infection starts in the epithelium of the posterior intestine and progresses towards a multifocal inflammatory response with sloughing and mucosal necrosis. Histologically, an acute inflammatory reaction in the intestine can be observed with proliferation of the connective tissue of the intestinal caeca and massive infiltration by the developing trophozoites and other developmental stages with subsequent spread to other organs. The occlusion and destruction of the lumen is considered to be the cause of the mortality among infected fish. A limited granulomatous inflammation may develop in the viscera with consequential peritonitis.

The actinosporean stage of *C. shasta* develops in a polychaete, *Manayunkia speciosa*. Fish become infected by coming into contact with water containing the infective

D.W. Bruno et al., *A Colour Atlas of Salmonid Diseases*,
DOI 10.1007/978-94-007-2010-7_9,

Table 9.1 Principal Metazoa parasites from salmonids

Name	Phyla or class	Common location	Environment
Ceratomyxa shasta	Myxozoa	Intestine	FW
Chloromyxum truttae	Myxozoa	Gall bladder	FW
Kudoa thyrsites	Myxozoa	Musculature	SW
Henneguya zschokkei	Myxozoa	Musculature	FW
Myxidium truttae	Myxozoa	Liver	FW, SW
Myxobolus cerebralis	Myxozoa	Cartilage, brain	FW
Parvicapsula pseudobranchicola	Myxozoa	Kidney, pseudobranch	SW
Sphaerospora truttae	Myxozoa	Kidney	FW
Tetracapsuloides bryosalmonae	Myxozoa	Many organs	FW
Philonema oncorhynchi	Nematoda	Abdominal cavity	FW
Cystidicola farionis	Nematoda	Swim bladder	FW, SW
Pseudoterranova decipiens	Nematoda	Musculature, liver	SW
Eustrongyloides sp.	Nematoda	Abdominal cavity	FW
Anisakis simplex	Nematoda	Musculature, abdominal cavity, vent	SW[a]
Eubothrium spp.	Cestoda	Intestine	FW, SW
Diphyllobothrium ditremum, D. dendriticum	Cestoda	On intestine, liver, abdominal cavity	FW[a]
Sanguinicola spp.	Trematoda	Heart, gills	FW
Cryptocotyle lingua	Trematoda	Gills, external surface	SW
Diplostomum spathaceum	Trematoda	Eye, brain	FW
Phyllodistomum umblae	Trematoda	Kidney, urinary bladder	FW
Apatemon gracilis	Trematoda	Pericardial and abdominal cavity	FW[a]
Cotylurus / Ichthyocotylurus spp.	Trematoda	Heart	FW
Stephanostomum tenue	Trematoda	Heart	SW
Nanophyetus salmincola	Trematoda	Many organs	FW
Gyrodactylus salaris	Monogenea	External surface	FW
Gyrodactyloides bychowskii	Monogenea	Gills	SW
Discocotyle sagittata	Monogenea	Gills	FW, BW
Acanthocephalus spp.	Acanthocephala	Intestine	FW, SW
Echinorhynchus spp.	Acanthocephala	Intestine	Normally SW[a]
Pomphorhynchus laevis	Acanthocephala	Intestine	FW
Lepeophtheirus salmonis	Maxillopoda	External surface	SW
Caligus elongatus	Maxillopoda	External surface	SW
Caligus rogercresseyi	Maxillopoda	External surface	SW
Argulus spp.	Maxillopoda	Skin, external surface	FW
Salmincola spp.	Maxillopoda	Gills, opercula	FW, SW
Margaritifera margaritifera	Bivalvia	Gills	FW
Piscicola geometra	Annelida	Gills, external surface	FW[a]

FW freshwater, SW sea water, BW brackish water
[a]Also found in anadromous fish

stage with subsequent migration from the gill epithelium into the gill blood vessels where replication and release of the parasite occurs. *C. shasta* spores are elongate and contain two polar capsules at the anterior margin (Fig. 9.1).

Diagnosis is carried out by the microscopic observation of typical spores in scrapes or stained sections prepared from the lower intestine, gall bladder or lesion within the body musculature. A polymerase chain reaction (PCR) is also available for diagnosis.

9.1.2 *Chloromyxum* spp.

Several myxosporeans are reported to infect salmonids, and *C. truttae* affecting farmed brown trout may result in a loss of appetite, emaciation and yellow discolouration of the skin and fins. Internally, the liver may be yellowish with hypertrophy of the gall bladder and associated enteritis. Infection can persist for several months and is fatal to some fish groups.

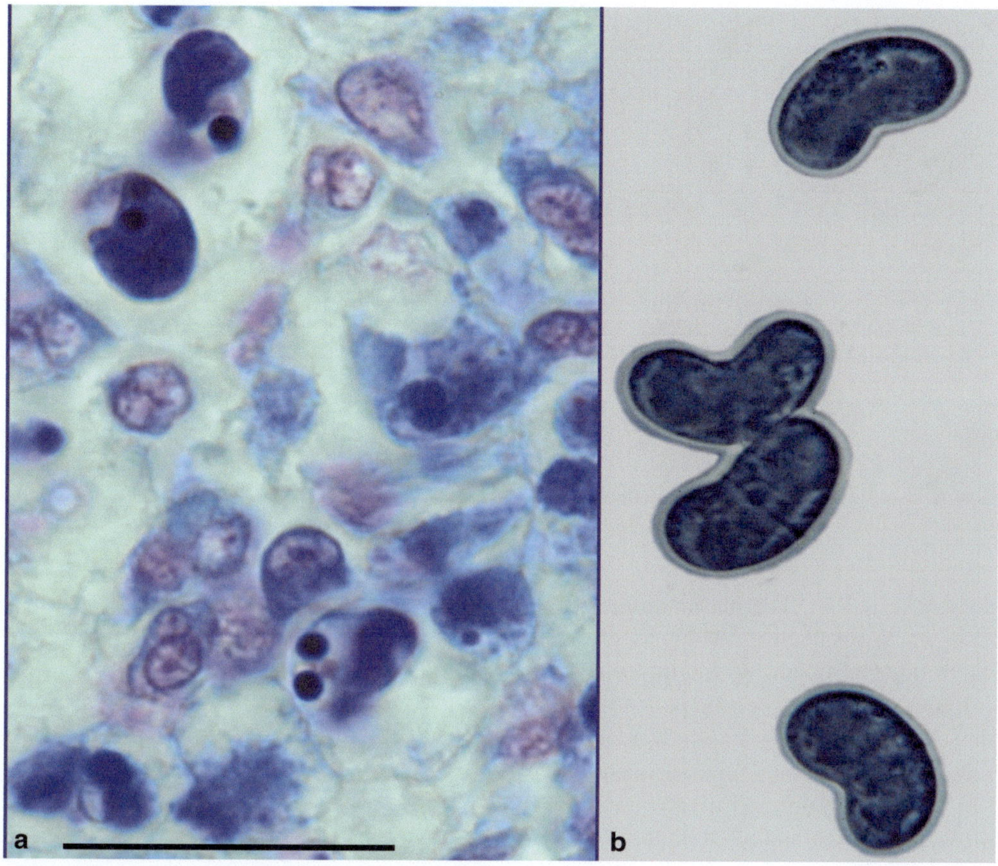

Fig. 9.1 Scrape showing characteristic elongate spores of *Ceratomyxa shasta* in steelhead trout. Bar = 20 μm. Giemsa stain (*left*). Fresh mount showing characteristic elongated crescent-shaped spores (*right*). High power

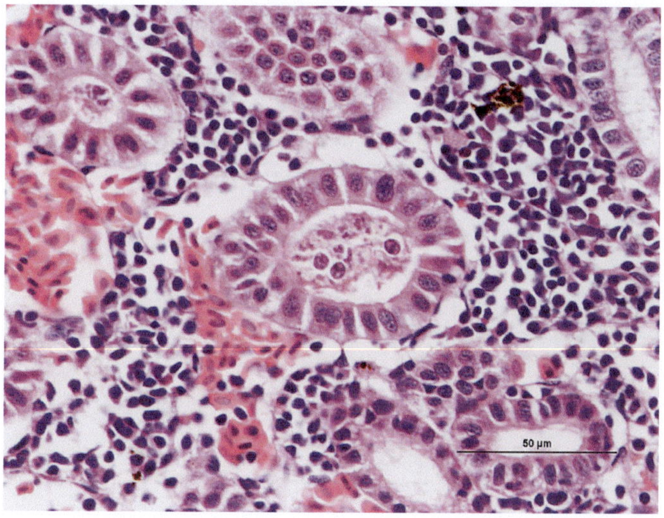

Fig. 9.2 *Chloromyxum* sp. plasmodia and spores in kidney tubules of brown trout. Medium power

C. schurovi in Atlantic salmon and brown trout sporulates in the kidney tubules and reported that the vascular system is used for transport with proliferation occurring remotely in target tissues (Fig. 9.2).

C. wardi is reported from the gall bladder of chum salmon. Infection occurs in fry during the fresh water phase, but the formation of the spherical or oval spores does not take place until the fish returns to the marine environment.

Species determination is considered difficult without the use of scanning electron microscopy or PCR. However, phylogenetic relationships based on ribosomal DNA data of the 18S rDNA provides significant taxonomic detail although not always agreeing with traditional taxonomic classification.

9.1.3 *Kudoa thyrsites*

Kudoa thyrsites primarily affects the trunk muscle but can also be observed in the heart of many marine fish. In North America its broad host range include Atlantic, Chinook, coho and pink salmon and rainbow trout. Infection causes typical focal lesions which are often the cause of the myoliquefaction condition known as 'milky flesh'.

Moribund fish are dark but generally no clinical signs are apparent until post-mortem. Internally, there is evidence of anaemia and the liver appears pale. The intramuscular stage starts with a single parasite in the muscle sacrolemma forming

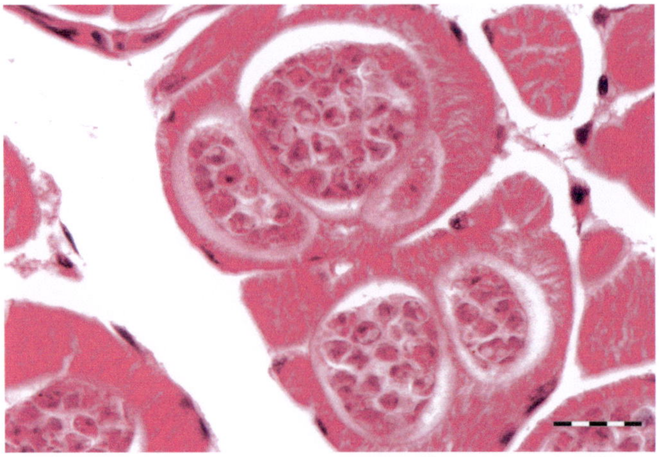

Fig. 9.3 Plasmodia of *Kudoa thyrsites* in transverse section of white muscle from Atlantic salmon. Bar = 20 μm

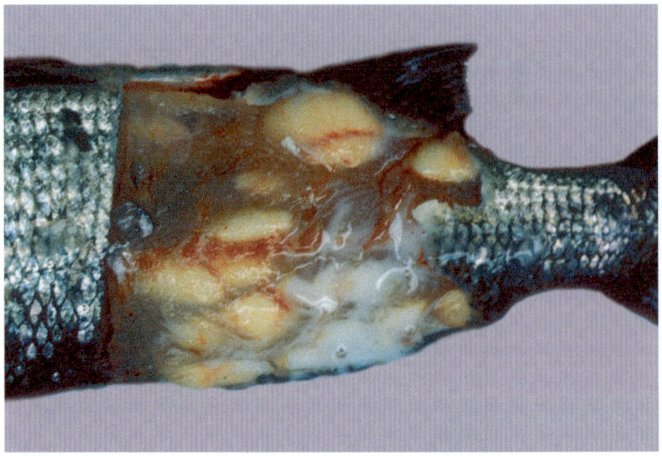

Fig. 9.4 *Henneguya zchokkei* cysts in white muscle of whitefish. Note punctured cyst with milk-like contents escaping

nodules or pseudocysts (Fig. 9.3). The plasmodium releases proteolytic enzymes which result in a histiolysis providing nutrients for growth of the plasmodia. Within the epicardium mononuclear cells infiltrate and an associated pericarditis may be evident. The dorsal musculature lesions show a characteristic multifocal intracellular infection with associated inflammatory response in the pericardium and myocardium. High numbers of *Kudoa* within the red and white muscle result in necrosis, fibrosis and inflammation followed by a chronic, active myositis with myolysis. In severely affected fish the kidney is swollen as a result of a markedly increased renal interstitium with occasional giant cells.

In Atlantic salmon a host response occurs after the polysporic plasmodia, containing the fully formed and the developing myxospores, rupture and the spores are released into the endomysium. The tissue damage and discolouration caused by *Kudoa* results in significant economic losses post-harvest as affected fish are rejected at processing.

Kudoa spores are stellate in shape and characterised by having four valves and four polar capsules, each containing a polar filament. Squash preparations, Gram or Giemsa stained imprints or sections of muscle tissue allows the spores to be observed.

9.1.4 *Henneguya* spp.

Henneguya zschokkei spores are found in the white muscle, cranial tissue and gills of different salmonids including Pacific salmon and whitefish from the Baltic Sea, while *H. cartilaginis* has been described from the head cartilage of wild masou salmon in Japan. Large, white-oval cysts with a creamy content occur and ruin the aesthetic appearance of the fillet (Fig. 9.4). The cysts mature and rupture through the integument and a large number of infective spores are discharged into the water. The open ulcers provide a port of entry for secondary pathogens. The diagnosis is based on the demonstration of the characteristic spores containing two polar capsules and two caudal projections.

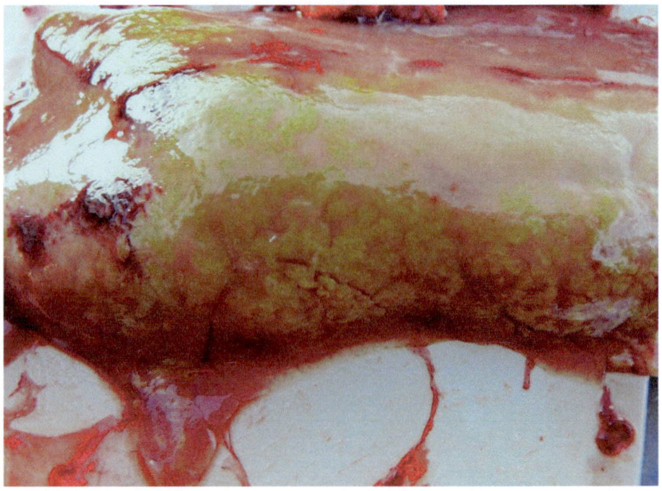

Fig. 9.5 Liver of adult wild Atlantic salmon with numerous plasmodia of *Myxidium truttae* protruding above the liver surface

9.1.5 *Myxidium* spp.

Myxidium truttae is a common fresh water parasite particularly of the liver of wild and farmed salmonids in Eurasia. *Myxidium salvelini* has been reported in Arctic char, but is probably apathogenic. Grossly, affected livers may have pale or yellowish surface protrusions. On incision, a thick yellowish or creamy coloured fluid can be recorded (Fig. 9.5). Similarly, *Myxidium minteri* has been reported in kidney of coho salmon (Fig. 9.6).

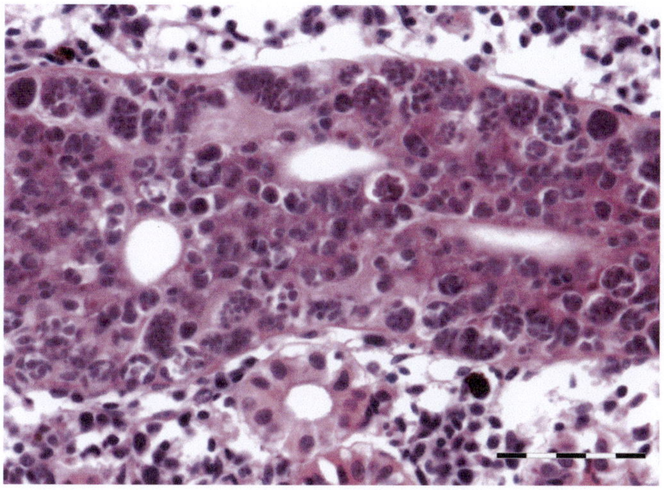

Fig. 9.6 *Myxidium minteri* in kidney of coho salmon. Bar = 50 μm

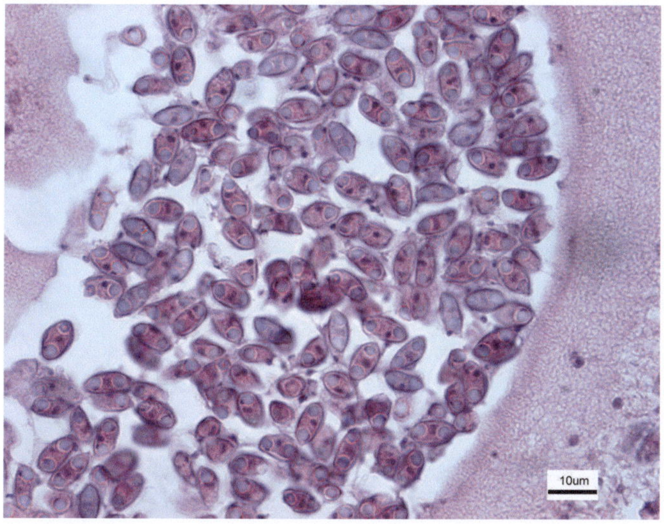

Fig. 9.7 *Myxidium truttae* spores in bile duct plasmodium from wild Atlantic salmon

During the life cycle, actinospores are discharged from an annelid host into the water and penetrate the skin of the fish intermediate host. The sporoplasm develops into presporogonic stages and eventually will locate in the bile ducts where the sporogonic stage, plasmodium develops. The cytoplasm of these large worm- or sac-like structures is filled with many cell types including sporogonic cells and pericytes. These develop into sporogony (spore formation), and subsequently into sporoblasts and myxospores.

Histological examination reveals dilated bile ducts and excretory canals with characteristic worm-like plasmodia filled with sporogonic cells and spores (Fig. 9.7). Diagnosis is based upon the examination of wet mounts or tissue sections, the identification of typical plasmodia and crescent shaped or fusiform spores and polar capsules at opposing ends.

9.1.6 *Myxobolus cerebralis*

Myxobolus cerebralis is the causal agent of a persistent and economical important condition termed 'whirling disease' (WD). The parasite has been identified from farmed and wild Pacific and Atlantic salmon, but within farmed species this is mainly a problem in trout reared in earthen ponds. Clinical signs include spiralling ('whirling behaviour'), darkening of the caudal region and severe skeletal deformities of the cranial area, jaw and opercula (Figs. 9.8 and 9.9) The whirling behaviour is often pronounced when the fish attempt to feed or when the fish are startled. Erosion of the cartilage surrounding the auditory organ is reported to contribute to the whirling. Infection of the cartilage of the spinal column causes pressure on the caudal nerves resulting in loss of control of caudal dermal melanophores. Infections cause a significant pathology usually in the first 3–4 months after fish start feeding and before ossification is complete. Maturing stages lyse and digest chondrocytes thus disrupting osteogenesis. Thereafter, older fish may be infected but infections are unlikely to result in a clinical complication.

Fig. 9.8 Darkening of the caudal region in brown trout due to infection with *Myxobolus cerebralis*

Fig. 9.9 Skeletal deformity in brook trout following infection with *Myxobolus cerebralis*

Spores are released from the cartilage after death of the fish where they are ingested by an intermediate host, the oligochaete *Tubifex tubiflex,* in which they localize in the intestinal epithelium and develop into the triactinomyxon stage. Re-infection in the fish host may occur through the skin and buccal cavity. Trophozoites migrate through the epidermis or gut lining and peripheral nerves to reach the cranial cartilage. In rainbow trout successful development of the parasite only occurs in sac fry which are older than 2 days. Trophozoites induce necrosis either directly or indirectly and lysis in the cartilage of the head and vertebrae. As ossification proceeds, there is an interruption of osteogenesis resulting in permanent cranial and other skeletal deformities. Trophozoites may also enter the labyrinth, causing damage and subsequent behavioural changes such as loss of balance. Although some fry die before any signs of WD are recorded, in most cases the course of the disease is chronic and spores remain for several years.

The classification of the group has been reanalysed and currently phylogenomic analyses of new genomic sequences of *Myxobolus cerebralis* firmly place Myxozoa as sister group to Medusozoa within Cnidaria. The morphological features of the spore which develop from the small amoeba-like trophozoite following infection of the fish are typically oval, measuring 8 × 10 μm and the two polar capsules are normally of equal size measuring 3 × 4 μm (Fig. 9.10). The multinucleate trophozoite grows and divides by nuclear division producing pansporoblasts, each of which will produce two spores which localise in the cartilage. Spore development in the fish has been linked to their acquisition of acid-fastness, and such spores have been referred to as 'mature'.

Diagnosis of whirling disease is made microscopically from stained tissue sections (e.g. Giemsa, Mallory-Heidenhain) of cartilaginous tissue or from wet mounts of macerated cartilage. Other species e.g. *M. arcticus* and *M. neurobius* infect the nervous tissue of salmonid species in fresh water in North America and Asia while *M. insidiosus* occurs in the striated musculature of Chinook salmon. Molecular genetic tests based on nuclear DNA are used to verify species.

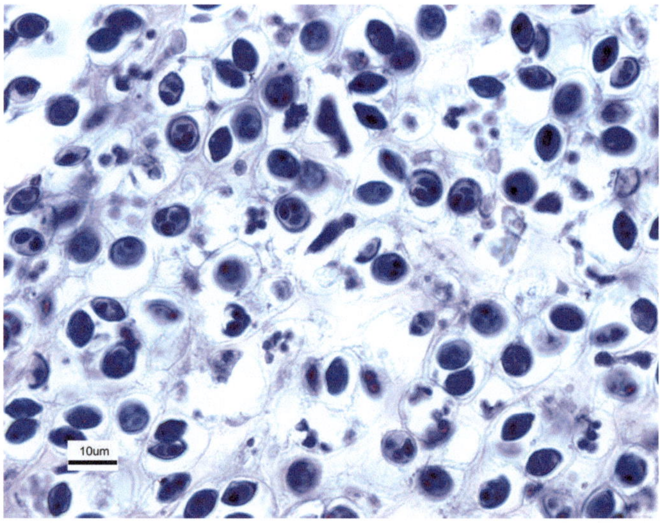

Fig. 9.10 *Myxobolus cerebralis* in cartilage of brook trout. Giemsa. High power

9.1.7 *Parvicapsula* spp.

Parvicapsula pseudobranchicola is a serious pathogen in sea-water farmed Atlantic salmon in Norway, particularly in northern areas. It primarily affects the pseudobranch where it may cause extensive inflammation and necrosis

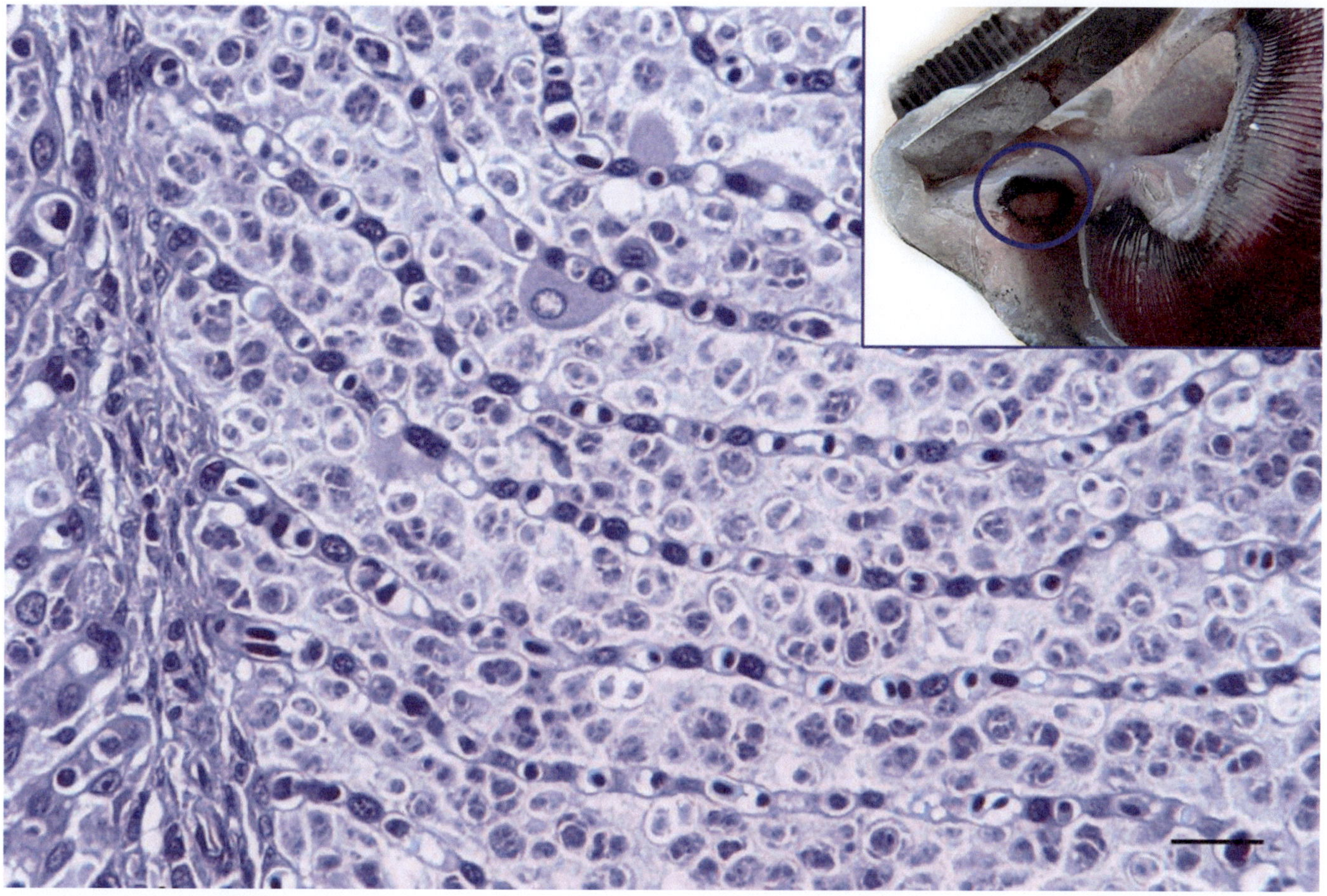

Fig. 9.11 *Parvicapsula pseudobranchicola* spores in pseudobranch of farmed Atlantic salmon. Low power. *Insert* Pseudobranch of farmed Atlantic salmon damaged by *Parvicapsula pseudobranchicola*

with up to 40 % mortality, although other organs may also be affected. Affected fish show unspecific clinical signs that may include cataract, snout ulcers and increased numbers of 'poor doing fish'. At autopsy, the pseudobranch is haemorrhagic or totally absent, leaving only a white pseudomembrane or a dark edge (Fig. 9.11). Histopathology reveals extensive infiltration of extrasporogonic stages, haemorrhage and necrosis of pseudobranchial tissue. The final host of this parasite are oligochaetes, while fish are the intermediate host. Infection occurs when actinospores are released from the final host and attach to the fish by means of the filaments from the polar capsules. The sporoplasm penetrates the epidermis and asexual reproduction follows, whereby new cells are infected. Mature spores are subsequently formed in the pseudobranch and are released into the environment allowing the final host infection. In this process, severe damage is inflicted on the pseudobranch resulting in considerable necrosis and loss of functional tissue. *Parvicapsula* can occur concurrently with other conditions and therefore the significance of the infection can be difficult to establish. *Parvicapsula* spores have a characteristic bean-or banana-shaped outline in smears

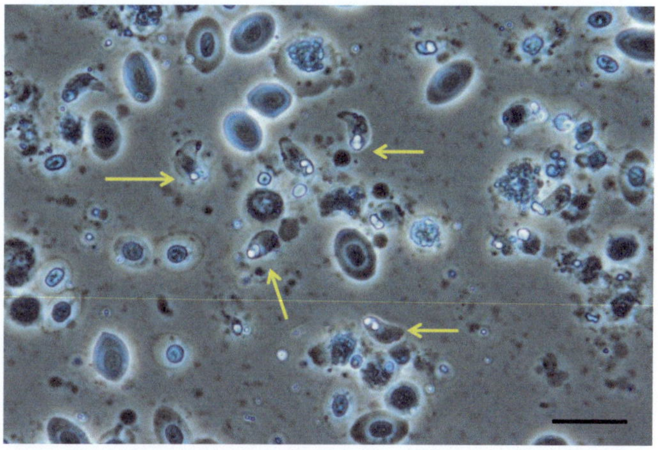

Fig. 9.12 Fresh squash mount from Atlantic salmon pseudobranch showing *Parvicapsula pseudobranchicola* spores (*arrows*). Dark field microscopy. High power

from the affected pseudobranch (Fig. 9.12). Diagnosis is based upon gross, histological lesions and RT-PCR.

Parvicapsula minibicornis may infect several species of wild and farmed Pacific species and farmed Atlantic salmon

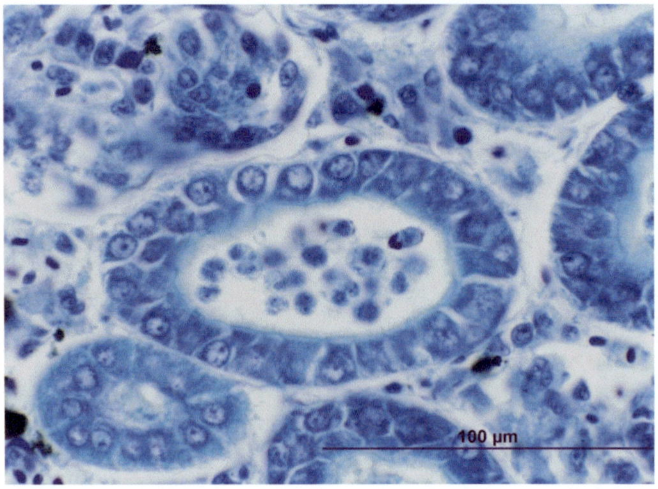

Fig. 9.13 *Parvicapsula minibicornis* spores in kidney tubules of juvenile coho salmon. Giemsa stain

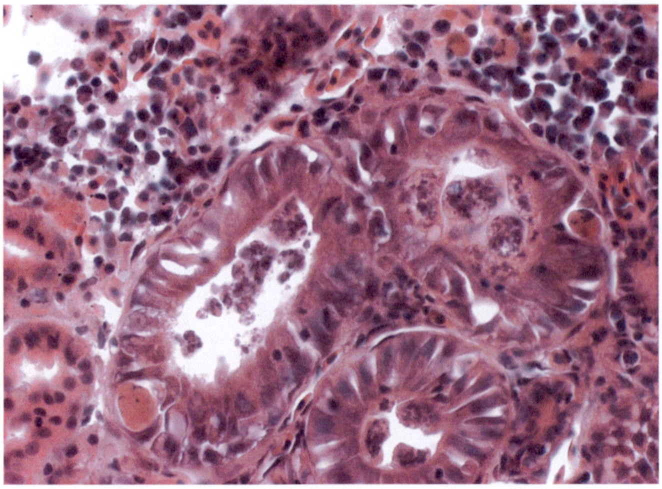

Fig. 9.14 Sporogonic stages of *Sphaerospora truttae* kidney tubules of farmed rainbow trout. High power

in the northwest coast of the USA (Puget Sound, Washington region). Elevated pre-spawning mortality in some of these species has been associated to infection with *P. minibicornis*. Clinical signs and pathological changes are unspecific and may include dark and lethargic fish with hypertrophied kidney (Fig. 9.13). Histologically, trophozoites and developing spores occur in the lumen and epithelium. A PCR assay has identified the myxospore in the freshwater polychaete, *Manayunkia speciosa*.

Parvicapsula kabatai has been described from renal tubules of pink salmon in British Columbia, Canada. The shape and size of the spores are similar to those of *P. pseudobranchicola*, but distinctly different from *P. minibicornis*. The significance of this parasite is currently unknown.

9.1.8 *Sphaerospora truttae*

Sphaerospora truttae was originally described from brown trout and grayling in Germany, but subsequently reported affecting and Atlantic salmon parr in Scotland. Brown trout are also proven susceptible. The gills have been identified as the predominant point of entry, which is followed by penetration of the vascular epithelia and thereafter, proliferation in the blood before exiting the vascular system through capillary walls. Subsequently, the kidney, as well as the spleen and the liver are infected. Parasites occur in the tubular lumen and sporogony takes place inside the renal tubules (Fig. 9.14). Histology can be used for presumptive identification but for early myxosporean stages and parasite specific identity, a DNA-based approach is appropriate.

9.1.9 *Tetracapsuloides bryosalmonae*

Tetracapsuloides bryosalmonae causes the condition proliferative kidney disease (PKD) which is a significant seasonal disease of young salmonids. PKD occurs both in farmed and wild fish and is associated with decline in wild populations in many countries. The endoparasitic myxozoan uses freshwater bryozoans as primary hosts. Environmental changes may play a role in the increased significance of PKD in wild populations. Bryozoans and *T. bryosalmonae* stages in bryozoans undergo temperature and nutrient-driven proliferation and above 15 °C are required for development of clinical disease. Infective spores enter the fish through the skin and gill epithelium.

Clinical signs include a dark body, bilateral exophthalmia, pale gills, and distended abdomen with pale visceral organs. Swelling due to extensive accumulation of ascites is recorded within the abdomen. The kidney, particularly the caudal region, is markedly swollen due to diffuse oedema (Fig. 9.15) and to a lesser extent the spleen.

Histozoic and extrasporogonic stages proliferate causing focal or multifocal granulomatous inflammation and renal interstitial tissue is replaced by mild haematopoietic hyperplasia during the early stages of infection, and followed by further granulomatous tissue with associated macrophages and mononuclear cells. Lymphoid cells and macrophages are frequently seen adherent to the PKX cells (Fig. 9.16) and hepatic lesions frequently include multinucleated giant cells scattered throughout (Fig. 9.17). There may also be extensive haemorrhage in the acute stages of the disease. The number of nephrons and melanomacrophage centres are severely reduced and an extensive chronic fibrosis occurs in the final stages.

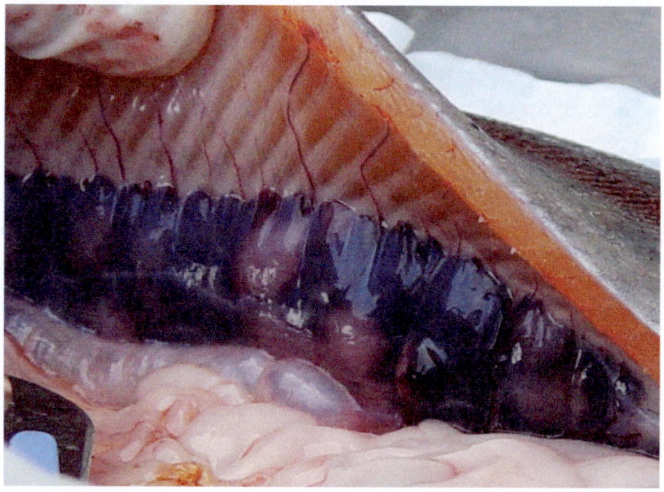

Fig. 9.15 Proliferative kidney disease in rainbow trout. Gross enlargement of the posterior kidney due to granulomatous response

Throughout the affected tissue characteristic PKX cells can be found, particularly in the endothelium of the portal vessels. These cells are large, eosinophilic to pale orange (often multinucleated), PAS-positive cells with a granular cytoplasm and typically surrounded by a clear halo. PKX cells may be seen in several other organs including the gills, spleen, liver and heart. Developmental stages (sporoblasts) of the parasite may also be recorded in the walls and lumen of the kidney tubules.

The diagnosis of PKD is based on clinical signs and the demonstration of the characteristic extrasporogonic stages in wet preparations, stained imprints or in histological sections. In addition, a specific PCR assessment can be used as a confirming test. *Sphaerospora truttae* is important from the viewpoint of differential diagnostics.

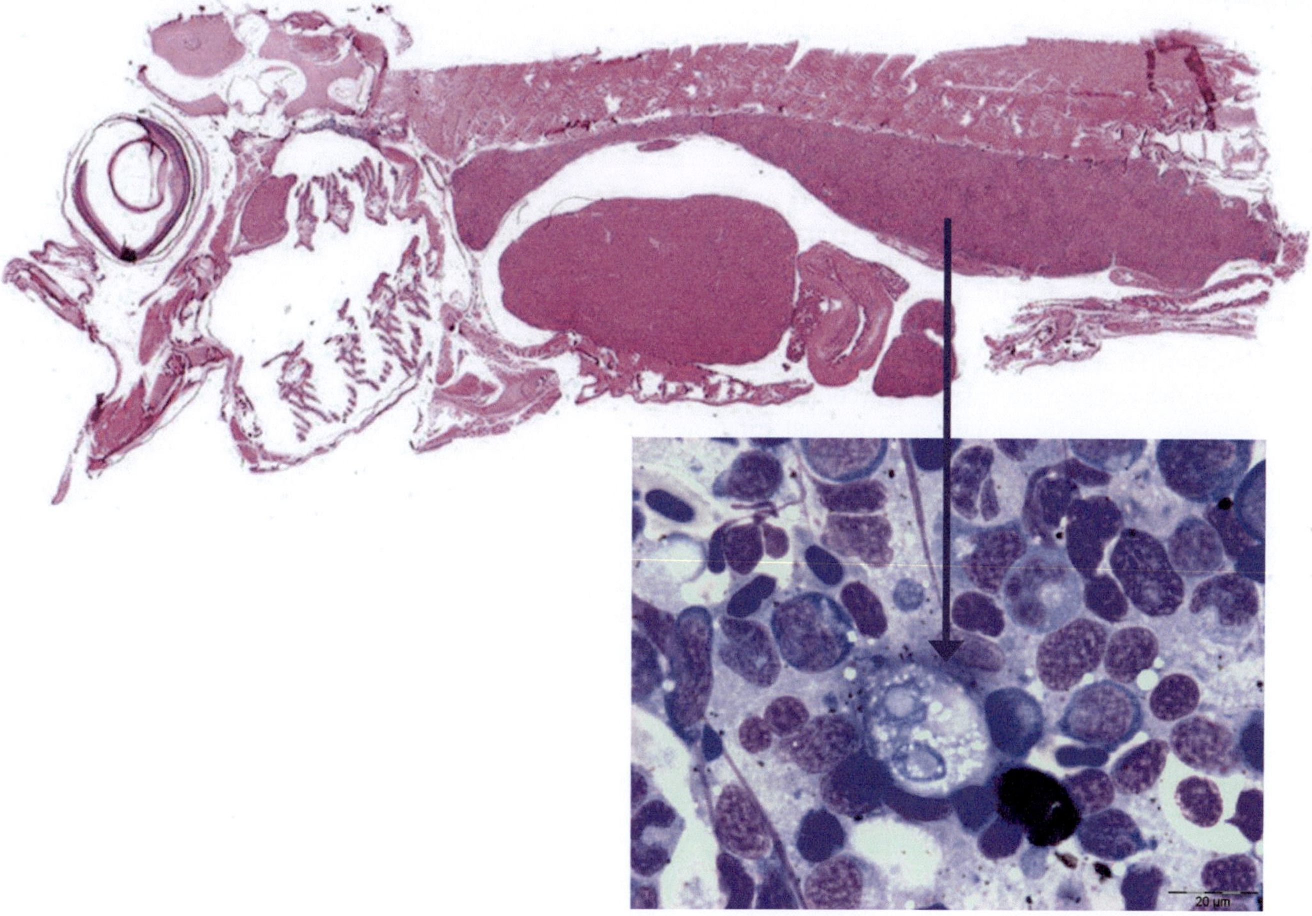

Fig. 9.16 Sagittal section of Atlantic salmon fry showing enlarged posterior kidney. *Insert* shows pre-sporogonic stage cell (*arrowed*). Insert Giemsa stained. High power

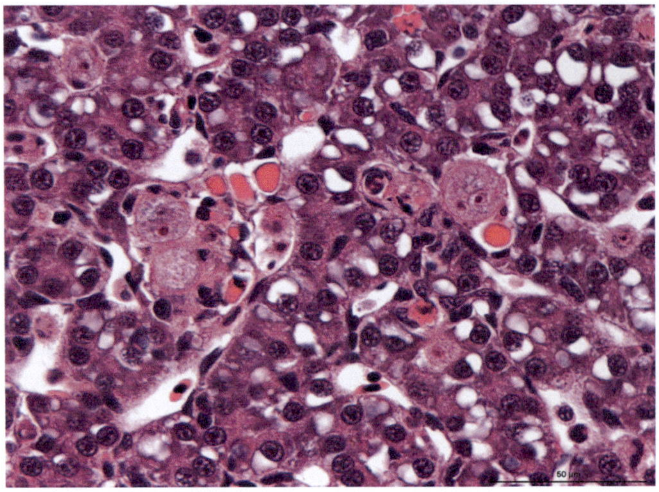

Fig. 9.17 Characteristic pre-sporogonic stage of *Tetracapsuloides bryosalmonae* in liver from Arctic char. High power

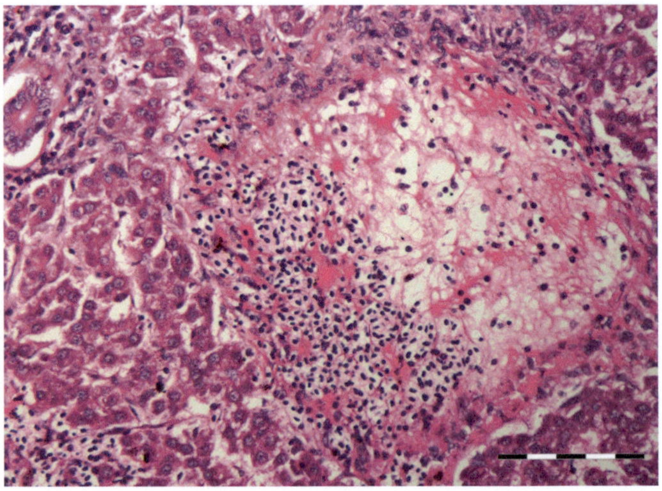

Fig. 9.18 Presumptive tract of migrating nematode in liver of Atlantic salmon. Low power

9.2 Nematoda

The Nematoda comprise numerous species of unsegmented roundworms that possess a pseudocoelom and a threadlike body that is circular in cross-section. Nematodes in fish can be found both as larvae or adult endoparasites. Nematode infection causes a variable but generally not extensive pathology, although more prominent where the parasite is migrating or degenerating (Fig. 9.18). Nematode digestive tract is divided into mouth (buccal) capsule, oesophagus, intestine and rectum. Histologically, a multi-layered cuticle which can be ornamented with ridges is recorded. The hypodermis is often thinner than the somatic musculature often with extensions called lateral chords protruding into the pseudocoelom. The musculature is composed of dense contractile elements and pale sarcoplasm. In section the oesophagus is recognised by the radial symmetry and triradiate lumen. The structure of the epithelial cells range from large multinucleate cells to cuboidal or tall columnar cells and can be been used to differentiate groups.

9.2.1 *Philonema oncorhynchi*

Philonema oncorhynchi (Philometridae) is noted in the visceral cavity of several fish species including rainbow trout, Arctic char, chum salmon and sockeye salmon in freshwater from various regions including Canada, Iceland, Japan and parts of Europe. *P. sibirica* is also described from European salmonids. Larva, sub-adult and adult worms can be present but are not usually fatal.

Juvenile fish acquire the nematode in freshwater by feeding on copepods, i.e. *Cyclops* spp. that act as the intermediate host. Larvae are released from gravid females present in the body cavity of affected hosts at spawning, and infect a copepod where they moult twice in their haemocoel into third stage larvae. The cycle is completed when the copepods are eaten by the final host. The orientation of seaward-migrating sockeye smolts is thought to be influenced by the presence of this parasite, which may have implications for smolt survival.

Fig. 9.19 *Philonema salvelini* in the peritoneal cavity of brook trout

The parasite lives in the gastrointestinal tract migrating into the visceral cavity where it matures; it rarely migrates to the musculature but it may induce distension of the abdomen and ascites (Fig. 9.19). Activation of peritoneal macrophages, neutrophils and eosinophilic granular cells (EGCs) result in the formation of visceral adhesions, characterised by the production of fibrous connective tissue from the host preventing the organs from hanging freely in the coelomic cavity. In some severe cases adhesions may cause atrophy of the gonads and prevent spawning.

Diagnosis of *Philonema* occurs at necropsy and is based upon the detection of a filiform worm up to 10 cm in length (female) with a rounded anterior end and a posterior tail tapering to a sharp point.

9.2.2 *Cystidicola farionis*

Cystidicola farionis is relatively common, particularly in wild salmonids, but also reported from farmed fish within Europe and North America. Adult males and females live in the swim bladder (Fig. 9.20) of fish including grayling, whitefish and rainbow trout. A pneumatic duct connected to the intestine is used by female worms to enter the intestine where eggs are deposited. Eggs are passed in the faeces and are eaten by amphipods, the first intermediate host. The L1 stage penetrates into the haemocoel and moults to the L3 stage, which is infective to the definitive host. After the infected intermediate hosts are consumed the L3 migrates up the pneumatic duct and completes its adult development.

Low numbers are not considered to affect the host, but fish with a severe infection may show evidence of anaemia and emaciation. Histologically, there is vascularisation and haemorrhage associated with the body cavity, stomach and swim bladder. In the latter there is an increase by histolytic inflammatory cells in the sub epithelium, with varying degrees of epithelial atrophy and the occurrence of round granular cells in the sub-mucosa connective tissue.

Diagnosis is made when adult worms are found in the swim bladder at necropsy, but not within the musculature. The adult worms are white and measure around 6 mm in length. The identification of *C. farionis* relies upon the examination of external anatomical features.

Fig. 9.20 *Cystidicola farionis* in swimbladder of wild Arctic char

9.2.3 *Pseudoterranova decipiens*

Pseudoterranova decipiens is a cosmopolitan parasite and occurs among many species of marine fish. Salmonids are considered as occasional hosts and serve as second intermediate hosts. This is a zoonotic parasitic condition and when ingested by humans the larvae may cause a disease condition similar to anisakiasis.

The parasite has an indirect life cycle with the larval stage developing in a benthic copepod, which acts as the first intermediate host, and the adult worm which matures within a seal. Adult worms may reach 60 mm in length.

Pseudoterranova encyst within the muscles and frequently cause a granulomatous encapsulation within the gut, with giant cell involvement. Mature capsules around the parasite consist of an inner layer of macrophages which undergo epitheloid transformation with gradual degeneration, and an outer layer, composed of collagen and fibroblasts.

9.2.4 *Eustrongylides* sp.

The larval stage of *Eustrongylides* are bright red and common in fresh water populations of feral trout in North America and Europe. The larvae are found in the fluid of thin-walled cysts within the abdominal cavity (Fig. 9.21). Soon after the host dies, the larvae will penetrate the cyst wall and migrate to adjacent organs and even penetrate the body wall. Piscivorous birds, such as mergansers and herons serve as the final host for the adult nematode where they are found in the stomach. Gross observations are sufficient to provisionally identify this group.

9.2.5 *Anisakis simplex*

Anisakis simplex is a very widespread parasite and found in almost all commercially exploited species in North Atlantic waters. Anisakiasis is also an important fish-borne zoonosis. Infection involves many marine fish species where it occurs in the visceral cavity and surrounding tissues. Adult worms live within the intestinal mucosa of marine mammals (especially cetaceans) as their final host, while fish, squid, as well as planktonic crustaceans act as paratenic or intermediate hosts that harbour the larval stages. The female worms produce eggs that are released through the faeces into the water where they develop into first stage larvae within the egg. The larva moults and hatches as the second and subsequent third stage (L3) which is infective to fish. The complex life cycle however, almost exclusively excludes *A. simplex* from farmed salmonids.

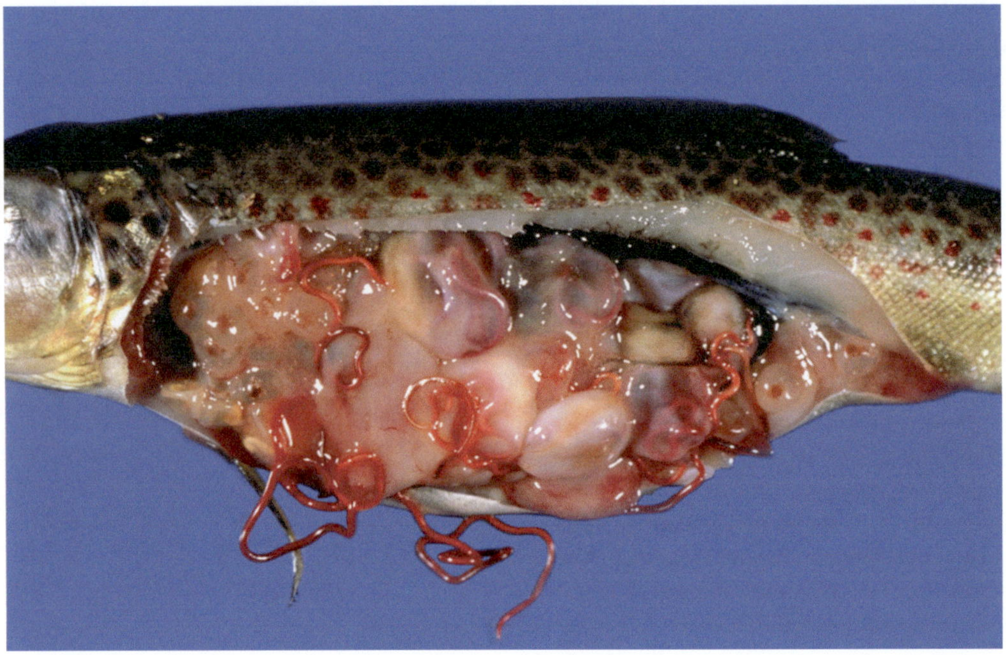

Fig. 9.21 Encysted *Eustrongylides mergorum* in the peritoneal cavity of brown trout

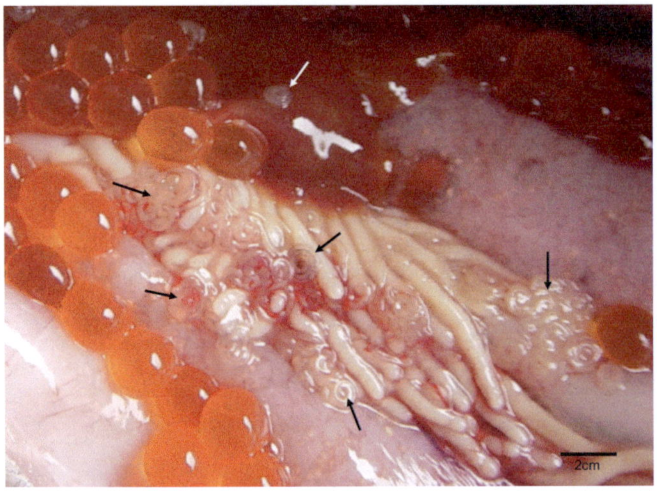

Fig. 9.22 *Anisakis simplex* (*arrows*) within the abdominal cavity of Atlantic salmon

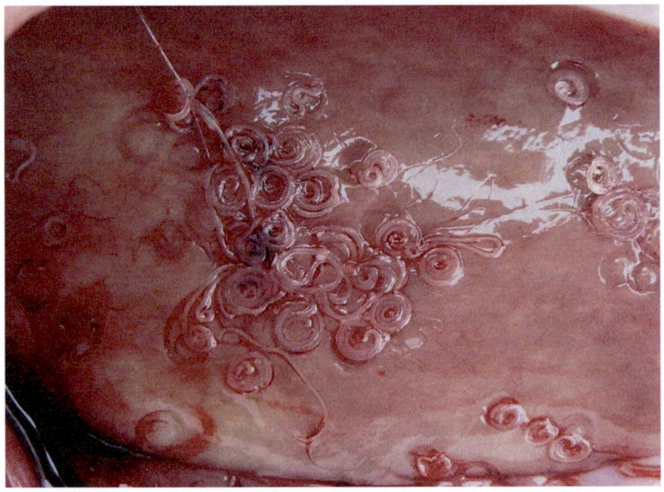

Fig. 9.23 *Anisakis simplex* larvae on the liver surface of wild Atlantic salmon

L3 worm penetrates the fish gut wall and normally encapsulates in the body cavity and the external surfaces of the gut, pyloric caeca, liver and fat tissue (Figs. 9.22 and 9.23), inducing a mild to moderate adhesion reaction in the fish host. In addition, they can also migrate and locate in the skeletal muscle thereby affecting the edible part of the fish.

Anisakis does not result in high losses, however, it is an important problem for the fishing industry because infection can reduce the quality of the flesh. Third stage larvae grow up to 2 cm in length, are almost colourless and are found tightly coiled and encased in the gut and flesh, particularly in the belly flaps. *A. simplex* is provisionally identified from morphological characteristics and confirmed through molecular techniques.

Recently it has been found that L3 larvae can invade in very high numbers the vent and urogenital region (red vent syndrome, RVS) an atypical location of the fish body of Atlantic salmon. RVS was initially described in the UK in returning wild Atlantic salmon which showed bloody, swollen, vents that gave the condition its popular name (Figs. 9.24 and 9.25). Although earlier sightings were suspected, simultaneous reports from geographically diverse rivers from all around the UK peaked in 2007, predominantly in one-sea winter 'grilse'. The condition was also recorded in two sea-winter

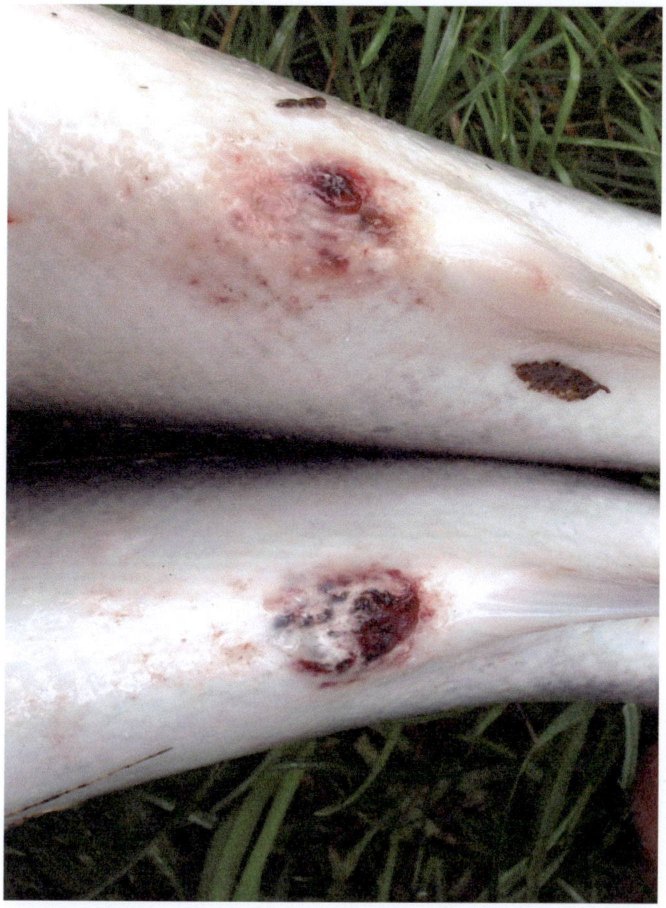

Fig. 9.24 Red vents in wild grilse caused by accumulation of *Anisakis simplex* larvae in the vent area

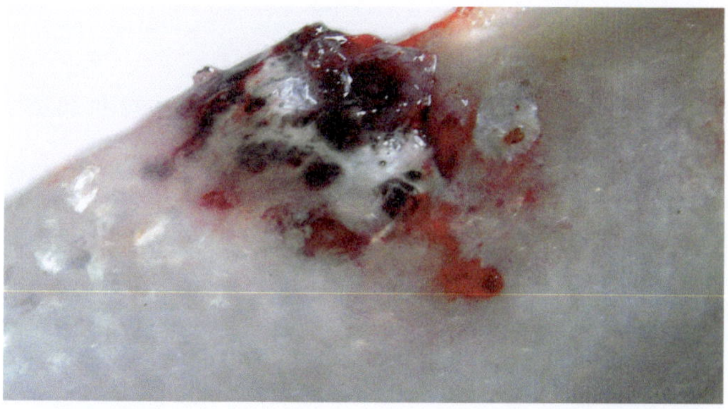

Fig. 9.25 Red vent in returning wild Atlantic salmon caused by accumulation of *Anisakis simplex* larvae in the vent area

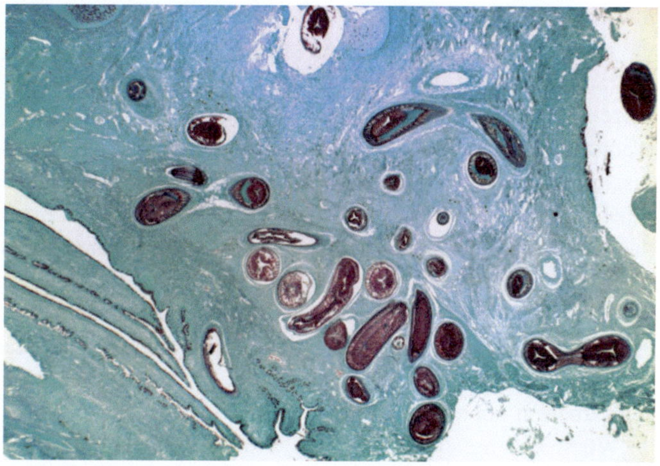

Fig. 9.26 Cross section through vent area of wild Atlantic salmon with red vent showing sections of *Anisakis simplex* larvae. PAS staining. Low power

salmon and in sea trout. External features of affected vents can be mild to severe and include protrusion and swelling, scale loss, skin breakdown, petechial or widespread haemorrhage, and bleeding in severe cases. Occasionally, larvae can be seen with the naked eye just beneath the skin surface in the vent region. At examination, both encapsulated and non-encapsulated larvae can be seen around the hind gut, within the discrete space towards the skin, between the hindgut and the genital cavity, and between this region and the urethra (Fig. 9.26). *Anisakis* have also been recorded deep within the skeletal musculature above the vent area and sometimes within the lumen of the genital cavity.

Histologically, severely affected tissues of the vent show scale loss and absence or a detached epidermal layer, however evidence of epidermal healing and scale regeneration has been observed after fish enter and remain in freshwater. Capillary dilation, blood congestion, haemorrhage and moderate to severe dermal inflammation associated to non-encapsulated migrating larvae has been reported. The inflammatory reaction is dominated by EGCs, but melanomacrophages and multinucleated giant cells have also been described.

RVS does not seem to prevent fish form spawning or is inducing mortality at least in those that have been examined or stripped artificially. The condition has also been reported from Canada, Iceland, Ireland and Norway, and is therefore considered to have a north Atlantic distribution.

9.3 Cestoda

The Cestoda or tapeworms have a wide geographical distribution, but usually a high degree of host specificity. Salmonids can act both as intermediate host for larval stages or final host of adult individuals, from infections by different species. Generally, when fish act as intermediate host the plerocercoids are found free or encysted on the viscera or in the musculature, where they can burrow deeply. Commonly, the plerocercoids trigger a granulomatous encystment on the viscera, with a dense fibrous coat incorporating fibroblasts, macrophages and collagenous tissue, and adhesions between viscera and the abdominal wall can be observed. In the muscle block plerocercoids may also cause granulomas, haemorrhage and necrosis with cavitation, negatively affecting the swimming ability of the fish. Plerocercoids are long lived in the fish intermediate host and will therefore accumulate in older fish that typically develop the most severe lesions. Piscivorous birds or mammals are final hosts where reproduction occurs in the gut.

When fish are the final host, adult worms are typically found attached by their scolex to the intestinal mucosa of the pyloric caeca and hindgut. Cestoda lack a digestive system and semi-digested nutrients from the host are absorbed through the whole body surface. The strobila is composed of individual proglottids where each segment contains a single set of reproductive organs. The proglottids with fertilized eggs are shed from the posterior end of the parasite as they mature. Eggs are shed with the faeces and hatch in water to release a motile coracidium which in turn is eaten by an intermediate invertebrate host (e.g. copepods) developing in a procercoid and continue the life cycle when eaten by a fish.

9.3.1 *Eubothrium* spp.

Eubothrium crassum and *E. salvelini* are the most abundant species of this genus and different species of salmonids may act as intermediate or final hosts. *E. salvelini* has been detected in rainbow trout, char, brook trout, Pacific and Atlantic salmon, while *E. crassum* has been found in brown trout, rainbow trout, Atlantic salmon, vendace and Danube salmon.

As adult parasites, the scolex is embedded in the pyloric caeca, while the conspicuous whitish, segmented strobila may extend through most of the intestine. Adult *E. crassum* may reach a length of 1 m, while *E. salvelini* seldom exceeds 30 cm. Both species can occur in considerable numbers in wild and farmed fish in fresh and sea water and may almost occlude the gut lumen (Fig. 9.27) with resultant loss of performance, emaciation and death. When present in large numbers, the parasites may also perforate the gut wall and end up in the peritoneal cavity (Fig. 9.28). Heavy infection has been shown to impair saltwater adaptation of migrant sockeye salmon. Fish are infected by ingesting infected copepods or small fish. *E. crassum* is a serious parasitic problem in farmed Atlantic salmon however successfully treated with medicated feeds containing praziquantel. Both species may occur simultaneously in the same fish and some fish may show multiple parasite infection (Fig. 9.29).

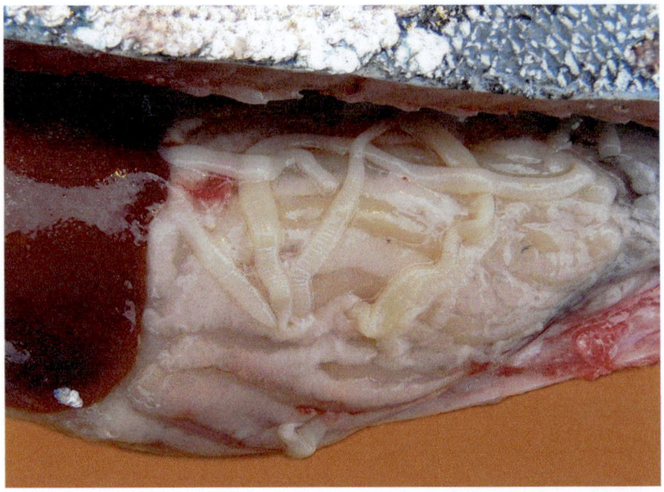

Fig. 9.27 *Eubothrium crassum* is usually located inside the intestine. In this case the parasites have penetrated the gut wall

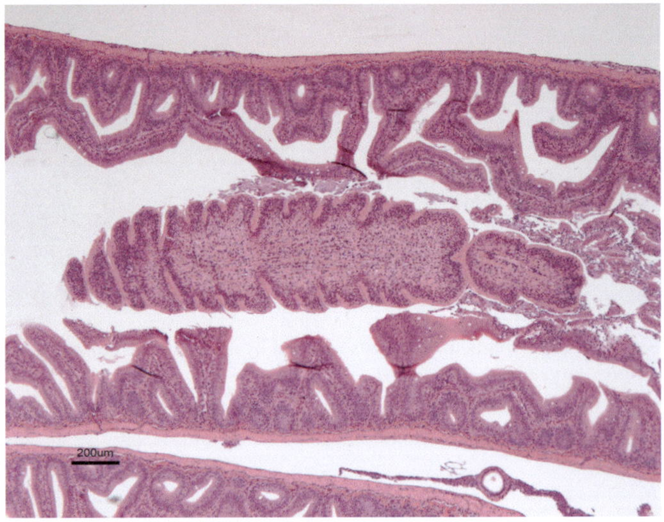

Fig. 9.28 Longitudinal section of *Eubothrium crassum* in pyloric caeca of farmed Atlantic salmon

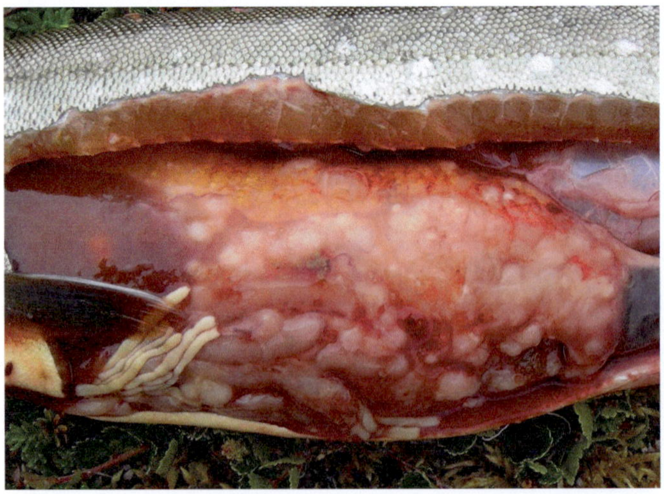

Fig. 9.29 *Eubothrium salvelini* in intestine, encysted *Diphyllobothrium ditremum* plerocercoids in the abdominal cavity and *Cystidicola farionis* in the swim bladder of wild Arctic char

9.3.2 *Diphyllobothrium* spp.

Several species within this group have fish as their intermediate host and plerocercoids of *D. dendriticum, D. latum* and *D. ditremum* are present in several salmonid species. *D. latum* has been found in whitefish and other salmonids in northern Scandinavia, Western Russia, the Baltic area and the Pacific north and south west of North and South America. Encysted plerocercoids are usually in the body cavity viscera (Figs. 9.30 and 9.31), but *D. dendriticum* have also been found in heart of farmed brown trout where they obliterate the cardiac lumen with subsequent circulatory failure and death. In European whitefish, plerocercoids cause a proliferation of mesenteric fibrous tissues of the gastric wall. Cysts are tri-layered and formed from a series of concentric whorls of fibroblast and collagen fibre-based connective elements. The extent of necrosis within each muscle layer and the serosa of the stomach differ, notably the latter is marked by a chronic inflammatory reaction and fibrosis. Within and around the cyst, many degranulating EGCs occur in addition to melanomacrophage centres.

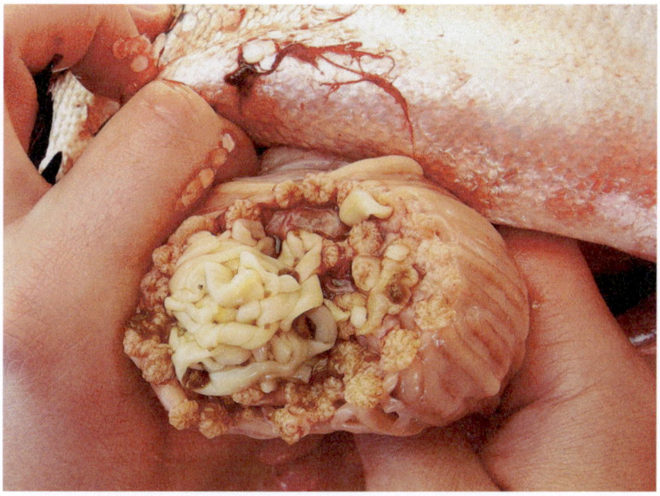

Fig. 9.30 *Diphyllobothrium* in the cut surface of pyloric caecae of farmed Atlantic salmon

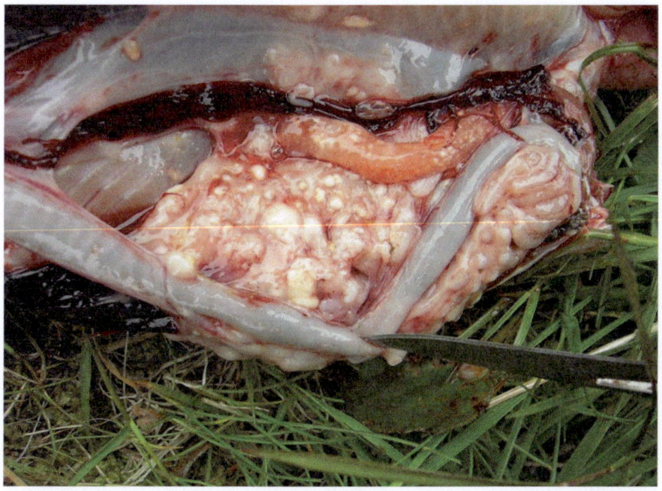

Fig. 9.31 Encysted plerocercoids of *Diphyllobothrium dendriticum* in the abdominal cavity of wild brown trout

D. latum (broad fish tapeworm) is of particular interest because of its zoonotic potential. Diphyllobothriasis in humans who have eaten raw or undercooked fish with infective plerocercoids may develop constipation, fatigue, abdominal pain and severe vitamin B_{12} deficiency.

Differential diagnosis of the visceral encysted forms include other parasites as the myxozoan *Henneguya zschokkei*, which need to be differentiated from the plerocercoid induced whitish nodules in the musculature, which can be done by dissection and microscopic examination.

9.4 Trematoda

Trematodes are flat and broad unsegmented helminth worms with two suckers, one anterior and a ventral adhesive sucker. Worms vary in size, from invisible to the naked eye to some fairly large individuals. With a single exception trematodes are hermaphroditic and oviparous, laying operculated eggs. They have a complex life cycle involving developmental stages (sporocysts and redias) in a mollusc (first intermediate host), a brief, free water living larval stage (cercaria) that requires a second intermediate host (different species including fish) where they develop into metacercaria, and when consumed by the vertebrate the final host (including fish) the adult stage develops. Once eggs are released by mature parasites back into the water, they hatch into a miracidia that will continue the cycle when it reencounters their mollusc host. The most frequently encountered stage in fish is the larval metacercaria but examples of adults stages are also found in salmonid fish.

9.4.1 *Sanguinicola* spp.

Sanguinicola spp. occur in the ventral aorta and branchial arteries of rainbow, cutthroat and brook trout. Eggs are carried through the blood stream to the gill capillaries where they become lodged (Fig. 9.32) causing rupture of pillar cells and vessel walls. Miracidia escaping from the gills can also cause severe mechanical damage, haemorrhage, and necrosis and calcification in the heart and kidney have been reported.

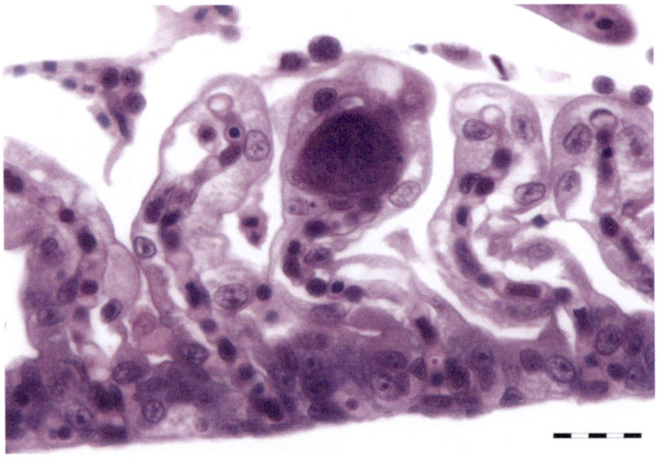

Fig. 9.32 *Sanguinicola* sp. on gills of rainbow trout. Bar = 20 μm

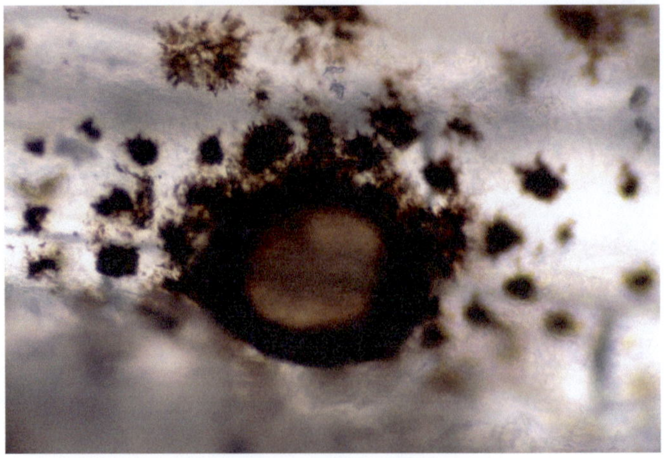

Fig. 9.34 *Cryptocotyle lingua* in skin of sea trout. There is heavy melanization around the encysted metacercaria

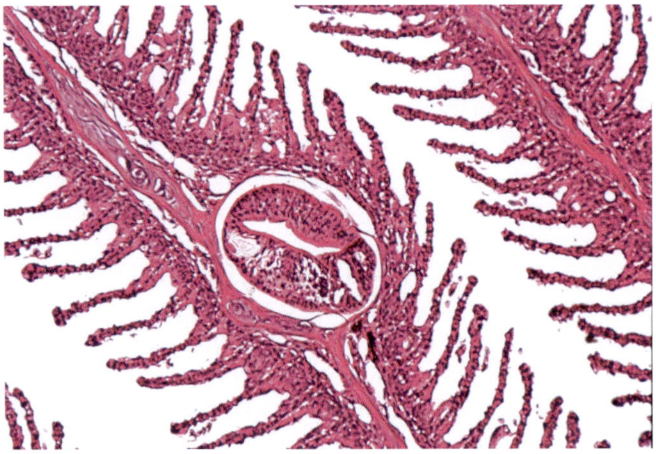

Fig. 9.33 Encysted metacercaria of *Cryptocotyle lingua* in the gill filament of farmed Atlantic salmon postsmolt. Low power

Fig. 9.35 Cataract resulting from the presence of metacercariae of *Diplostomum spathaceum* in rainbow trout

Serological or molecular techniques for differential diagnosis are not currently available and diagnosis therefore determined by morphological identification of adults.

9.4.2 *Cryptocotyle lingua*

Cryptocotyle lingua metacercaria infect several marine fish including both wild and farmed salmonids during the sea water phase. The condition is known as 'black spot disease' and has been diagnosed in salmon, sea-run brown and brook trout, and sea-run Arctic char. These parasites may cause mortality in heavily infected fish with a loss of condition factor and an increased susceptibility to other diseases.

Metacercariae encapsulate in the skin, fins and gills (Fig. 9.33), and evoke a melanomacrophage reaction generally visible macroscopically, but may also found elsewhere such as the eyes causing exophthalmia and blindness. In the heart, cercariae become surrounded by a focal myocarditis. The presence of circular black spots (0.2 mm diameter) in the skin (Fig. 9.34) and other organs, assists with diagnosis.

9.4.3 *Diplostomum spathaceum*

Diplostomum spathaceum cercariae cause diplostomiasis, generally a seasonal disease of wild and farmed fresh water fish. Under farming conditions, *D. spathaceum* is primarily a problem with rainbow trout reared in earth ponds or cages in shallow waters. Infected fish usually show cataract and exophthalmia (Fig. 9.35), skin petechiae is particularly obvious on the ventral surface but can also be seen in the internal organs where haemorrhage may also be observed. Reduced growth and emaciation is reported for farmed fish. In

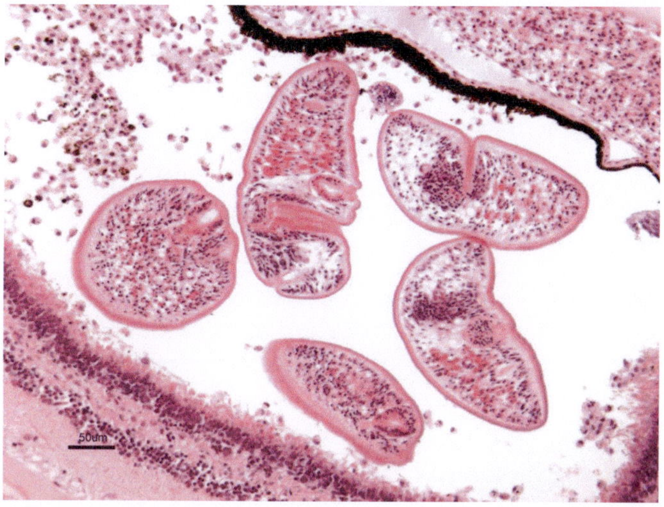

Fig. 9.36 Metacercariae of *Diplostomum spathaceum* in the eye chamber of wild Atlantic salmon parr

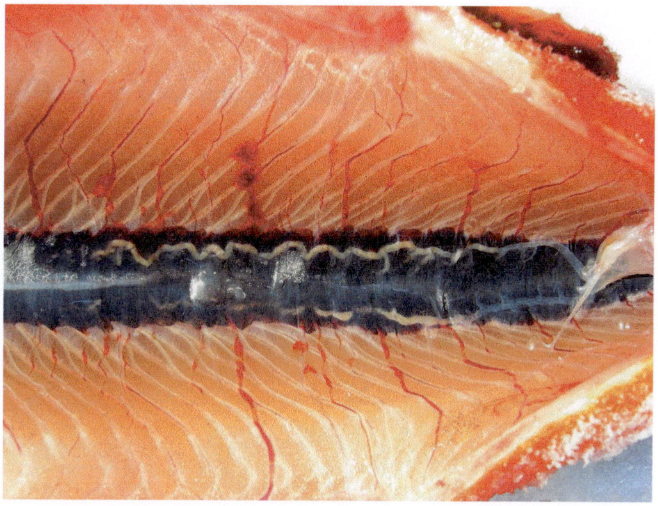

Fig. 9.37 Kidney flukes, *Phyllodistomum umblae* in the ureters of wild Arctic char. Note urinary bladder to the *right*

chronically infected individuals the normal transparent lens becomes whitish due to proliferation of lens epithelium; capsular rupture and detachment of the retina may impair host vision. This occurs as results of the active migration of cercaria to the anterior chamber, retina, vitreous body and lens, mostly through subcutaneous connective tissue and skeletal muscle of the trunk (Fig. 9.36). The site of entry is marked by tiny capsular perforations through which cortical lens fibres exude. These perforations can lead to lens rupture and severe endophthalmitis, and a generalised cortical liquefaction as the flukes migrate to the anterior cortex with the consequent proliferation of the lens epithelium. Predation of fish by birds completes the life cycle of the parasite.

The diagnosis of diplostomiasis is based on clinical observations, demonstration of metacercariae in fresh mounts of vitreous or aqueous humours smears, or in histological sections of the eye.

9.4.4 *Phyllodistomum umblae*

Phyllodistomum umblae is known as the kidney fluke and commonly found in the collecting ducts, ureters and urinary bladder of Arctic char and others including chum salmon, rainbow and brook trout and Atlantic salmon in freshwater. The parasite has a holarctic distribution. Fish that are heavily parasitized develop grossly distended and whitish ureters resulting in osmotic imbalance (Fig. 9.37). The condition may have some resemblance to early stages of nephrocalcinosis in all fish species, but can be readily differentiated from the latter by microscopy of the urethra contents. Adult flukes are up to 3 mm long and can be readily recognized.

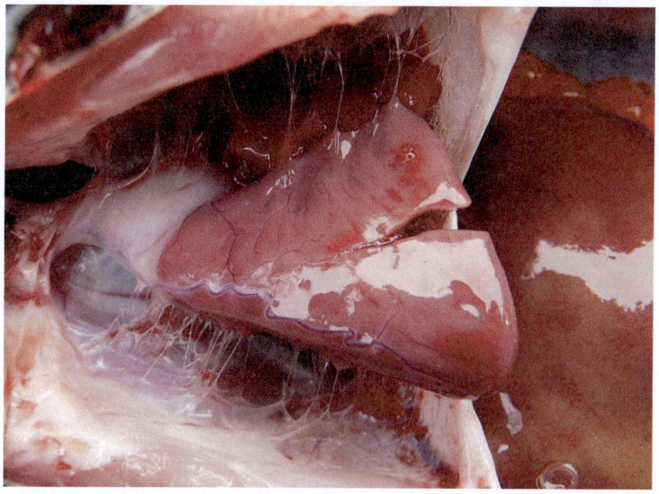

Fig. 9.38 Restrictive pericarditis caused by metacercariae of *Apatemon gracilis* in wild adult Atlantic salmon

9.4.5 *Apatemon gracilis*

Apatemon gracilis commonly occurs in the pericardial cavity of many freshwater fish including farmed rainbow trout in Scotland. Fish carrying this infection may have a low food intake, low growth rates and reduced activity levels. The adult form is found in the small intestine of waterfowl from where their eggs are passed in the host's faeces.

The severity of pericardial lesions seem to progress with encystment of the metacercariae within the epicardium. It has been shown experimentally that the stroke volume in severely affected rainbow trout may be reduced by up to 50 %. Severe constrictive pericarditis (Fig. 9.38) and degenerated encysted metacercaria (Fig. 9.39) in wild

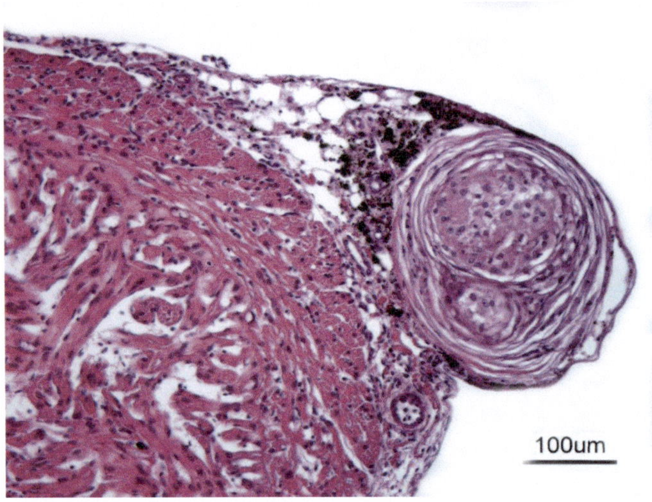

Fig. 9.39 Degenerating and encysted metacercaria of *Apatemon gracilis* near the apex of the ventricle in wild adult Atlantic salmon. Note melanin around the parasite

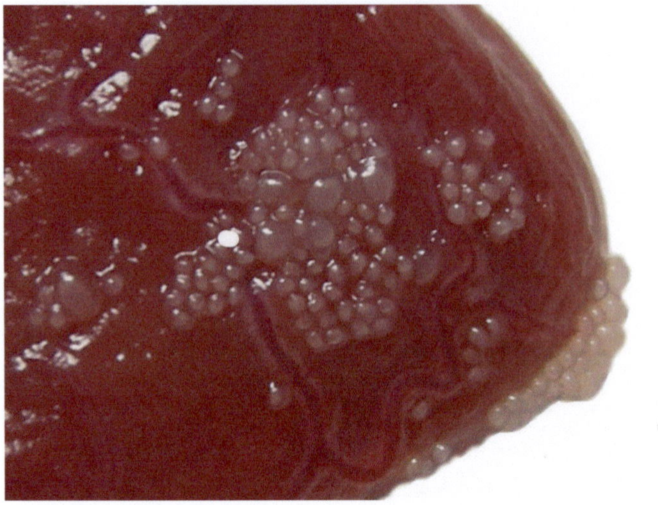

Fig. 9.40 Encysted metacercariae of *Ichthyocotylurus erraticus* near the apex of the ventricle of Atlantic salmon

spawning Atlantic salmon and sea-trout, has been recorded in Norway. Infected trout may show fibrogranulomatosis of the epicardium. Diagnosis can be made by clinical signs and light microscopy.

9.4.6 *Cotylurus* spp.

There are several species of *Cotylurus* that infect fish. The eggs are excreted by fish-eating birds including cormorants, herons and gulls, and hatch after around 15–16 days. The circulatory system and loose connective tissue serve as migratory routes of the metacercaria into the pericardium, where it encysts in the pericardial cavity of rainbow trout and Arctic grayling (Fig. 9.40). Overall, there is typically little tissue response around the encysted metacercariae although for some species, there may be severe haemorrhagic enteritis. In heavily infected fish *Cotylurus* cause a loss of condition factor and an increased susceptibility to other diseases.

9.4.7 *Ichthyocotylurus erraticus*

Ichthyocotylurus has a similar life cycle to *Apatemon* and *Cotylurus*. Aggregates of encysted metacercariae of *I. erraticus* occur around the bulbus arteriosus and to some extent also in the epicardium and myocardium forming a reactive fibro-connective capsule around the parasite in fish such as grayling and whitefish (Fig. 9.41).

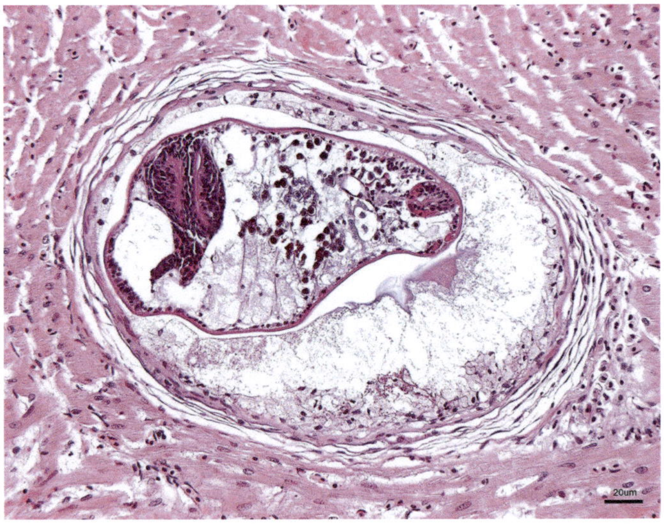

Fig. 9.41 Encysted metacercariae of *Ichthyocotylurus erraticus* in the ventricle of wild whitefish

Fish may show skin haemorrhage, especially on the ventral side with raised scales. The pericardium can be covered with several layers of metacercarial cysts and the mass of parasites extending deep into the compact layer of the heart; a PAS-positive reaction in tissue sections helps identify the cysts walls. A heavy inflammatory reaction occurs around the parasites. Mononuclear cells (presumably lymphocytes) and EGCs are the most prevalent cell types in the reaction. Chronic granuloma formation occurs around the parasites. The diagnosis is based upon gross, histological findings and PCR identification of the parasite.

9.4.8 *Stephanostomum tenue*

Metacercariae of the trematode *Stephanostomum tenue* occur in the heart of sea water-farmed rainbow trout and a good example of a parasite causing disease in a farmed species outside its normal range. In this case, the rainbow trout serves as an accidental host and associated mortality may be high.

9.5 Monogenea

Parasitic haptorworms represent a large group mainly affecting, but not exclusively, the gills and skin of many fish species including salmonids. They attach to the host by their characteristic anchoring structure, the posterior opisthaptor, which bares hooks, clamps and/or suckers. Monogenea have no true body cavity and the digestive tract has only one opening; relevant species are seldom longer than 1 mm. All species are hermaphroditic with both male and female reproductive organs in the same individual. They have no intermediate hosts and most species are oviparous, but some, including *Gyrodactylus salaris*, is viviparous and the offspring when released, can attach directly to the host and therefore the number of parasites on a single fish increases rapidly.

9.5.1 *Dactylogyrus* sp.

Dactylogyrus, commonly known as gill flukes, occur on the gills and buccal cavity of affected fish. These flukes are very common on cyprinids with a reduced occurrence on rainbow trout (Fig. 9.42). Inflamed gills, excessive mucous and increased respiratory rate can be observed as clinical signs. The hermaphroditic adults produce eggs directly into the water which after hatching, new parasites re-attach to the gills of a fish. The parasite is diagnosed by clinical signs and identification of flukes at necropsy.

9.5.2 *Gyrodactylus salaris*

Gyrodactylus salaris is a small, viviparous ectoparasite of ~0.5 mm in length which mainly occurs on the skin and gills of freshwater fish (Fig. 9.43), Atlantic salmon is especially affected. Fish may become greyish as a result of increased mucous and as the numbers increase, the dorsal and pectoral fins may become whitish from epidermal hypertrophy. Brown trout, rainbow trout and Arctic char may also harbour the parasite but seldom develop lesions. Infestation has caused serious losses to salmon in Norwegian rivers following its introduction with infected smolts in the 1970s from Swedish hatcheries. The latter, were located in watercourses draining into the Baltic Sea where the fish are generally resistant to the parasite. Unfortunately, widespread fish movements took place before the parasites pathogenic potential was fully understood and more than 40 rivers became seriously affected. Controlling and attempts to eradicate the parasite remain a high priority. Affected rivers' natural smolt production has been reduced to approximately 15 % of the original population due to the devastating effect on the fry and parr. Corresponding low figures for returning fish occurred as a low number of young fish manage to survive through to smolts. Severely infected fish may harbour thousands of parasites, particularly on the dorsal and pectoral fins.

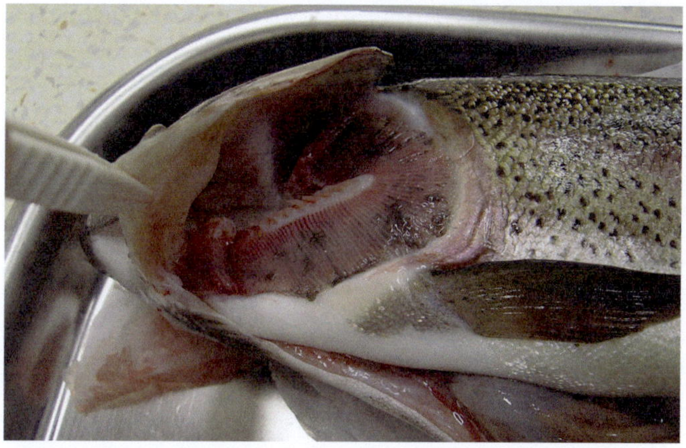

Fig. 9.42 *Dactylogyrous* spp. on the gills of a farmed rainbow trout

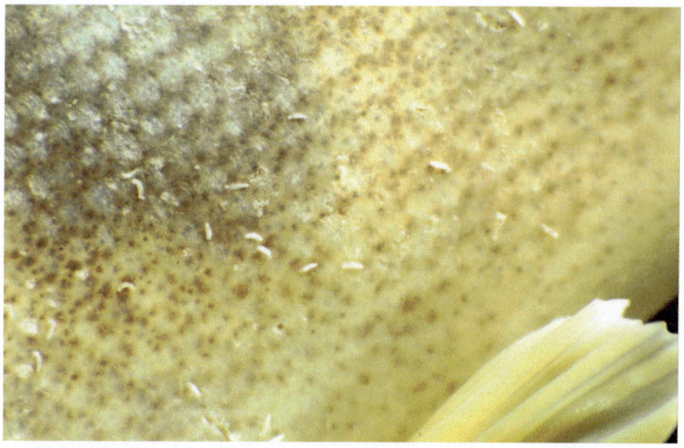

Fig. 9.43 *Gyrodactylus salaris* on the skin of Atlantic salmon parr

The lesions inflicted on the host epidermis arise both from attachment by the opisthaptor, and the feeding activity of the mouth in the anterior part of the body (Fig. 9.44). The loss of osmotic integrity causes fatal water imbalance and parasite activity on the surface indirectly leads to reduction in mucous cells through disruption of cell dynamics within the epidermis. Skin ulcers may also be invaded by secondary pathogens such as *Saprolegnia* and bacteria, e.g. pseudomonads and aeromonads.

G. salaris is transmitted horizontally between fish or via the river-stream substrate, where the parasite can survive for some days depending on temperature. Infected out-migrant smolts may transmit the parasite to other rivers in fjord systems with low surface salinity (less than 20 mg l^{-1}) as these fish may move back and forth briefly after leaving their home river. *G. salaris* does not survive in sea water.

Other gyrodactylids occurring in salmonids include *G. truttae, G. derjavinoides* and G. *teuchis* in Europe, and *G. salmonis* in North America. They may all have some significance when occurring in large numbers. *Gyrodactyloides bychowskii* may be found on the gills of both farmed and wild Atlantic salmon in the northwest Atlantic and may cause lesions when present in large numbers, including epidermal gill hyperplasia and hypertrophy and a decline in overall condition of the fish.

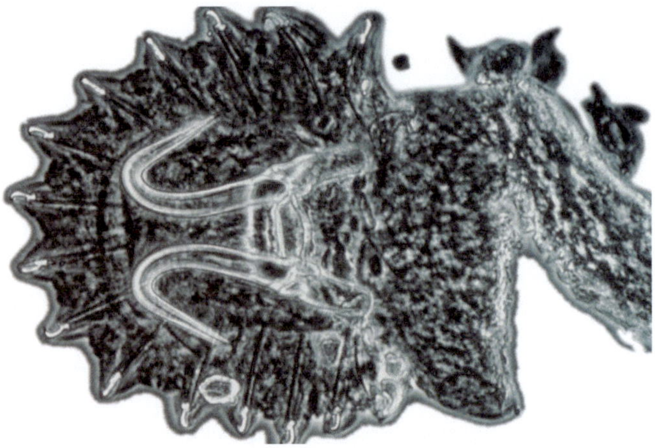

Fig. 9.44 Attachment organ (opisthaptor) of *Gyrodactylus salaris*

Diagnosis of this group is mainly based upon the size and morphology of structures in the opisthaptor and molecular techniques (PCR) are applied to identify species.

9.5.3 *Discocotyle sagittata*

Discocotyle sagittata larvae become attached to the gills of rainbow and brown trout by an adhesive apparatus consisting of four pairs of clamps on the opisthohaptor, with new infections taking place during the summer and autumn. This oviparous parasite survives transfer to sea water and may be found on returning sea-trout and salmon but overall, it is generally of little significance to the host. When present in high numbers however, they are associated with pale gills, decreased body condition and host mortality.

9.6 Acanthocephala

Acanthocephalans (spiny-headed worms) are characterised by the presence of an evertable proboscis in the anterior part of the body, armed with rows of chitinous spines, which they use to hold the gut wall of their host (Fig. 9.45). Acanthocephalans have complex life cycles involving at least two hosts, including invertebrates, fish, amphibians, birds and mammals. Infection can alter the intermediate host behaviour, morphology and other features that enhances the probability of transmission to the definitive hosts. The adaptation to a parasitic life style resulted in a drastic simplification of their morphology, they have a cylindrical non segmented body and their general cavity is a pseudocœlom. The muscular, excretory and nervous systems are greatly reduced and they lack respiratory and circulatory systems as well as an alimentary canal, thus nutrient uptake occurs directly through their body wall.

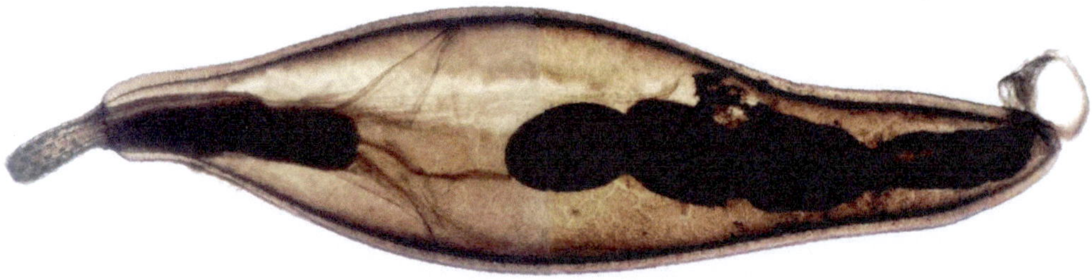

Fig. 9.45 Whole mount of adult *Acanthocephalus tumescens* from the intestine of rainbow trout

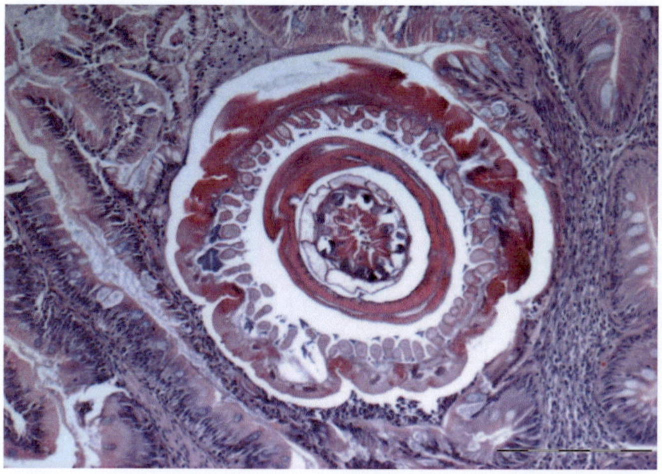

Fig. 9.46 *Acanthocephalus* sp. in the intestine of grayling. Low power

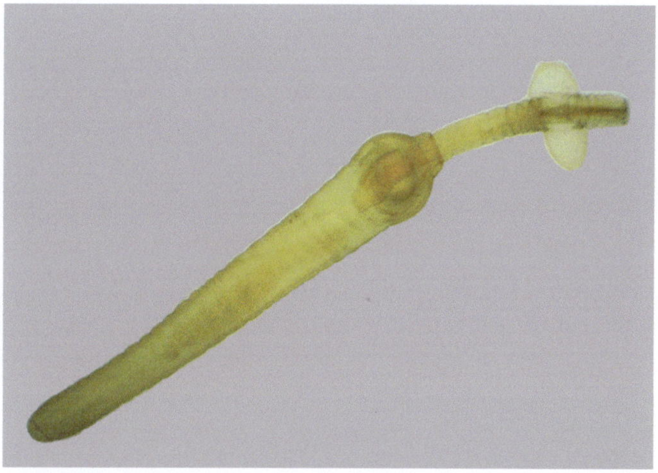

Fig. 9.47 Whole mount of *Pomphorhynchus patagonicus* from rainbow trout

9.6.1 *Acanthocephalus* spp.

Several species of *Acanthocephalus* have been described from salmonids from Japan, North and South America. Fish may tolerate heavy infections in the intestine, but this results in chronic catarrhal inflammation, haemorrhage, compression of intestinal folds and loss of columnar appearance of epithelial cells, and consequently poor growth. The host's epithelial lining is eroded through the action of the armed retractable proboscis. Spindle-shaped eggs are produced in the intestinal tract and pass into the water where they are ingested by various isopods and amphipods. The representatives of the genus *Acanthocephalus* are primarily recognised by their morphology and the number of hooks on the proboscis. In stained histological sections the distinguishing features include a thin, non-rigid acellular cuticle and thick hypodermis consisting of layers of fibres overlying thin bands of circular and longitudinal smooth muscle. In cross section, distinctive lacunar channels are seen as clear oval or circles in the hypodermis. Consequently raised subserosal nodules occur in the gut mucosa with a severe granulomatous reaction (Fig. 9.46).

9.6.2 *Pomphorhynchus* spp.

Pomphorhynchus laevis is a frequent parasite of European and North American fresh water fish. Grayling, Atlantic salmon, brown and rainbow trout are reported susceptible hosts. *P. patagonicus* is reported affecting several fresh water species including rainbow trout in fresh water lakes in the Patagonia region of South America (Fig. 9.47). A hooked proboscis penetrating the intestine wall results in a chronic proliferative host response with the formation of whitish, fibrous capsule of inflammatory cells including EGCs and fibroblasts. Mechanical damage to the intestinal epithelium is the principal lesion with severe infections blocking the lumen and depriving the host of nourishment. In some cases the parasite may penetrate through the intestine and cause a proliferative reaction in other organs such as the liver and pancreas.

Diagnosis is based upon the finding of adult worms in the intestine or invasive larvae within the body cavity. The shape of the proboscis, the number and arrangement of the hooks are also diagnostic features.

9.7 Maxillopoda

Sea lice are the most economical important parasitic copepod affecting salmon culture worldwide, and typically infest the external surface of marine and brackish-water fish.

Infestation can result in significant morbidity and mortality of hosts in addition to being expensive to control (Table 9.2).

9.7.1 *Lepeophtheirus salmonis*

Lepeophtheirus salmonis is a parasitic marine copepod belonging to the family Caligidae referred to as sea lice. These occur on wild and farmed salmonids in both the North Atlantic and North Pacific oceans and is considered the most important ectoparasite in salmon farming. Affected species include Atlantic and Pacific salmon, sea-run strains of rainbow trout, brown trout and Arctic char. Recently *L. salmonis* have been shown to develop, although not to complete a full life-cycle, on the three-spined stickleback in coastal areas of British Columbia.

Sea lice are seldom reported in large numbers from wild fish. However, farmed salmon may become heavily infested

Table 9.2 'Laxe-Luus'

Peder Clausson Friis (1545–1614) and Erik Pontoppidan (1698–1764) were a great scholar and bishop respectively in Norway, and both showed a great interest in biology and natural history. They are probably the first to describe sea-lice in migrating salmon. Friis (ca.1600) accurately describes the parasite without naming it: 'søge strax op i Elffuen oc Fosser at de kunde afftoe i Fosser og paa Steen affscraabe store Lus aff sig, som sider i hans Nache' ('it moves quickly up in the river and waterfalls to wash off in the falls, and to scrape off on stones, large lice attached to its neck'). Furthermore: 'De Lus som eer paa Laxen er som stuore Edderkopper, dog lenger, haffuer en lang Neb, sidder haart paa hannom, bider igjennom den sterche oc sej Hud indtill Blod gaar ut' ('the lice on the salmon are the size of great spiders, but longer, with a long beak, firmly attached to it (i.e. the salmon), it bites through the strong and tough skin until blood emerges'.

Later, Pontoppidan in 1753 states that 'da den i store Flokke kommer fra Havet og søger op i Elverne, deels for at forfriske sig i det ferske Vand, deels for at afgnie og afskylde, ved skarpe Strømmes og Fossers Fald, et slags grønagtig Utøy, kaldet Laxe-Luus, som sette sig imellem Finnerne og plage den i Foraars Varme'. ('great schools of salmon moving from the sea into fresh water, partly to refresh themselves, and partly to rid themselves by rubbing and washing in the swift currents and waterfalls, of a kind of greenish vermin called Laxe-Luus, attached between the fins, plaguing it in the heat of spring').

From: Berland and Margolis (1983)

but the number of lice is usually low due to intensive surveillance and prophylactic treatments. Maximum agreed counts of parasite numbers trigger a treatment, to help diminish infestations and this is supplemented with cleaner fish to remove lice (wrasse species and lumpsucker). However, atypical strains of *Aeromonas salmonicida* have been isolated from wrasse, and recent testing has established that wrasse species can harbour viral haemorrhagic septicaemia virus.

Sea lice undergo ten developmental stages on the fish with a moult between each stage. The infective, free-living stage is called the copepodid and newly attached chalimus larvae are typically found on the ventral surface and the fins, while older stages can occur anywhere on the body. Adult individuals typically locate in scale-less areas such as the neck and head, around the anal fin and on the ventral side of the caudal peduncle (Figs. 9.48 and 9.49). If untreated, both chalimus and the adult lice may occur in substantial numbers and cause severe skin lesions. Early gross signs are seen as light grey patches on the head, neck and the perianal area. Much of the damage caused by copepods is associated with attachment to the host and feeding behaviour. Lice graze on

Fig. 9.48 Severe skin lesions caused by sea lice, *Lepeophtheirus salmonis*, on farmed Atlantic salmon

host tissues ranging from mucus, epidermal, dermal or subcutaneous tissues, causing skin ulceration, petechiae and resultant hyperpigmentation. Microscopically, lesions may be variable depending on fish species but include mechanical

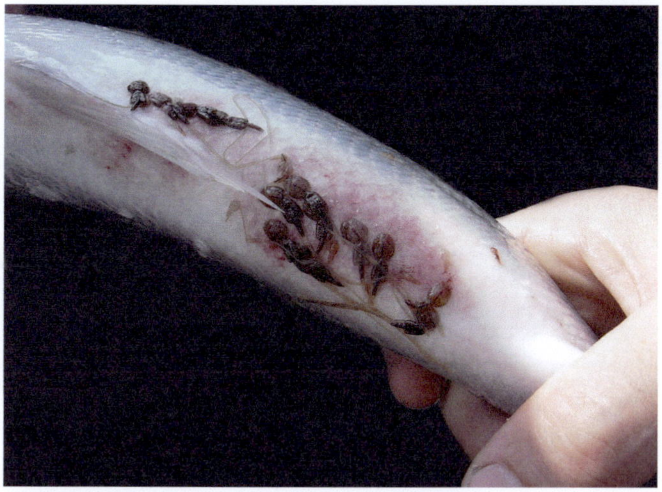

Fig. 9.49 Adult female *Lepeophtheirus salmonis* and *Caligus elongatus* on the ventral side of the peduncle of a farmed Atlantic salmon

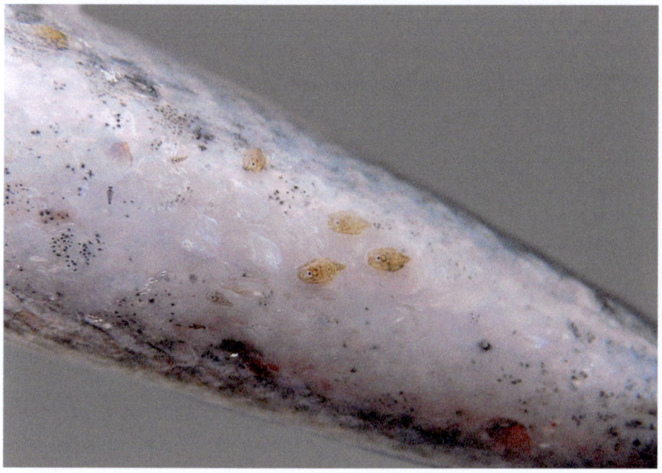

Fig. 9.50 Chalimus larvae of *Lepeophtheirus salmonis* on the ventral side of the peduncle in farmed Atlantic salmon

disruption with scale loss, followed by epidermal hyperplasia, mucus cell hypertrophy, macrophage infiltration with some fibrosis, sometimes extending into the cranium. The resultant ulcers will break the osmotic barrier and are consequently sites for secondary infection.

9.7.2 *Caligus elongatus*

Caligus elongatus is a non-host specific parasitic copepod that may infest salmonids in aquaculture in Europe and eastern Canada. It is distinctly smaller and lighter than *L. salmonis* (Fig. 9.50). Lesions are similar to those caused by *L. salmonis*, but seldom as deep and extensive. The life cycle comprises eight stages and a separate pre-adult stage is not present.

Fig. 9.51 Skin wounds inflicted by *Caligus rogercresseyi* in farmed Atlantic salmon

Fig. 9.52 Whole mount of *Caligus rogercresseyi*

9.7.3 *Caligus rogercresseyii*

Caligus rogercresseyii is the predominant parasitic problem in the salmonid industry in Chile, and is also present on sea-run brown trout on the Atlantic side of Patagonia (Figs. 9.51 and 9.52). Coho salmon is not considered susceptible due to a well-developed inflammatory response. A range of non-salmonids are natural hosts and reservoir for the parasites. Adult males and females are approximately 5 mm in length and the life cycle is similar to *C. elongatus*. Affected fish show multifocal skin abrasion and petechiae with microscopic lesions similar to those described for *L. salmonis*. It has also been shown that the parasite may act as a vector for *Piscirickettsia salmonis* and infectious salmon anaemia virus, and predispose the fish to other diseases. Other caligids that may parasitize salmonids include *C. flexispina* and *C. teres* in the southern hemisphere and *C. clemensi* in the North Pacific region.

9.7.4 *Argulus coregoni, A. foliaceus* and *A. japonicus*

Members of the Argulidae represent the freshwater counterparts of sea lice and infest the skin and fins of numerous fish worldwide, including Atlantic salmon, trout, whitefish and grayling. These parasites have a preoral stylet that causes local mechanical injury and particularly the release of digestive enzymes. They are common on wild fish and can cause severe losses in pond culture and inland sport fisheries.

Argulus coregoni is the largest of the three species and measures up to 13 mm in length (Fig. 9.53). The anterior part of the body is covered by an oval semi-transparent carapace. The feeding activity causes intense irritation and affected fish typically jump repeatedly, scrub against surfaces, stop feeding and appear dark. Affected fish are also prone to secondary infections e.g. *Pseudomonas* spp. and *Aeromonas* spp. Histologically, necrosis and epidermal proliferation occurs around entry wounds and haemorrhage with a severe inflammatory response involving lymphocytic infiltration, typically follow. Due to their size and characteristic shape, these parasites are easily diagnosed on gross examination of fins and skin.

Argulus foliaceus can have a significant impact on yield in recreational trout fisheries, partly by increasing losses but also by reducing the appetite of infected fish, making them less likely to react to bait.

A. japonicus has spread worldwide from its original Far East habitats through the movement of fish. These obligate ectoparasites feed on mucus, epidermal cells, blood and tissue fluids through a proboscis-like mouth inserted into the skin. The tissue is partly pre-digested by enzymes inserted via the stylet. The parasites attach to the fish surface with a pair of round suckers and able to move freely on the fish and to swim from host to host.

Fig. 9.53 Whole mount of *Argulus coregoni* from the skin of a wild grayling

9.7.5 *Salmincola* spp.

The genus *Salmincola* has a circumpolar distribution and has been found in many species of salmonid fish (Fig. 9.54). Some parasites may survive on their host during their sea water migration and occur in large numbers in returning adult salmon broodstock. It has been hypothesized that heavily infected salmon may suffer reduced growth and survival at sea, potentially reducing the abundance of repeat spawners. These conspicuous parasites may cause a severe local inflammatory response, particularly on the distal ends of the gill filaments or on gill arch, where their mouth parts are deeply embedded.

Fig. 9.54 *Salmincola salmoneus* from the gills of wild adult Atlantic salmon

A severe infestation may result in significant blood loss. *Salmincola* may also attach to the gills, oral cavity and to the inside of the operculum (Figs. 9.55 and 9.56). At histological examination, gills with attached adults show hyperplasia and hypertrophy, although in the long term may induce atrophy or growth inhibition of affected lamellae. Eosinophilia and absence of mucous cells is also reported. The identification of the different species is based on examination of the females, as the males are small, difficult to distinguish and die after fertilization.

9.8 Bilvalvia

The Bivalvia represent the class of marine and freshwater that includes molluscs, clams, oysters, mussels and scallops.

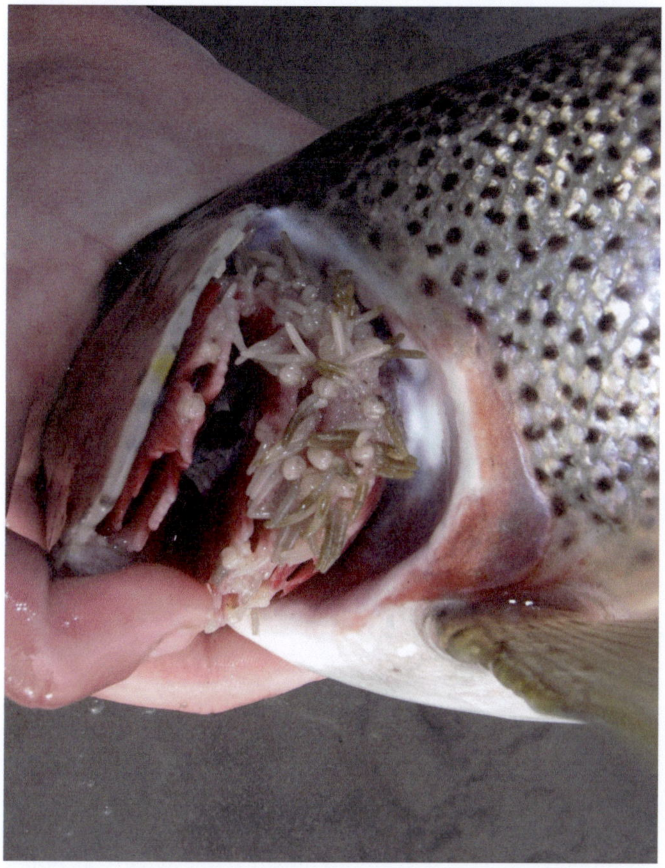

Fig. 9.55 Severe infestation of *Salmincola* sp. in the gills of wild rainbow trout. Note necrotic gill tips

9.8.1 Glochidiosis

Margaritifera margaritifera, the fresh water pearl mussel is an aquatic bivalve resident in fast flowing cool waters low in calcium, and their distribution is restricted and threatened in Europe. Glochidia, the microscopic larval stage of the mussel, are released from adult animals into the water and attach to the gill lamellae of resident trout and salmon fingerlings (or parr) for a period, before it detaches and establish in the bottom substrate. On contact with the gills, the glochidia clamps to the lamellae using sharp teeth and enclose a portion of the lamellae in the mantle cavity. Both wild and farmed stock including Atlantic salmon in Scotland and Norway, and Chinook and coho parr in America have been affected by glochidia (Fig. 9.57). In Japan, the glochidia of *M. laevis* have occasionally been reported on salmonids. Following attachment to the gill epithelium, a localised hyperplasia and fusion of lamellae occurs and may surround the developing larvae, with the area becoming thinner as the glochidia develop. Encysted parasites occur to a lesser degree in the gill rakers and occasionally, the pseudobranch. When large glochidia are located distally, this results in clubbing of the filaments (Fig. 9.58). Stricture of gill capillaries and hyperplasia is

Fig. 9.56 Salmincola thymalli on the gills of wild grayling

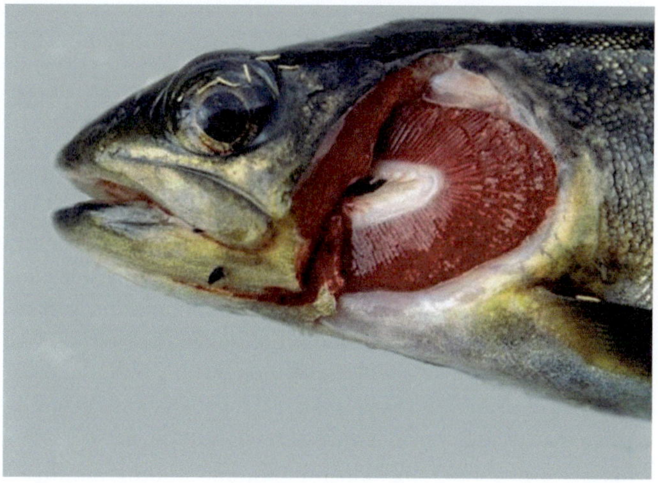

Fig. 9.57 Glochidia of *Margartifera margartifera* attached to the gills of wild brown trout

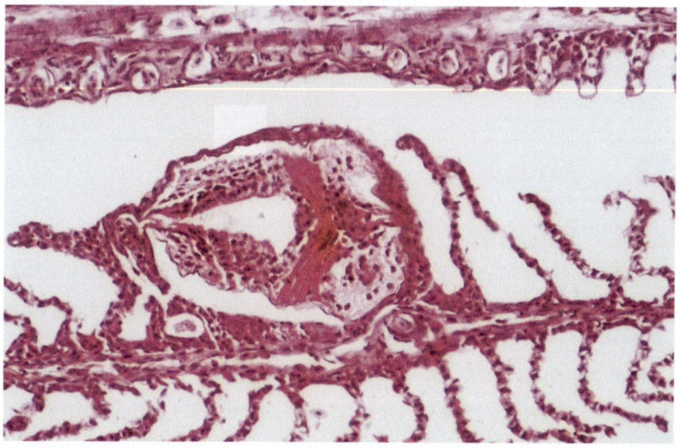

Fig. 9.58 Glochidia of *Mytilus edulis* (spat) on gills of Atlantic salmon. Low power

Fig. 9.59 *Piscicola* spp. on operculum of rainbow trout. *Insert* leech on Arctic char

linked with reduced or absent functional respiration. Fish swimming upstream help to maintain the distribution of the mussel. Attached glochidia excyst as juvenile mussels and their release from their branchial cysts results in open lesions which are subject to secondary infections. Preliminary diagnosis can be achieved by microscopic examination of fresh smears and confirmed by tissue sections.

Net cages provide attachment sites for planktonic larvae and other organisms and this is referred to as 'fouling'. Mussels are among the most important fouling molluscs and their colonisation of salmon nets can lead to a reduction in water flow. However, despite the close proximity to the farmed fish, there are only a few reports of larval settlement to the gill filaments by post-veliger larvae. Gross and subsequent examination can confirm the presence of larvae with associated hyperplasia and fusion.

9.9 Annelida

9.9.1 Leeches

Leeches (family Piscicolidae) are parasitic annelid worms that are predominant in freshwater environments (Fig. 9.59). Some leeches live on oligochaete worms and crustaceans and others are hematophagous, and therefore predominantly feeding on blood from vertebrate and invertebrate animals. Fish can become infested by leeches and the presence of red or white circular bite marks on their body surface are evidence of feeding. In heavily infected fish, the host may suffer from anaemia and the affected area offer opportunities for secondary infections. Leeches are equipped with suckers at either end of the body and also possess a clitellum, hence are hermaphrodite. Leeches play a role in the transmission of certain agents for example, *Piscicola geometra* acts as mechanical vectors of spring viraemia of carp virus (SVCV), while *P. salmositica* can transmit *Cryptobia salmositica*. The parasite multiplies in the crop and is then transmitted to new fish at the next feed. The leech, *Myzobdella lugubris* have also proven positive by cell culture for viral haemorrhagic septicaemia virus, particularly in the Lake Erie watershed (North America).

Further Reading

Adams A, Richards RH, Marin de Mateo M (1992) Development of monoclonal antibodies to PKX, the causative agent of proliferative kidney disease. J Fish Dis 15:515–552

Bakke TA, MacKenzie K (1993) Comparative susceptibility of native Scottish and Norwegian stocks of Atlantic salmon, *Salmo salar* L., to *Gyrodactylus salaris* Malmberg: laboratory experiments. Fish Res 17:69–86

Bartholomew JL, Atkinson SD, Hallett SL (2006) Involvement of *Manayunkia speciosa* (Annelida: Polychaeta: Sabellidae) in the life cycle of *Parvicapsula minibicornis*, a myxozoan parasite of Pacific salmon. J Parasitol 92:742–748

Berland B, Margolis L (1983) The early history of 'Lakselus' and some nomenclatural questions relating to copepod parasites of salmon. Sarsia 68:281–288

Bower SM (1985) *Ceratomyxa shasta* (Myxozoa: Myxosporea) in juvenile chinook salmon (*Oncorhynchus tshawytscha*): experimental transmission and natural infections in the Fraser river, British Columbia. Can J Zool 63:1737–1740

Boyce NP (1979) Effects of *Eubothrium salvelini* (Cestoda: Pseudophyllidae) on the growth and vitality of sockeye salmon, *Oncorhynchus nerka*. Can J Zool 57:97–602

Boyce NP, Clarke WC (1983) *Eubothrium salvelini* (Cestoda: Pseudophyllidae) impairs salt water adaptation of migrant sockeye yearlings (*Oncorhynchus nerka*) from Babine lake, British Columbia. Can J Fish Aquat Sci 40:821–824

Bravo S, Perroni M, Torres E, Silva MT (2006) Report of *Caligus rogercresseyi* in the anadromous brown trout (*Salmo trutta*) in the Río Gallegos Estuary, Argentina. Bull Eur Assoc Fish Pathol 26:186–193

Bristow GA, Berland B (1991) The effect of long term, low level Eubothrium sp. (Cestoda: Pseudophyllidae) infection on growth of farmed salmon (Salmo salar L.). Aquaculture 98:325–330

Bruno DW, Stone J (1990) The role of saithe, *Pollachius virens* L., as a host for the sea lice, *Lepeophtheirus salmonis* and *Caligus elongatus*. Aquaculture 89:201–207

Bruno DW, McVicar AH, Waddell IF (1988) Natural infection of farmed Atlantic salmon, *Salmo salar* L., parr by glochidia of the freshwater pearl mussel, *Margaritifera margaritifera* L. Bull Eur Assoc Fish Pathol 8:23–26

Bruno DW, Collins CM, Cunningham CO, Mackenzie K (2001) *Gyrodactyloides bychowskii* (Monogenea: Gyrodactylidae) from sea-caged Atlantic salmon *Salmo salar* in Scotland: occurrence and ribosomal RNA sequence analysis. Dis Aquat Org 45:191–196

Bullock WL (1963) Intestinal histology of some salmonid fishes with particular reference to the histopathology of acanthocephalan infections. J Morphol 112:23–44

Campbell AD (1971) The occurrence of *Argulus* (Crustacea: Branchiura) in Scotland. J Fish Biol 3:145–146

Chappell LH, Hardie LJ, Secombes CJ (1994) Diplostomiasis: the disease and host-parasite interactions. In: Pike AW, Lewis JW (eds) Parasitic diseases of fish. Samara Publishing, Dyfed, pp 59–86

Clifton-Hadley RS, Bucke D, Richards RH (1984) Proliferative kidney disease of salmonid fish: a review. J Fish Dis 7:363–377

Fast MD, Ross NW, Mustafa A, Sims DE, Johnson SC, Conboy GA, Speare DJ, Johnson G, Burka JF (2002) Susceptibility of rainbow trout, *Oncorhynchus mykiss*, Atlantic salmon, *Salmo salar* and coho salmon *Oncorhynchus kisutch* to experimental infection with sea lice *Lepeophtheirus salmonis*. Dis Aquat Org 52:57–68

Freeman MA, Sommerville C (2009) *Desmozoon lepeophtheiri* n. sp., (Microsporidia: Enterocytozoonidae) infecting the salmon louse *Lepeophtheirus salmonis* (Copepoda: Caligidae). Parasite Vectors 2:58

Freeman MA, Bell AS, Sommerville C (2003) A hyperparasitic microsporidean infecting the salmon louse, *Lepeophtheirus salmonis,* an rDNA-based molecular phylogenetic study. J Fish Dis 26:667–676

Garnick E, Margolis L (1990) Influence of four species of helminth parasites on orientation of seaward migrating sockeye salmon (*Oncorhynchus nerka*) smolts. Can J Fish Aquat Sci 47:2380–2389

Gilbert MA, Granath WO (2003) Whirling disease of salmonid fish: life cycle, biology, and disease. J Parasitol 89:658–667

Gonzáles L, Carvajal J, George-Nascimento M (2000) Differential infectivity of *Caligus flexispina* (Copepoda, Caligidae) in three farmed salmonids in Chile. Aquaculture 183:13–23

Hallett SL, Ray RA, Hurst CN, Holt RA, Buckles GR, Stephen D, Atkinson SD, Bartholomew JL (2012) Density of the waterborne parasite *Ceratomyxa shasta* and its biological effects on salmon. Appl Environ Microbiol 78:3724–3731

Harrod C, Griffiths D (2005) *Ichthyocotylurus erraticus* (Digena: Strigeidae): factors affecting infection intensity and the effects of infection on powan (*Coregonus autumnalis*); a glacial relict fish. Parasitology 131:511–519

Higgins MJ, Margolis L, Kent ML (1993) Arrested development in a freshwater Myxosporean, Myxidium salvelini, following transfer of its host, the sockeye salmon (*Oncorhynchus nerka)*, to salt water. J Parasitol 79:403–407

Hoffman GL (1990) *Myxobolus cerebralis,* a worldwide cause of salmonid whirling disease. J Aquat Anim Health 2:30–37

Hoffman RW, El-Matbouli M (1994) Proliferative kidney disease (PKD) as an important myxosporean infection in salmonid fish. In: Pike AW, Lewis JW (eds) Parasitic diseases of fish. Samara Publishing, Dyfed, pp 3–15

Hoffmann R, Kennedy CR, Meder J (1986) Effects of *Eubothrium salvelini* Schrank, 1790 on Arctic charr, *Salvelinus alpinus* (L.), in an alpine lake. J Fish Dis 9:153–157

Johnson SC, Albright LJ (1992) Comparative susceptibility and histopathology of the response of naive Atlantic, chinook and coho salmon to experimental infection with *Lepeophtheirus salmonis* (Copepoda: Caligidae). Dis Aquat Org 14:179–193

Johnson SC, Treasurer JW, Bravo S, Nagasawa K, Kabata Z (2004) A review of the impact of parasitic copepods on marine aquaculture. Zool Stud 43:8–19

Jones MW, Sommerville C, Bron J (1990) The histopathology associated with the juvenile stages of *Lepeophtheirus salmonis* on the Atlantic salmon, *Salmo salar* L. J Fish Dis 13:303–310

Jones SRM, Prosperi-Porta G, Dawe SC, Barnes DP (2003) Distribution, prevalence and severity of *Parvicapsula minibicornis* infections among anadromous salmonids in the Fraser River, British Columbia, Canada. Dis Aquat Org 54:49–54

Jones SRM, Prosperi-Porta G, Kim E (2012) The diversity of Microsporidia in parasitic copepods (Caligidae: Siphonostomatoida) in the Northeast Pacific Ocean with description of *Facilispora margolisi* n. g., n. sp. and a new family Facilisporidae n. fam. J Eukaryot Microbiol 59:26–217

Jørgensen A, Nylund A, Nikolaisen V, Alexandersen S, Karlsbakk E (2011) Real-time PCR detection of *Parvicapsula pseudobranchicola* (Myxozoa: Myxosporea) in wild salmonids in Norway. J Fish Dis 34:365–371

Karlsbakk E, Saether PA, Høstlund C, Fjellsoy KR, Nylund A (2002) *Parvicapsula pseudobranchicol*a n. sp. (Myxozoa), a myxosporidian infecting the pseudobranch of cultured Atlantic salmon (*Salmo salar*) in Norway. Bull Eur Assoc Fish Pathol 22:381–387

Karna DW, Millemann RE (1978) Glochidiosis of salmonid fishes. III. Comparative susceptibility to natural infection with *Margaritifera margaritifera* (L.) (Pelecypoda: Margaritanidae) and associated histopathology. J Parasitol 64:528–537

Ko RC, Anderson RC (1979) A revision of the genus *Cystidicola* Fischer, 1798 (Nematoda: Spiruroidea) of the swim bladder of fishes. J Fish Res Board Can 26:849–864

Lom J, Dykova I (2006) Myxozoan genera: definition and notes on taxonomy, life cycle terminology and pathogenic species. Folia Parasitol 53:1–36

Lyndon AR (2001) Low intensity infestation with the heartfluke *Apatemon gracilis* does not affect short-term growth performance in rainbow trout. Bull Eur Assoc Fish Pathol 21:263–265

Mackinnon BM (1993) Host response of Atlantic salmon (*Salmo salar*) to infection by sea lice (*Caligus elongatus*). Can J Fish Aquat Sci 50:789–792

Meyers TR, Millemann RE, Fustish CA (1980) Glochidiosis of salmonid fishes. IV. Humoral and tissue response of coho and chinook salmon to experimental infection with *Margaritifera margaritifera* (L.) (Pelecypoda: Margaritanidae). J Parasitol 66:274–281

Mo TA (1994) Status of *Gyrodactylus salaris* problems and research in Norway. In: Pike AW, Lewis JW (eds) Parasitic diseases of fish. Samara Publishing, Dyfed, pp 43–56

Mo TA, Poppe TT, Vik G, Valheim M (1992) Occurrence of *Myxobolus aeglefini* in salt-water reared Atlantic salmon (*Salmo salar*). Bull Eur Assoc Fish Pathol 12:104–106

Mo TA, Senos MR, Hansen H, Poppe TT (2010) Red vent syndrome associated with *Anisakis simplex* diagnosed in Norway. Bull Eur Assoc Fish Pathol 30:197–201

Nesnidal MP, Helmkampf M, Bruchhaus T, El-Matbouli M, Hausdorf B (2013) Agent of whirling disease meets orphan worm: Phylogenomic analyses firmly place Myxozoa in Cnidaria. PLoS One 8(1):e54576. doi:10.1371/journal.pone.0054576

Nezlin LP, Cunjak RA, Zotin AA, Ziuganov VV (1994) Glochidium morphology of the freshwater pearl mussel (*Margaritifera margaritifera*) and glochidiosis of Atlantic salmon (*Salmo salar*): a study by scanning electron microscopy. Can J Zool 72:15–21

Noguera P, Collins C, Bruno D, Pert C, Turnbull A, McIntosh A, Lester K, Bricknell I, Wallace S, Cook P (2009) *Anisakis simplex* sensu stricto (Nematoda: Anisakidae) in red vent syndrome affected wild Atlantic salmon *Salmo salar* in Scotland. Dis Aquat Org 87:199–215

Nowak BF, Hayward CJ, González L, Bott NJ, Lester RJG (2011) Sea lice infections of salmonids farmed in Australia. Aquaculture 320:171–177

Nylund A, Karlsbakk E, Sæther PA, Koren C, Larsen T, Nielsen BD, Brøderup AE, Høstlund C, Fjellsøy KR, Leirvik K, Rosnes L (2005) *Parvicapsula pseudobranchicola* (Myxosporea) in farmed Atlantic salmon *Salmo salar*; tissue distribution, diagnosis and phylogeny. Dis Aquat Org 63:197–204

Nylund S, Andersen L, Sævareid I, Plarre H, Watanabe K, Arnesen CE, Karlsbakk E, Nylund A (2011) Diseases of farmed Atlantic salmon *Salmo salar* associated with infections by the microsporidian *Paranucleospora theridion*. Dis Aquat Org 41:41–57

Öxer A, Wootten R (2000) The life cycle of *Sphaerospora truttae* (Myxozoa: Myxosporea) and some features of the biology of both the actinosporean and myxosporean stages. Dis Aquat Org 40:33–39

Pettersen RA, Hytterød S, Vøllestad LA, Mo TA (2013) Osmoregulatory disturbances in Atlantic salmon, *Salmo salar* L., caused by the monogenean *Gyrodactylus salaris*. J Fish Dis 36:67–70

Rahkonen R, Aalto J, Koski P, Särkkä J, Juntunen K (1996) Cestode larvae *Diphyllobothrium dendriticum* as a cause of heart disease leading to mortality in hatchery-reared sea and brown trout. Dis Aquat Org 25:15–22

Ramakrishna NR, Burt MDB (1991) Tissue response of fish to invasion by larval *Pseudoterranova decipiens* (Nematoda; Ascaridoidea). Can J Fish Aquat Sci 48:1623–1628

Rand TG, Cone DG (1990) Effects of *Ichthyophonus hoferi* on condition indices and blood chemistry of experimentally infected rainbow trout (*Oncorhynchus mykiss*). J Wildl Dis 26:323–328

Ratanarat-Brockelman C (1974) Migration of Diplostomum spathaceum (trematoda) in the fish intermediate host. Parasitol Res 43:123–134

Roberts RJ, Johnson KA, Casten MT (2004) Control of *Salmincola californiensis* (Copepoda: Lernaeapodidae) in rainbow trout, *Oncorhynchus mykiss* (Walbaum): a clinical and histopathological study. J Fish Dis 27:73–79

Rodger HD (1991) *Diphyllobothrium* sp. infections in freshwater-reared Atlantic salmon (*Salmo salar* L.). Aquaculture 95:7–14

Rubio-Godoy M, Tinsley RC (2008) Recruitment and effects of *Discocotyle sagittata* (Monogenea) infection on farmed trout. Aquaculture 274:15–23

Saksida SM, Marty GD, Jones SRM, Manchester HA, Diamond CL, Bidulka J, St-Hilaire S (2012) Parasites and hepatic lesions among pink salmon, *Oncorhynchus gorbuscha* (Walbaum), during early salt water residence. J Fish Dis 35:137–151

Sharp GJE, Pike AW, Secombes CJ (1992) Sequential development of the immune response in rainbow trout (*Oncorhynchus mykiss* (Walbaum, 1792)) to experimental plerocercoid infections of *Diphyllobothrium dendriticum* (Nitzsch, 1924). Parasitology 104:169–178

Shulman BS, Ieshko EP (2003) *Chloromyxum schurovi* sp. n. – a new myxosporidian species (Myxosporea: Sphaerosporidae) of salmonids (Salmonidae). Parasitology 37:246–247 (in Russian)

Sterud E, Forseth T, Ugedal O, Poppe TT, Jørgensen A, Bruheim T, Fjeldstad H-P, Mo TA (2007) Severe mortality in wild Atlantic salmon *Salmo salar* due to proliferative kidney disease (PKD) caused by *Tetracapsuloides bryosalmonae* (Myxozoa). Dis Aquat Org 77:91–198

Tort L, Watson JJ, Priede IG (1987) Changes in in vitro heart performance in rainbow trout, *Salmo gairdneri* Richardson, infected with *Apatemon gracilis* (Digenea). J Fish Biol 30:341–347

True K, Purcell M, Foott JS (2009) Development and validation of a quantitative PCR to detect *Parvicapsula minibicornis* and comparison to histologically ranked infection of juvenile Chinook salmon, *Oncorhynchus tshawytscha* (Walbaum), from the Klamath River, USA. J Fish Dis 32:183–192

Watson JJ, Pike AW, Priede IG (1992) Cardiac pathology associated with the infection of *Oncorhynchus mykiss* Walbaum with *Apatemon gracilis* Rud.1819. J Fish Biol 41:163–167

Weiland KA, Meyers TR (1991) Histopathology of *Diphyllobothrium ditremum* plerocercoids in coho salmon *Oncorhynchus kisutch*. Dis Aquat Org 6:175–178

Whitaker DJ, Kent ML (1991) Myxosporean *Kudoa thyrsites*: a cause of soft flesh in farm-reared Atlantic salmon. J Aquat Anim Health 3:291–294

10 Production Diseases and Other Disorders

Abstract

Production related diseases and disorders cover a large area that the authors believe deserves increased attention, although for some areas we cannot cover the subject in great depth. These refer to a wide range of conditions that may or not be attributed to a biological agent with examples discussed in this chapter covering environmental related conditions, vaccination, developmental and congenital abnormalities and disorders, dietary imbalance, disorders affecting the heart and the eye, general skeletal abnormalities, and deformities among eggs and fry and predators.

Keywords

Production diseases • Abnormality • Salmon • Trout

Production related diseases and disorders refer to a wide range of conditions with often multifactorial origin, and therefore they require a multidisciplinary approach in their management. Unspecific mortality varies from rather obscure causes like 'loser syndrome' or 'failed smolt' (Fig. 10.1), to other more specific causes such as poor smoltification, handling, transport, negative treatment effects and also certain infectious agents. A significant proportion of the mortalities can occur early post transfer to sea water and in some cases, might be linked to the quality of the smolt at the time of transfer, something that can and should be managed.

Overall, production related diseases and disorders cover a large area that the authors believe deserves increased attention, although for some areas we cannot cover the subject in depth. Examples discussed in this chapter cover environmental related conditions, vaccination, developmental and congenital abnormalities and disorders, dietary imbalance, disorders affecting the heart and the eye, general skeletal abnormalities, deformities among eggs and fry and predators.

10.1 Environment

10.1.1 Water Quality and Husbandry

Poor water quality is an obvious detrimental factor for any aquatic species including wild and farmed fish, with induced hypoxia being one of the major associated risks. Gill neuroepithelial cells (NECs) provide the sensory mechanisms for low oxygen detection, conversely, hyperoxia causes a decline in NECs number. These changes are concurrent with marked vascular distension, increased gill mucous, hyperplasia, elongation of respiratory lamellae and eventually, metabolic acidosis and death.

Fish excrete ammonia, and to a lesser amount, urea into the water as waste. Ammonia is highly toxic and exposure is a hazardous issue in fish farming resulting in a significant increase in oxygen consumption, with higher ventilation volume and respiratory distress. This can lead to acute mortality while the long term exposure results in reduced growth. Histologically, a severe branchial hyperplasia and associated widespread lamellar fusion, particularly at the tips, can be

D.W. Bruno et al., *A Colour Atlas of Salmonid Diseases*,
DOI 10.1007/978-94-007-2010-7_10,

Fig. 10.1 Production Atlantic salmon 'smolts' recovered from well-boat during sea-water transfer

observed. The tissue may become oedematous, with a mild inflammatory response and occasional aneurysms.

In the farm environment, husbandry induced fin and skin damage can occur, although measures are generally in place such that these factors are kept to a minimum. Examples of production related damage in Atlantic salmon include frayed or damaged fins, scale loss and scrapes and are shown in Fig. 10.2

10.1.2 Jellyfish

Fish reared in the marine environment can come into contact with true jellyfish. Depending on prevailing environmental conditions single animals or large swarms can be moved against the side of sea cages. Small jellyfish can be washed into cages whereas larger jellyfish tend to be break into pieces and tentacles or parts of tentacles enter the cages. In some cases the quantity of jellyfish can result in anoxia in the cages or obstruct respiration causing respiratory distress. Traumatic enucleation of the eye and marks on the side of the fish are also reported.

Several species of jellyfish have been associated with major loss of farmed Atlantic salmon, e.g. *Aurelia aurita*, *Cyaneae capillata*, *Pelagia noctiluca* and *Phialella quadrata*. Throughout European waters *Muggiaea atlantica*, *P. noctiluca* and *Solmaris corona* have contributed to significant loss in farmed fish. Blooms of *C. capillata* represent serious welfare issues due the irritant whiplash-like injuries inflicted by nematocysts of broken tentacles passing over the surface of the fish. *P. quadrata* (>15 mm in diameter) can pass through the mesh of sea cages and consequently sucked into the mouth of fish during respiration. At autopsy, as many as 40 jellyfish have been reported in the stomach of individual fish. In addition this species probably acts as a vector for the bacterium, *Tenacibaculum maritimum*.

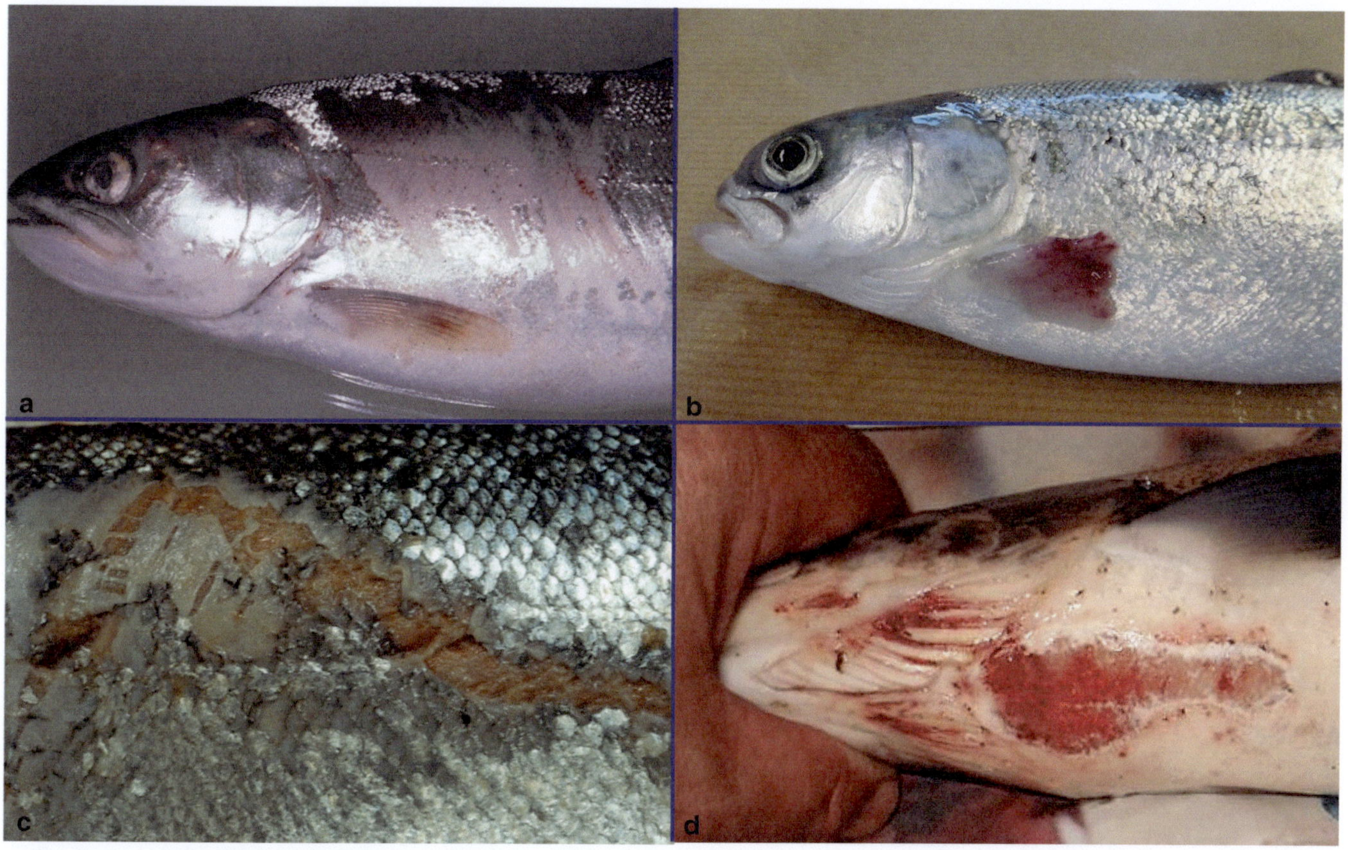

Fig. 10.2 (**a**) Net damage with scale loss in farmed Atlantic salmon. (**b**) Farmed Atlantic salmon with severe pectoral fin erosion with haemorrhage. (**c**) Farmed Atlantic salmon with skin wound after being scrapped on metal walkway. (**d**) Net scrape

Another species, *Bolinopsis infundibulum* has been reported to be associated with farmed fish mortality during the autumn in northern Norway. This species is fragile and ruptures on contact with the cages liberating a jelly-like substance which interferes with oxygen uptake by the fish gills.

Histologically, in general affected gills show sloughed lamella and necrosis, with oedema and inflammation of the filaments as emphasised by the presence of granulocytes. After approximately 48 h gill lesions show large areas of epithelial sloughing, haemorrhage and lysis of erythrocytes.

Cnidaria vary from a few millimetres to a few metres in size, and may be solitary (i.e. medusae of Hydrozoa, Scyphozoa and Cubozoa) or colonial (i.e. hydrozoan siphonophores) organisms.

10.1.3 Phytoplankton and Algal Blooms

Blooms occur naturally and can adversely affect human health as well as fish stocks. The blue-green algae, Cyanobacteria are among the most damaging blooms that impact on water quality. Impact depends upon the type, size and frequency of the blooms but reported as seasonal changes rather than influenced by fish farming. Fish stocks can suffer directly from toxins or indirectly through damage to the gill epithelia, resulting in acute necrosis, swelling, pyknosis and congestion.

10.1.4 Gas Bubble Disease

Gas bubble disease (GBD) is a non-infectious physically induced process that is caused by uncompensated hyperbaric pressure of total dissolved gases within the fish vascular system. GBD can either occur as a natural phenomenon in lakes and rivers (e.g. heating of water, photosynthesis) or artificially when supersaturated water is drawn into fish tanks without adequate aeration (e.g. pumping and heating of water, hydroelectric plants or leaking pumps). When pressure compensation is inadequate, a sudden decompression in the external environment (water) leads to blood dissolved gases (initially nitrogen) to form emboli in several tissues as it abruptly comes out of solution when trying to balance with the decreased external pressure. Highly vascularised tissues suffer most and severe exophthalmia due to the physical accumulation of gas bubbles in the choroid gland of the posterior uvea, is a frequent finding, as well as corneal degeneration and haemorrhage. Macroscopically bubbles may also

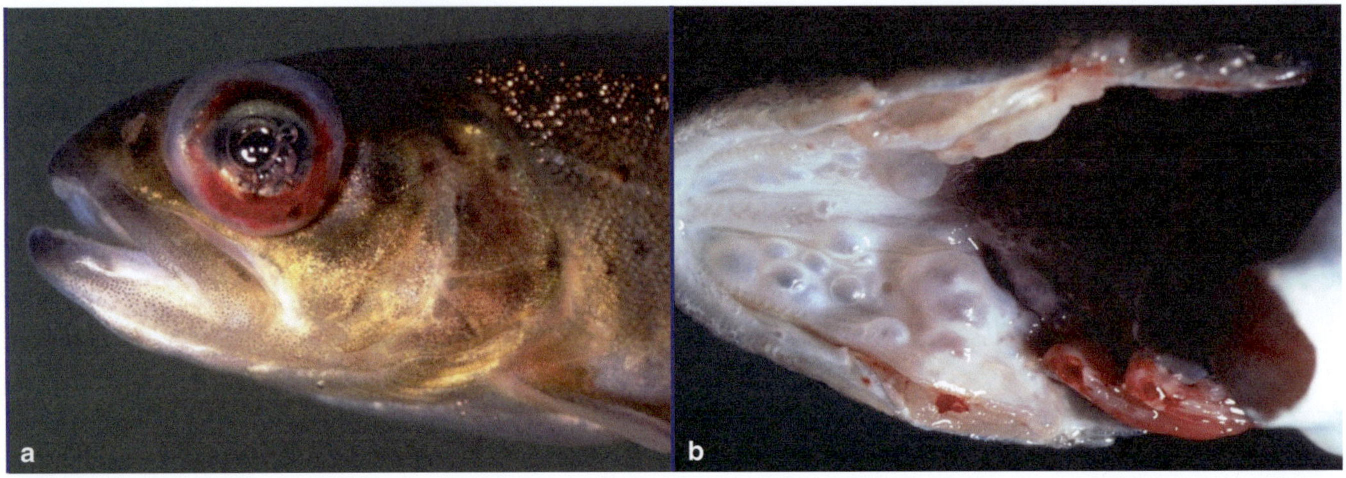

Fig. 10.3 (**a**) Exophthalmia, haemorrhage and ocular gas bubbles in brown trout with gas bubble disease. (**b**) Gas bubbles in the palate of brown trout with gas bubble disease

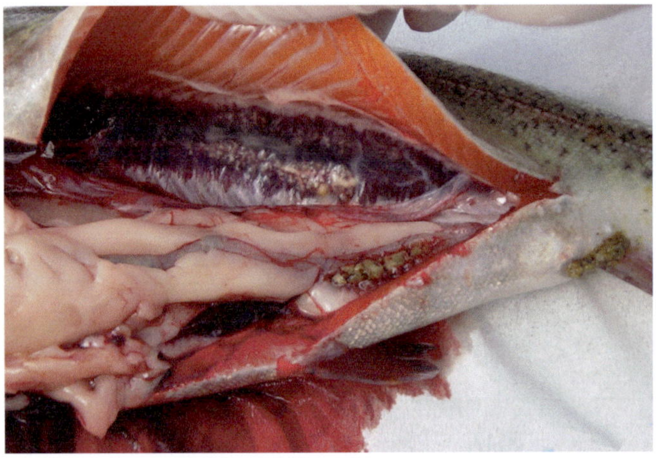

Fig. 10.4 Dilated ureters and granulomatous inflammation in the posterior kidney of farmed rainbow trout with nephrocalcinosis

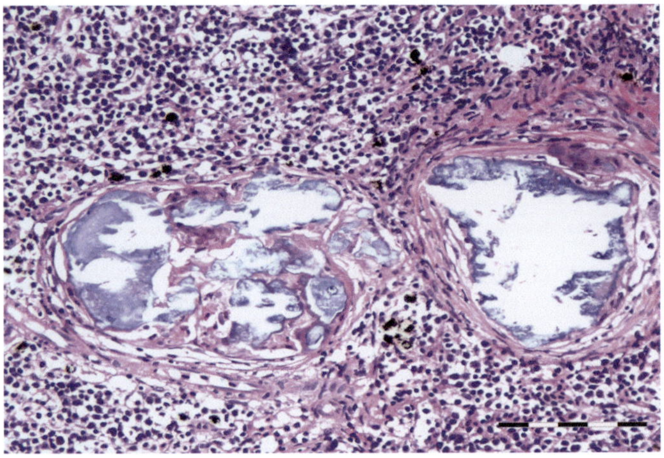

Fig. 10.5 Nephrocalcinosis in farmed rainbow trout. Dilated ureters are filled with amorphous basophilic material. Bar = 100μm

be seen in the mucous membranes lining the oral cavity, the gills and fins (Fig. 10.3). Gas bubbles in the heart cavities can disrupt the blood flow and acutely affected fish subsequently die from asphyxiation. GBD can also lead to indirect problems and injuries when the air vesicles rupture (as in the skin and gills) leads to haemorrhage, open small wounds and secondary infections. Oedema of the lamellae with degeneration of the covering respiratory epithelium occurs with tissue necrosis and ischemia of the capillary beds. Safe limits for gas supersaturation depend upon the size of the fish, species, degree of super-saturation and water temperature.

10.1.5 Nephrocalcinosis

Nephrocalcinosis occurs in intensively farmed salmonids, in particular rainbow trout and brook trout, but has also been seen in wild fish. The aetiology appears to be complex, but is often associated with high ambient free CO_2 levels and/or nutritional aspects involving magnesium deficiency or selenium toxicity. Common signs in affected fish are abdominal swelling, exophthalmia and ventral haemorrhage which may continue to develop after transfer to sea water. At necropsy, ascites, splenomegaly and thickening of the ureters with white, chalky caseous deposits are frequent findings (Fig. 10.4). The kidney may become swollen, grey and urinary cysts can develop. Histologically, mineral deposits occur in distended distal tubules and ureters (Fig. 10.5). Deposits in adjacent parenchyma provoke a granulomatous inflammation with fibrosis and severe distortion of normal tissue. Mortality is generally low, but food conversion ratio in affected fish is impaired and the carcass quality is reduced. Diagnosis is based on gross lesions and the histopathological changes showing deposits in the collecting ducts which typically stain dark blue (basophilic) in H&E sections and black with von Kossa stain.

10.2 Vaccination Side-Effects

With the exception of Tasmania, virtually all salt-water farmed salmonids are vaccinated. Most vaccines in current use are multivalent, i.e. they contain antigens from several different bacteria and/or viruses. In order to enhance and extend the duration and/or directing the nature of the immune response, an oil based adjuvant is mixed with the antigens. The composition of the vaccines may vary depending on the zoosanitary situation. After a few days of starvation fish are typically immunized prior to smoltification and sea-water transfer. The fish are anesthetized and the vaccine is given intraperitoneally, either by vaccination teams or automated machines. The injection site is typically in the ventral midline, 1.5 fin lengths cranial to the base of the pelvic fins, and the injection volume ranges from 0.05 to 0.1 ml.

The combination of antigens and adjuvant provokes a strong localized inflammatory response. This chronic inflammation is macroscopically visible as fibrinous strands between the peritoneal wall and internal organs (Fig. 10.6). Typically, the caudal parts of the pyloric caeca and spleen are involved in the lesions. Thrombosis and granulomatous inflammation is also reported in the liver. Moderate lesions are considered acceptable, both from an animal welfare and from the consumers' points of view.

Histologically, the affected region between the wall of the pyloric caeca and pancreatic tissue and around the spleen, show clear, empty spaces within the granulomatous response. These correspond to the oil droplets in the vaccine which are dissolved during the tissue processing. The inflammatory response is dominated by macrophages, lymphocytes, fibroblasts and multinucleated giant cells. Eosinophilic granular cells and melanomacrophages are

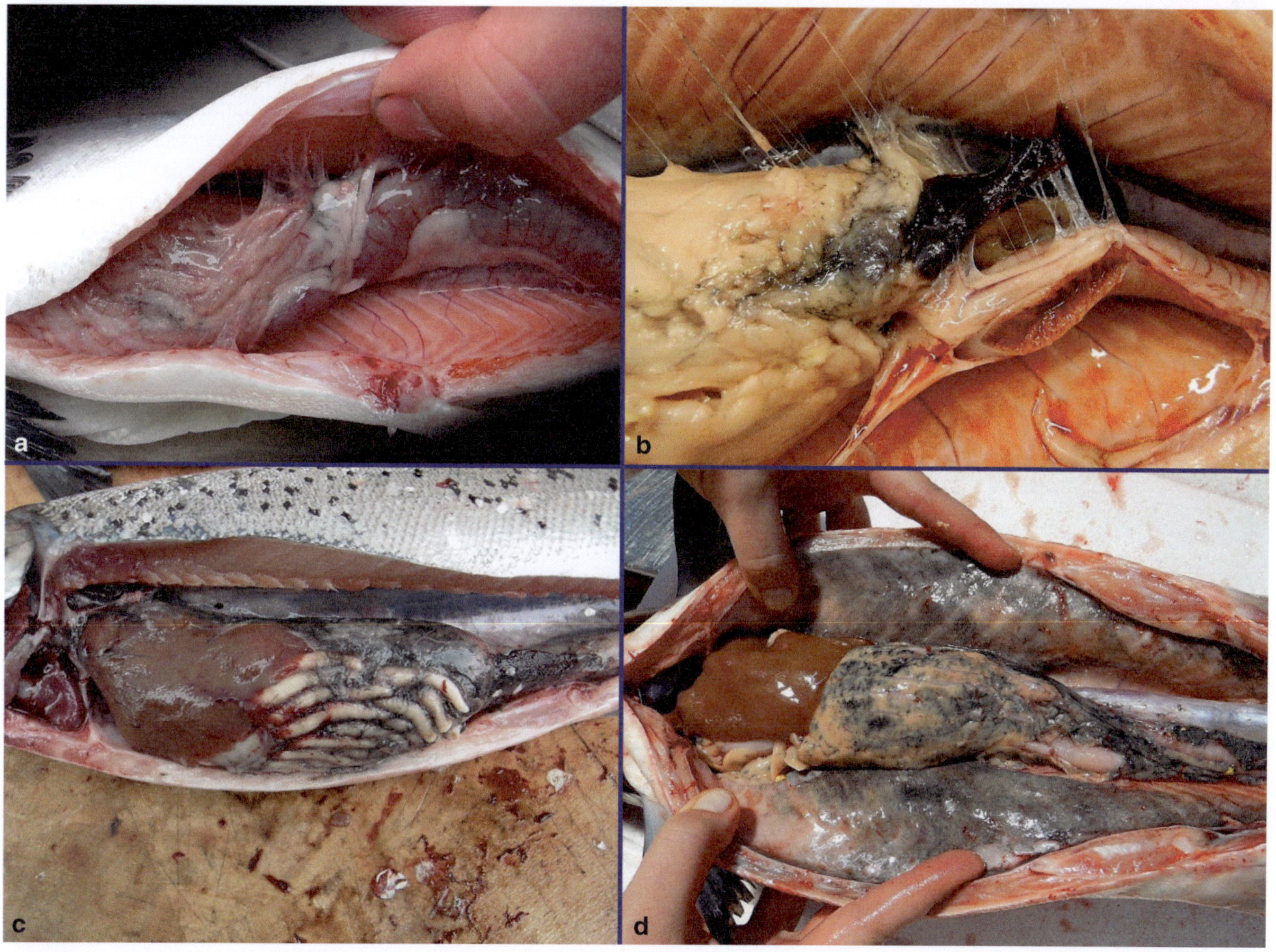

Fig. 10.6 Vaccinated Atlantic salmon showing. (**a**) Mild adhesions between visceral organs and body wall near the injection site. (**b**) Moderate adhesions between visceral organs and body wall; and (**c–d**). Severe melanisation of internal organs and peritoneal surface

usually present. In some cases, lesions become extensive and can involve most of the abdominal cavity with fusion of most organs to the body wall (Fig. 10.7). Extensive melanisation is sometimes evident. Granulomatous responses and location of oil droplets have also been recorded in organs such as gill, liver and muscle.

Vaccination may also induce autoimmunity with marked glomerulonephritis (see Fig. 4.16) characterized by deposition of immune complexes in the mesangial cells of the glomerulus. Inflammatory responses may also be found in the uveal tract and in the heart (epicarditis and multifocal myocarditis). Occurrence of characteristic Splendore-Hoeppli reactions (asteroid bodies) in lesions is typical in hypersensitivity reactions.

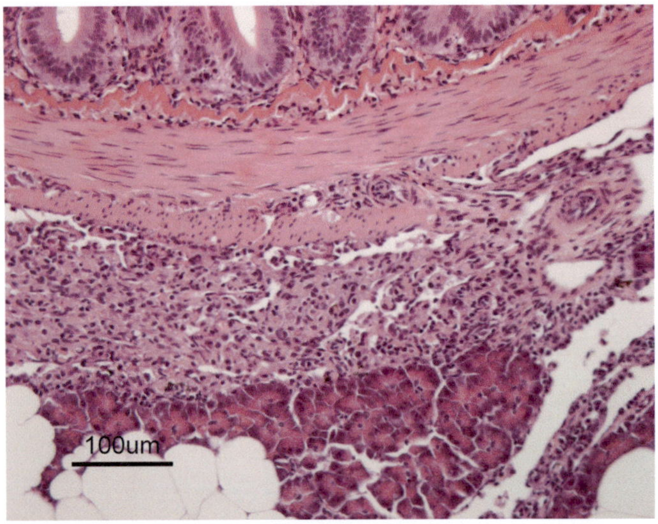

Fig. 10.7 Granulomatous peritonitis in Atlantic salmon vaccinated with oil-adjuvanted vaccine

10.3 Dietary Imbalance

Salmonids are, in principle, carnivores with a relatively short digestive tract and essentially identical, although slight modifications exist between species. Growth, food intake, stomach evacuation rate and feed efficiency ratio of Atlantic salmon are influenced by temperature and fish size. However, farmed and wild fish face very different challenges in relation to their nutrition. The former have a fairly continuous access to a dry pelleted feed throughout their life cycle and overfeeding results in overweight fish (Figs. 10.8 and 10.9). Conversely wild fish experience extreme variations in availability, quantity, quality and composition throughout the year and overweight wild fish are not observed. The composition and quality of pelleted feed is generally high, but scarcity of fish meal protein has resulted in their gradual replacement by plant components imposing a challenge to the feed manufacturers and to the fish. Quality control and customer demand have resulted in appropriate nutritional balanced diets for farmed salmonids and only rarely are deficiencies reported. Nevertheless, prolonged storage of feed under

Fig. 10.8 Extremely obese seawater-farmed rainbow trout. The fish have frayed fins and the bottom fish an abnormality of the caudal part of the vertebral column

Fig. 10.9 Adipose tissue in the abdominal cavity of a sea-water farmed rainbow trout. Only the liver is visible; all other organs are hidden in fat

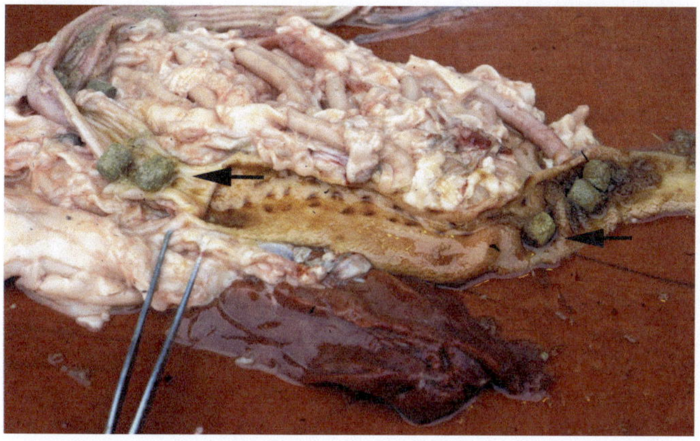

Fig. 10.10 Undigested pellets (*arrow*) in the anterior part of the rear intestine in farmed Atlantic salmon

unfavourable conditions (e.g. heat, light, moisture) may compromise the quality and inconsistency may occasionally occur. In welfare terms there is also increasing evidence that intensively reared fish significantly alters some aspect of cardiac anatomy and physiology. In this section the effect of poor or deficient diets on the health of farmed fish is discussed.

10.3.1 Inadequate Digestion of Pellet

Dry pellets will normally decompose to a porridge-like consistency in the stomach before passing to the pyloric region for further digestion, but at low temperatures, the feed is not always dissolved in the stomach and may become lodged as a hard plug in the sphincter area with resultant accumulation of dry and hard pellets in the stomach (Fig. 10.10). The mucosa is typically hyperaemic and irritated, which is also reflected in histopathological lesions.

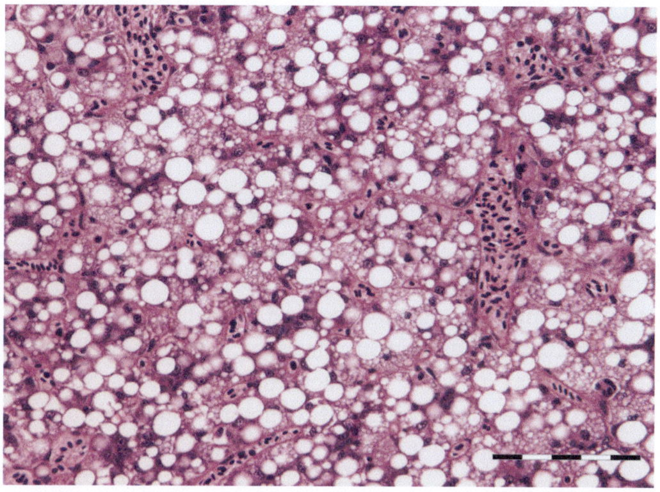

Fig. 10.11 Vacuolation of hepatocytes in farmed rainbow trout with fatty liver degeneration. Bar = 100μm

Another condition that occurs under similar circumstances is where faeces of farmed salmon turn yellowish and foamy and float to the surface. The posterior gut is hyperaemic and the vent opening may be swollen. However, the direct causes of both conditions are unknown, but are believed to be associated to the technical and biological qualities of the pellet at variable ambient temperatures.

10.3.2 Hepatic Lipidosis

Farmed salmonids are susceptible to hepatic lipidosis or lipoid liver disease (LLD) when fed polyunsaturated fatty acids combined with insufficient amounts of antioxidant protection, such as vitamin E. Polyunsaturated fatty acids are prone to auto-oxidation on exposure to atmospheric oxygen.

Clinical signs include reduced appetite, impaired equilibrium and increased mortality. Fish become anaemic with pale gills and ascites, loose scales and develop fin necrosis. Internally, a yellow-orange discolouration and slight enlargement of the liver with a friable fatty consistency has been observed. The haematocrit is usually severely depressed as a result of a microcytic anaemia with increased blood cell fragility. Mortalities result from the cumulative effects of anaemia and hepatocellular and biliary tract dysfunction. Typical histological lesions show accumulation of excessive lipid in hepatocytes, macrophages and fat-storing cells (Fig. 10.11). Degenerating hepatocytes with pyknotic nuclei may occur over large areas. Red cell breakdown frequently results in haemosiderosis in the spleen. Diagnosis is based on internal appearance and evidence of a generalised fatty infiltration of the liver in stained tissue sections.

10.3.3 Steatitis

Steatitis or pansteatitis is an inflammation of the visceral adipose tissue that is normally abundant around the pyloric caeca in salmonids. The inflamed tissue is firm, yellowish or greyish and the swim bladder may be opaque with streaks of greyish yellow. Histological findings include thickening of the fat cell walls, infiltration, aggregation of ceroid-containing macrophages and pigment deposit (Fig. 10.12). A granulomatous response centred on multinucleate giant cells and eosinophilic inclusions may also be seen. Steatitis may occur as a sequel to pancreatitis, but also as an idiopathic disease where vitamin E deficiency and/or rancid fat in the diet are possible causes.

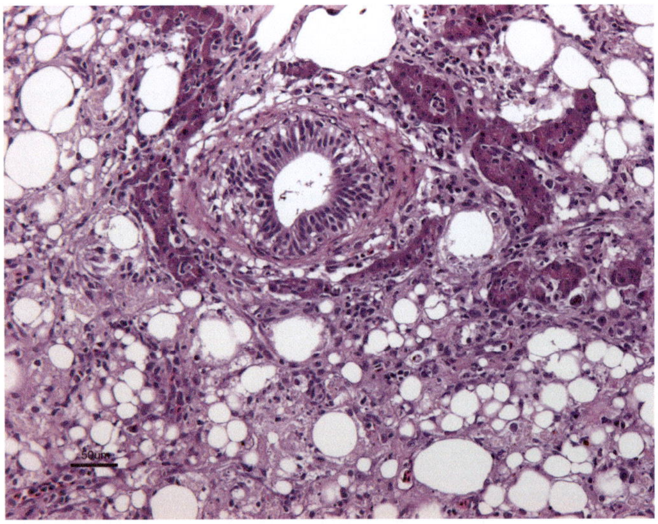

Fig. 10.12 Steatitis of peripancreatic fat in farmed Atlantic salmon

10.3.4 Heart Fat Infiltration

Fat infiltration in the epicardium can be encountered in both farmed and wild fish (Fig. 10.13). For example, wild whitefish may accumulate large amounts of fat, particularly between the ventricle and bulbus arteriosus and along the edges of the ventricle. In farmed salmon, the amount of epicardial fat is highly variable and is generally considered a negative trait.

10.3.5 Soybean-Induced Enteritis

Salmon fed on diets rich in proteins from soybean can develop a non-infectious sub-acute enteritis affecting the distal intestine. Lesions include thickening of the lamina propria due to infiltration of macrophages, neutrophils, eosinophilic granular cells, loss of the normal supranuclear vacuolization of the absorptive cells in the intestinal epithelium and widening and reduction in height of the intestinal folds (Fig. 10.14). Absorption from the gut is reduced due to loss of apical vacuoles in epithelial cells. Lesions develop faster at 12 °C than at 8 °C and are generally resolved within a few weeks after removal of the soybean components from the diet.

10.3.6 Vitamin Deficiencies

Ascorbic acid (vitamin C) is an essential dietary requirement which deficiency results in an impaired biosynthesis of collagen and chronic reduction of the supporting tissues, giving rise to a typical kyphosis, lordosis or scoliosis (Figs. 10.15 and 10.16). In some scorbutic fish, fractures in the axial

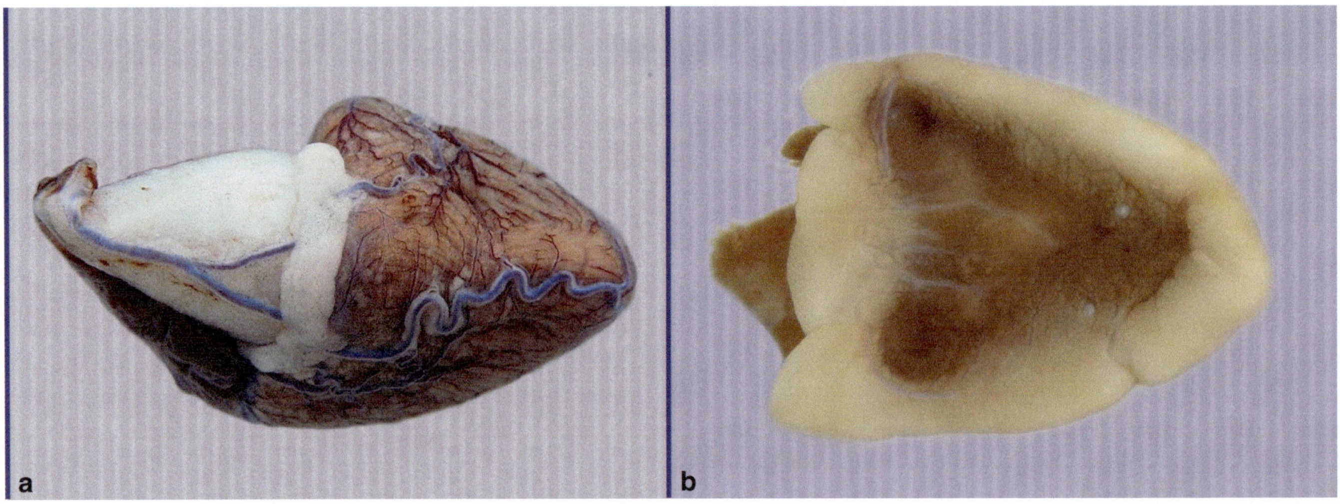

Fig. 10.13 (**a**) Congested coronary vessels and epicardial fat accumulation in the heart of farmed Atlantic salmon. (**b**) Extensive fat accumulation along the margins of the ventricle in wild whitefish (*caudal view*)

skeleton with dislocations, focal haemorrhage and atrophy are noted (Fig. 10.17). A reduction in collagen and irregular hyalinisation of the remaining tissue results in a loose structure and dysplasia, including osteoporosis. Within the gill lamellae marked cell degeneration of the cartilaginous rod with large vacuoles accompanies deformation and distension (Fig. 10.18). Fish may show a hypoplastic anaemia and a decrease in haematopoietic tissue can be seen in the anterior kidney. Macroscopically, a shortened operculum is a common feature (Fig. 10.19). Gross appearance of the fish and the examination of histological sections provide provisional information for a diagnosis. An analysis of the vitamin content of the feed should be carried out to support a diagnosis.

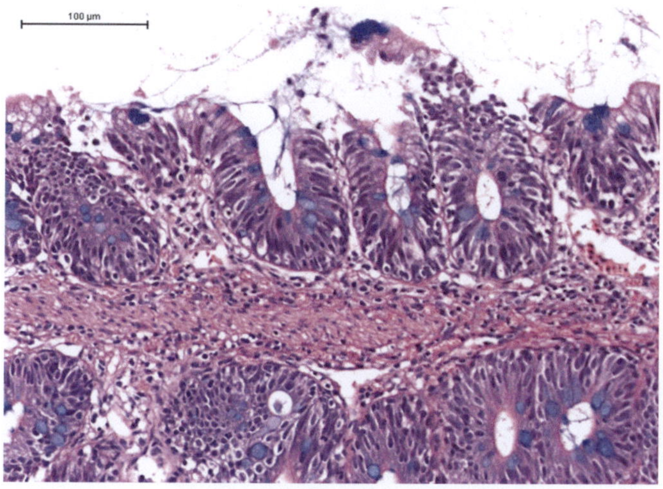

Fig. 10.14 Soybean enteritis in farmed Atlantic salmon

Vitamin E is an essential nutrient for many species of fish and deficiencies can occur where diets are partly oxidised before use. Several factors influence the dietary requirement for vitamin E, including the level of polyunsaturated fatty acids and other oxidants in the diet and fish tissues, as well as the level of selenium. A deficiency has an impact on several tissues including observations of white muscle degeneration (Fig. 10.20) and under experimental conditions, it is associated with calcification of the pseudobranch (Fig. 10.21).

Pathological changes include lordosis, exophthalmia, splenomegaly, mottled liver, pale kidney, poor growth and ascites, and can be correlated with marked microcytic anaemia, reduced haematocrit and increased haemolysis. Incomplete maturation of these cells is consistent with increased fragility due the recognized role in cell membrane integrity and growth. In addition, steatitis has also been reported as well as small areas of degeneration with dilation of the liver sinusoids often with ceroid deposition. Splenic haemosiderosis is prevalent and occurs as a fine, stippled appearance to the tissue in Perl's stained sections (see Fig. 4.22) and this correlates with the accelerated

Fig. 10.15 Farmed rainbow trout fingerlings with vertebral abnormalities resulting from vitamin C deficiency

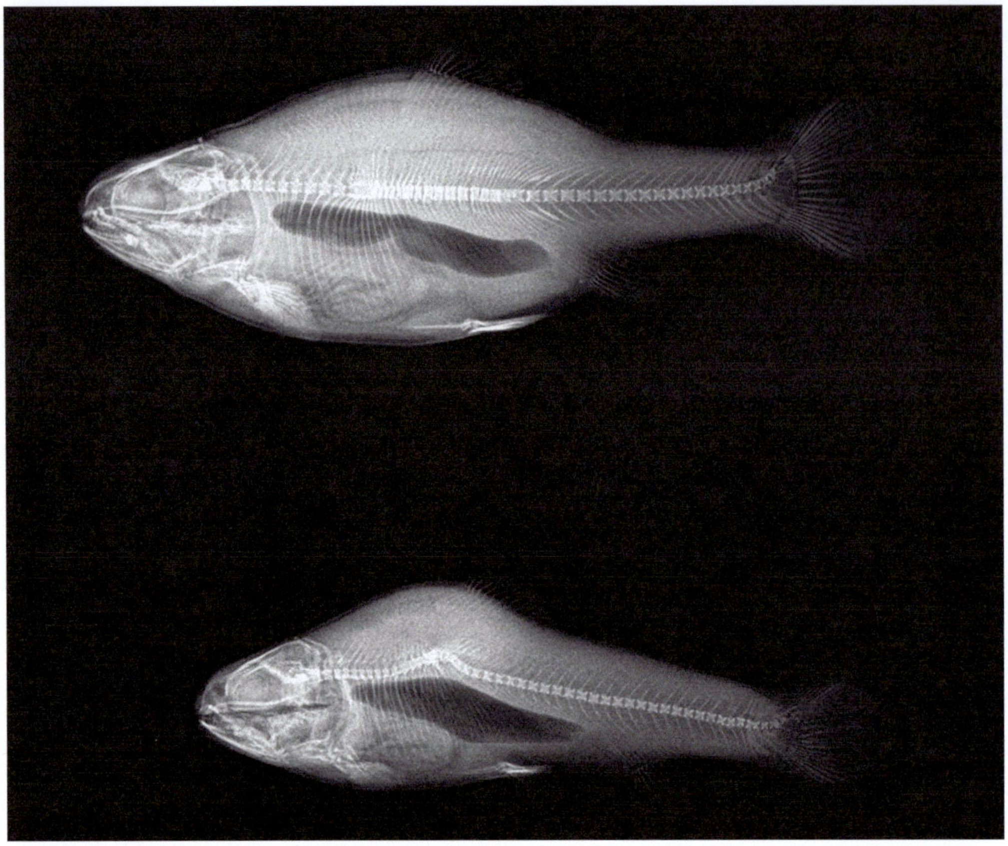

Fig. 10.16 X ray showing two different types of spinal deformity in farmed rainbow trout with vitamin C deficiency

Fig. 10.17 Abnormality of the caudal part of the vertebral column in farmed rainbow trout suffering from vitamin C deficiency

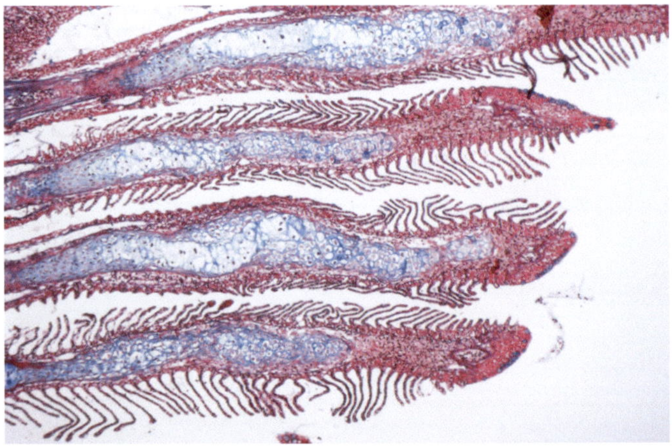

Fig. 10.18 Vitamin C deficiency in rainbow trout. The cartilage of the lamellae is curved and show irregularities. The cartilage cells are hydropic and vacuolated. Azan and toluidine blue. Low power

erythrocyte lysis. Vitamin E deficient salmonids are recognised from gross observations and the examination of stained tissue sections.

Histological abnormalities associated with pantothenic acid-deficient fish (vitamin B5) exhibit dermatitis and exudate-covered gill lamellae, with extensive hyperplasia (Fig. 10.22). Differential diagnosis include changes attributed to infestation with *Paramoeba perurans* (see Fig. 8.2).

Vitamin B_6 deficiency in rainbow trout includes anorexia, listlessness, frantic erratic swimming, and ataxia. Deficiency can be readily determined by measuring pyridoxine-enhanced liver aspartate aminotransferase (ASAT) activity before clinical signs of deficiency are apparent. Histological

Fig. 10.19 Farmed rainbow trout with shortened operculum. Note exposed gill tissue

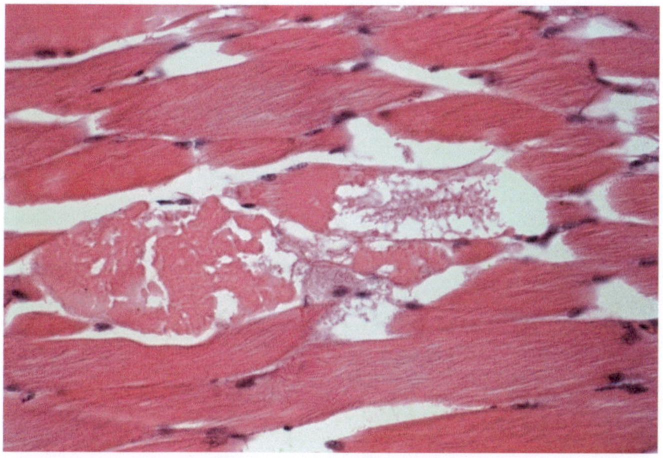

Fig. 10.20 Experimentally induced white muscle degeneration in Atlantic salmon with deficiency of vitamins C and E. Low power

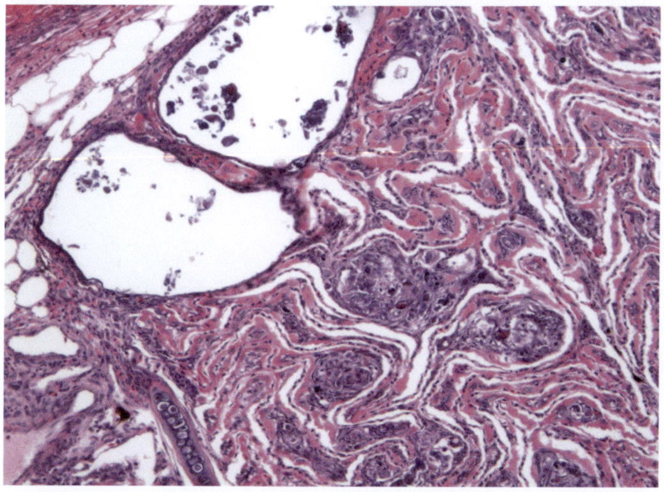

Fig. 10.21 Experimentally induced calcification of the pseudobranch in Atlantic salmon with vitamin E deficiency. Low power

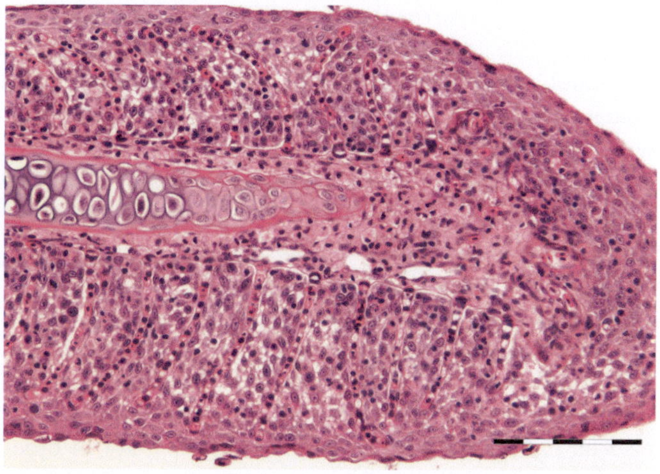

Fig. 10.22 Experimentally induced severe hyperplasia of gill epithelium and fusion of lamellae in rainbow trout with vitamin B5 deficiency. Bar = 100μm

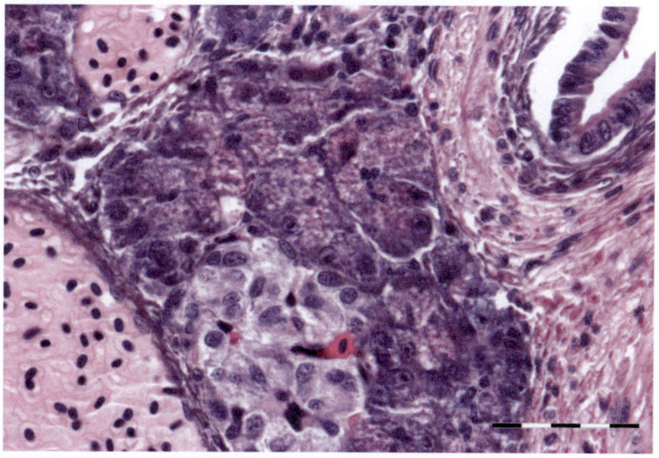

Fig. 10.23 Acinar cell atrophy in pancreas of Atlantic salmon with vitamin B6 deficiency. Bar = 50μm

changes include acinar cell atrophy (Fig. 10.23) and hyperplasia of renal haematopoietic tissue.

Alack of thiamine (vitamin B1) is linked to the condition known as 'M74' in eyed eggs of wild Atlantic salmon and sea-trout in the Baltic area. The deficiency continues through the fry stages. Early mortality syndrome (EMS) and Cayuga syndrome are recognised as more or less identical conditions in several wild salmonids in the Great lakes area in North America. These conditions are characterized by up to 100 % mortality in progeny from certain female fish. Clinical signs include spiral swimming, loss of equilibrium and hyperexcitability, lethargy, dark body and subcutaneous oedema. Affected fry go off the feed and develop hydrocephalus, yolk-sac precipitate and haemorrhage (Fig. 10.24). Histologically, characteristic lesions are found in the molecular layer of the cerebellum developing cellular degeneration and necrosis, nuclear pyknosis and karyorrhexis, and sometimes haemorrhage. It is generally accepted that these lesions are the consequence of thiamine deficiency in the female. Salmonids are top predators and some of their prey may include forage fish with high levels of thiaminase, e.g. alewife in the Great Lakes region and sprat and herring in the Baltic Sea. The composition of the forage fish diet may be variable from 1 year to the other and individual fish may have different feeding strategies explaining the female dependant factor. High thiaminase levels results in low levels of thiamine in eggs and progeny with resultant characteristic signs of thiamine deficiency. These conditions can be prevented or reversed by exposing eggs or fry to thiamine. Females can also be injected with thiamine prior to stripping in order to avoid the condition. White spot disease and blue-sac disease are both differential diagnosis.

10.3.7 Mineral Deficiencies

In aquatic ecosystems zinc is an essential micronutrient as well as a toxicant of considerable significance. Consequently, zinc deficiency leads to physiological perturbation of growth, reproduction, vision and immunity. Cataracts attributed to zinc deficiency are usually bilateral and characterised by diffuse or focal opacity of the lens. Other gross signs of this deficiency include a reduced growth rate with skin and fin erosion. Histologically, lesions consist of vacuolation, lysis of fibres and proliferation of capsular cells. Keratitis and panophthalmitis may sometimes occur concurrently. Cataracts occurring through zinc deficiency are recognised by a diffuse or focal opacity of the lens and the examination of stained sections. Diagnosis is based on histological assessment of eye sections.

Magnesium deficiency in rainbow trout include evidence of poor growth, loss of appetite, calcinosis of kidney and muscle and a dramatic increase in the muscle extracellular fluid volume. Histologically, there are degenerative changes in kidney, ovary and liver as well as hyperplasia of the renal haematopoietic tissue.

Fig. 10.24 M74 in Baltic salmon fry. Note haemorrhage in gill area

10.4 Diseases Associated with the Eye

A range of production related eye diseases are identified ranging from sub-acute lesions include cataract, anterior synechia and suppurative panophthalmitis (Figs. 10.25, 10.26, 10.27, 10.28 and 10.29). Lesions may be the result of haematogenous (endogenous) spread of pathogens, physical or mechanical (e.g. handling) with subsequent infection, chemical (e.g. medical treatments), thermal damage or nutritional factors.

In farmed fish, numerous nutritional imbalances are known to cause cataract, i.e. deficiency of zinc, vitamin A, or the amino acids thiamine, riboflavin, methionine and tryptophane. Cataracts are often bilateral and the result of denatured proteins in the lens or the result from infection, e.g. parasites (see Fig. 9.36). Posterior cortical, bilateral cataracts were documented in the 1990s following the removal of blood meal from diets and this was attributed to

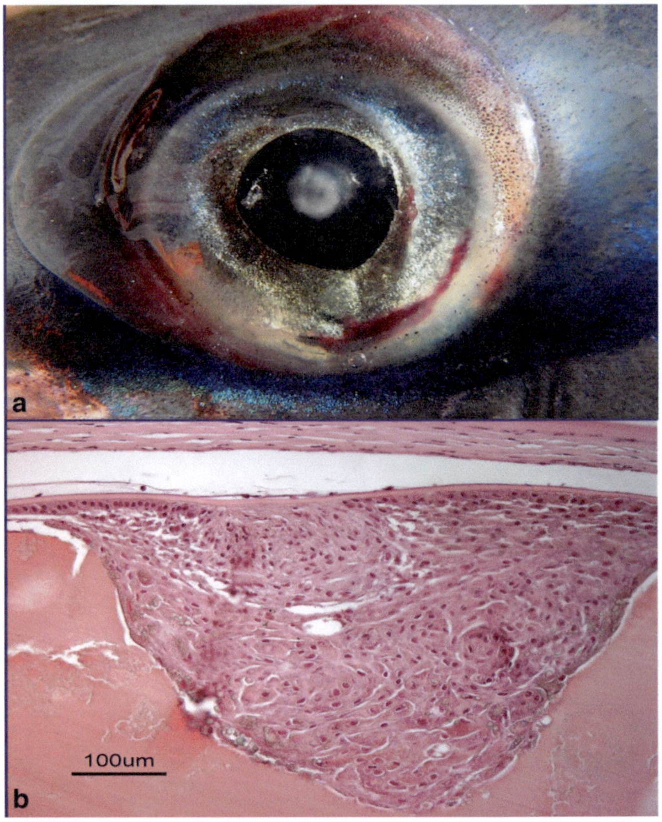

Fig. 10.25 (**a**) Central cataract in farmed Atlantic salmon. (**b**) Section through cataract in farmed Atlantic salmon

Fig. 10.26 Keratitis and exophthalmia in farmed rainbow trout

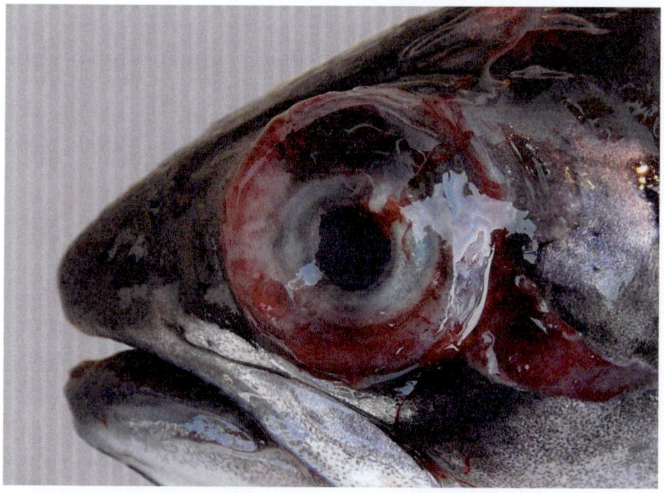

Fig. 10.27 Panopthalmitis in farmed Atlantic salmon

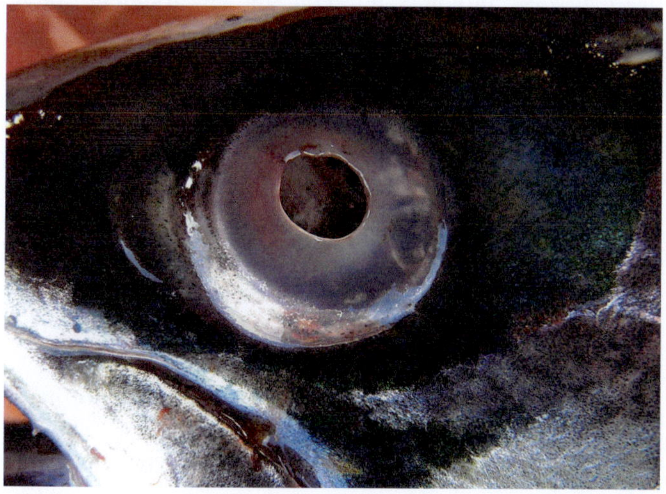

Fig. 10.28 Punctured eyeball in farmed Atlantic salmon

Fig. 10.29 Eye haemorrhage in farmed Atlantic salmon with infectious salmon anemia

Fig. 10.30 Panopthalmitis and cranial erosion in farmed Atlantic salmon with *Tenacibaculum* infection

a lack of dietary histidine. Histologically, cataracts typically manifest as vacuolation, lysis of fibres and proliferation of capsular cells. Depending on severity, it may cause impaired vision or total blindness.

Osmotic cataract is the result of fluctuating osmolality where the fish is unable to adjust to the environmental salinity and may be reversible. Temperature fluctuations, gas supersaturation and exposure to UV light are other examples.

Keratitis may result from physical, nutritional, chemical or thermal damage to the cornea, or it may be caused by an infectious agent. This is fairly common with systemic infectious diseases as for example bacterial conditions, where scleritis is also reported. For instance, *Tenacibaculum* spp. infection may completely destroy the eye (Fig. 10.30).

In water supersaturated with gas, gas bubble disease (GBD) involving the eyes, may occur (see Fig. 10.3). Eye lesions include gas bubbles in the anterior chamber, cataract, synechia and panophthalmitis.

Physical damage with perforation of the cornea or endogenous spread of bacteria to the ocular structures may result in endopthalmitis or panopthalmitis, which may be accompanied by haemorrhage and subsequent puncture of the eye. Damage to the cornea may be superficial or deep to penetrating (perforating). Superficial damage will heal by increased mitotic activity and migration of epithelial cells from the periphery of the lesions. During the healing process, the eye and the open socket may completely fill with scar tissue.

Another source of eye damage is some parasitic infections as for example infections by metacercaria of *Diplostomum spathaceum* in wild fish or those reared in earth ponds. The parasite affects the lens or in the anterior eye chamber (see Fig. 9.36). Similar eye changes have been seen in fish infected with the trematode *Tylodelphys clavata*.

10.5 Developmental Disorders

Deformities, skeletal disorders or pigment abnormalities are an unwelcome but inherent aspect in salmonid farming, fortunately, with low prevalence. The similarity of malformations across fish species and culture systems implies that there is a general causal effect within the environment. Particular groups or batches of salmon fry may show a range of deformities which are evident within the egg, or are apparent shortly after hatching. Such deformities include a general discolouration or pigment abnormality of the egg, 'Siamese twin' fry, kyphosis, lordosis and others associated with the cranium or jaw. Many of these conditions are attributed to genetic factors, nutritional imbalance, hypoxia, poor husbandry, or contact with parasites such as *Myxobolus cerebralis*. Furthermore, fish can be the recipients of numerous injuries that are potentially deleterious to aquaculture production performance and welfare. For many deformities the proposed causes in natural populations remain speculative.

10.5.1 Deformities Among Eggs and Fry

High mortality during early life is common and may be caused by numerous aetiologies, but often developmental (Fig. 10.31). All salmonids can be affected but most reports are from widely farmed species, i.e. salmon and rainbow trout. Hereditary and environmental factors are the most important causes. Moderate abnormalities may become evident as the fish grows, and provided these do not interfere significantly with normal function, such fish can generally survive. However, as a farmed animal they will not perform well, they will be removed during grading and consequently rejected at market.

10.5.2 Yolk Sac Constrictions

Yolk sac constrictions affect the posterior part of the yolk-sac which is separated and becomes unavailable as an energy source for developing fry. The cause(s) are not known, but may be related to density and presence of inadequate substrate. Insufficient water flow and/or accumulation of waste metabolites are also possible causes, alone or in combination.

10.5.3 Soft Egg Disease

Soft egg disease is a condition where the eggs become soft and flaccid and collapse, probably as a result of altered water-flow through the membranes of the egg. Possible causes include high concentration of ammonia and amoeba infestation.

10.5.4 Blue Sac Disease

Blue sac disease or dropsy (hydrocoele embryonalis) is characterized by abnormal accumulation of fluid with a bluish tint between the yolk-sac and the outer membranes of the fry (Fig. 10.32). This can be observed shortly after hatching and become more apparent within a few days. Affected individuals often display humpback, a wide open mouth and focal haemorrhage. Affected fry are lethargic with reduced respiration and heart rate. Fry may recover if the condition has not had a long duration or if the lesions are moderate. The causes are believed to be accumulation of toxic waste products, unfavourable environment, rough handling and transport. Xenobiotic chemicals may act directly or following enzymatic catalysis to intermediate products, becoming more or less toxic to the organism. The condition has been reported to be induced by dioxin-inducing the syndrome characterized by increased mortality at early life stages and spinal defects, oedema and haemorrhage.

10.5.5 White Spot Disease

White spot disease can occur in eggs in late stages of incubation and the yolk-sac of newly hatched fry. The white or greyish spots in the yolk are coagulated proteins following a leak in the foetal membrane with consequent mortality (Fig. 10.33). Unfavourable environmental factors including

Fig. 10.31 Examples of malformations in yolk-sac fry

Fig. 10.32 Blue sac disease in Atlantic salmon yolk-sac fry. Note periocular and gill haemorrhage

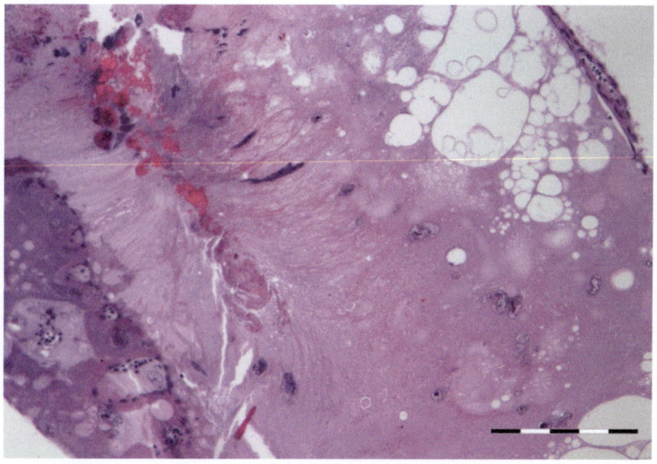

Fig. 10.33 Coagulated yolk in Chinook salmon fry with white spot disease. Bar = 200μm

heavy metals, fluctuating pH, ammonia, low temperature, chemical treatments and rough handling during critical phases of egg development, are likely causes.

10.6 Abnormalities of the Opercula

Shortened or frayed opercula with scars and indentations result in direct physical irritation and damage to the gill epithelium and are common in intensively farmed salmonids. The condition may be uni- or bilateral, although difficult to record until larvae reach a certain size (~>12 mm in length). In severe cases, most of the gill tissue is visible and exposed (Fig. 10.34). As with other production-related problems, the causes appear to be multifactorial and complex, involving aggression, irritation from chronic inflammation and wear and tear. This is considered distinct from eroded opercula where an infectious cause is likely.

Egg incubation temperature has been linked to shortened opercula in salmonids. Shortened opercula are thickened with excess mucus along the trailing edge and implies reduced capacity of the buccal pump and consequently interferes severely with normal breathing. In severe cases the fish is obliged to swim constantly in order to move water over the gills. Furthermore, the fish will be unable to 'cough', i.e. reverse the water flow over the gills in order to remove debris, particles or parasites. Exposed gills are also more vulnerable to damage and infection by bacteria, fungi and parasites.

Mechanical damage to the opercula may result from fish jumping and hitting the bird net over the cages (particularly salmon). The fish struggle to get loose resulting in severe opercula damage (Fig. 10.35). In wild fish, damage from gill nets will typically cause vertical wounds and chronic ulceration around the entire fish in front of the dorsal fin and in the opercula area.

Diagnosis is based on gross observation of characteristic opercula lesions. As a result of the indirect effect on the gill tissue, shortening and thickening of filaments with epithelial hyperplasia and lamellar fusion are diagnosed by light microscopy.

10.7 Jaw and Head Deformities

Jaw and head deformities have been described from Atlantic salmon fry. For example conspicuous, cream coloured unilateral nodules are reported. These occur on the cranium above the optic tectum and located posterior and obliquely to the eye (Fig. 10.36). The nodules are ovoid, but

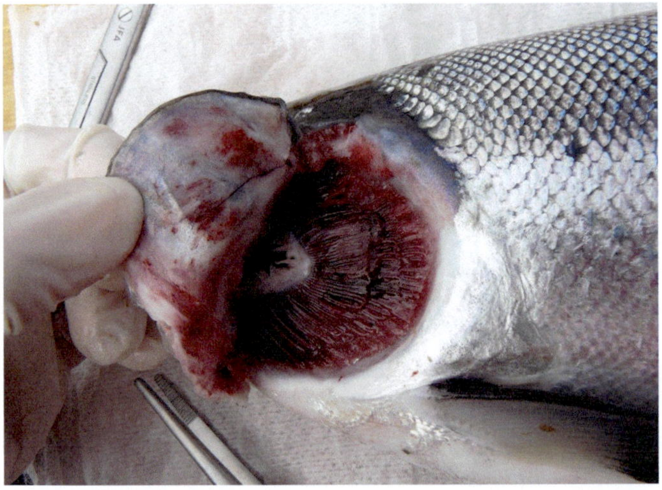

Fig. 10.35 Haemorrhage on the inside of the operculum of a farmed Atlantic salmon resulting from the fish being caught in the bird net while jumping

Fig. 10.34 Farmed Atlantic salmon with inverted operculum

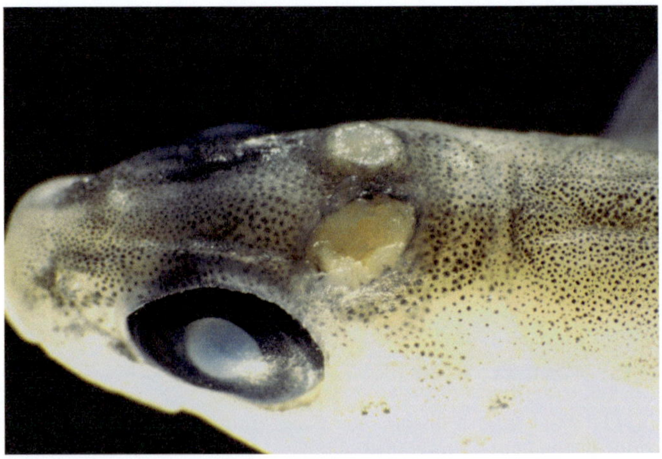

Fig. 10.36 Cranial nodules in Atlantic salmon parr

occasionally round with a smooth, non-pigmented surface and measuring up to 1.5 mm in diameter. Mortality attributed to this defect has ranged from 10 to 15 % in certain egg batches. Growth of the surviving fry is not impaired and behaviour appears normal.

Histologically the cerebellum appears normal, but displaced dorsally. Karyorhectic Malpighian cells are absent from the epidermis and the meninges appear normal. The molecular and granular layers of the cerebellum are displaced upwards towards the frontal plate with no inflammatory reaction or changes in the Purkinje cell layer. There is no apparent contact of the cerebellum with the water, or evidence of infection. The eye has been shown to be affected, with the retina appearing excessively folded with a decrease in the volume of the vitreous chamber. The aetiology of this condition is currently unknown. Visual observation and histology are used to identify these nodules.

Jaw (mandible) deformities have been documented in wild fish but frequencies are greater in hatchery populations. Deformities can affect both the maxilla and/or the mandible, resulting in a short or long lower jaw deformity. The latter can also be displaced laterally. These abnormalities can be induced during embryonic and post-embryonic periods of life and are often lethal with over 80 % of the affected larvae dying.

Further causes of deformities have been attributed to a genetic or environmental origin, the result of adverse environmental changes, phosphorous deficiency, excessive cartilage deposition and physical injury.

Fig. 10.37 Mandibular ankylosis (ventral deviation) of the mandible in farmed Atlantic salmon

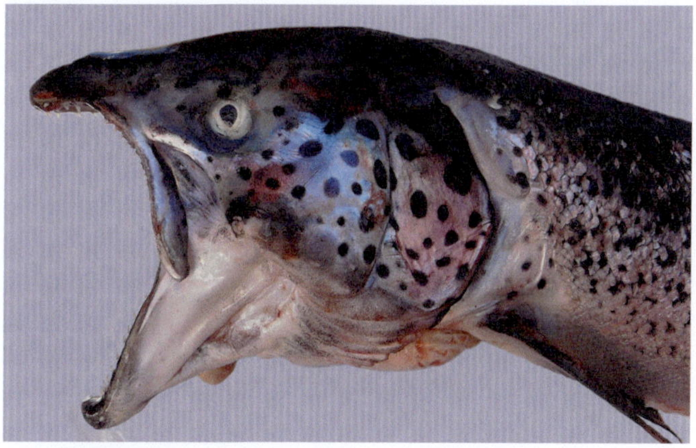

Fig. 10.38 'Screamer disease' in farmed Atlantic salmon

10.7.1 Mandibular Ankylosis

Mandibular ankylosis 'gape jaw' in farmed Atlantic salmon results in a permanently fixed wide open mouth and flared opercula due to ankylosis of the mandibular articulation. Consequently fish cannot close the mouth and have to swim continuously to irrigate the gills (ram ventilation). The condition is associated with the tooth-bearing, dentary bone and glosso-hyal (lingual plate) which curves downward, in a region approximately two-thirds from the anterior end of the jaw (Fig. 10.37). A localized dysplastic reaction involves the Meckel's cartilage. The bones of the upper jaw including the premaxilla, lacrymal and maxilla appear normal. In some fish there is a lateral displacement and twisting of the articular bone on one side of the fish. This results in the quadrate bone pushing against the body wall and a separation of the branchiostegal rays. The absence of supporting cartilage results in a downward displacement of the jaw which became more apparent as fish grow larger. The hypertrophic reaction is considered to be a compensatory mechanism for the deformity. Deformities of the lower jaw generally impede swimming and feeding activity with consequent reduction in their mean weight. X-ray images show that this jaw deformity results from incomplete ossification within the Meckel's cartilage and displacement of the angular bone as such represent a serious welfare issue and believed to be linked to a phosphorus imbalance (Fig. 10.38).

10.7.2 Pug Head

'Pug head' results from an under development or hypoplasia of the upper jaw (maxilla) and consequently the mandible appears over developed. Incubation temperature appears to be a contributing factor (Fig. 10.39).

10.7.3 Microstomia

Microstomia or small mouth is seen infrequently in particular groups of fish and thought to be congenital in origin (Fig. 10.40).

Fig. 10.39 Pug-head condition of farmed Atlantic salmon

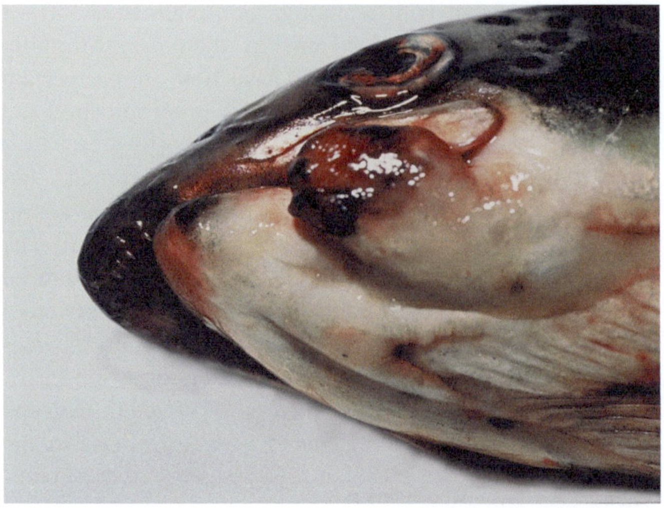

Fig. 10.41 Double jaw deformity in farmed Atlantic salmon resulting from displacement of the lower end of the hyoid arch

Fig. 10.40 Microstomia in farmed Atlantic salmon

Fig. 10.42 Exposed fused vertebrae in farmed Atlantic salmon

10.7.4 Double Mouth Deformity

Double mouth deformity results from displacement of the lower end of the hyoid arch downwards and backwards through a gap in the mouth floor (Fig. 10.41). This occurs because the protractor muscles of the arch are not functioning correctly, usually, as the result of accidental injury. The retractor muscles, thus left unopposed, pull the lower end of the arch into the deformed position. The cause is speculative but could be due to air being trapped in the floor of the mouth following obstruction of the pneumatic duct. Gross observation is sufficient to confirm these deformities.

10.8 Deformities of the Vertebral Column

Skeletal anomalies and deformities, particularly those of the vertebral column, have been observed in a number of species of wild and farmed fish. These deformities can take many forms such as compression and ossification of the vertebral joints, occurrence of compressed and fused vertebrae caudal to the dorsal fin and excessive proliferation of collagen, resulting in grossly evident increased thickness of the spinal column, often where no single cause can be established (Fig. 10.42). A linkage with pollution exists for some spinal deformities in feral fish and in farmed fish, spinal deformities are commonly reported production-related diseases.

The vertebral column is composed of bone, cartilage and connective tissue and constantly remodelled and adjusted to the environment, physical challenges and physiology. Specific terminology may be used to describe different forms and variations of vertebral abnormalities and include kyphosis, lordosis and scoliosis. Ankylosing lesions can affect the entire spine or individual sections, giving the fish a characteristic appearance (e.g. humpback, hunchback; anterior part, short-tails; posterior part or the entire spine ('short fish')). Combinations of the above may also occur. The impact of these abnormalities will always be impaired swimming capability, hence more energy is used for locomotion and feeding and therefore vertebral column abnormalities lesions are often protracted. Such fish have reduced stress tolerance, are more susceptible to physiological imbalance and are consequently always inferior to fish with normal function. Wild fish with such lesions will typically be poorer in the competition for food and territory and will also be readily predated.

The aetiology may be complex including congenital (e.g. inbreeding, genetic factors), idiopathic or acquired (e.g. infectious, nutritional or physical). Under farming conditions, vertebral column abnormalities has been associated alone or in combination, to different factors such as parasitic infections, nutritional deficiency and imbalance (e.g. phosphorous and vitamin C), vaccination, elevated temperature (during egg incubation and yolk-sac period), adverse environmental conditions such as hypoxia, exposure to toxicants, fast growth, or obstruction of the pneumatic duct.

Vertebral column malformations are commonly diagnosed by gross signs and X-ray, while the exact aetiology may be a greater challenge to solve. An exemption might be the parasite aetiology, where examination of homogenized cartilage and gill arches are needed to exclude *Myxobolus* spores (see Fig. 9.10)

10.9 Congenital Malformations

Congenital malformations occur in both wild and farmed progeny of salmonid fish and different types may occur simultaneously in the same fish. Examples include hypoplasia of the swim bladder, absence of pyloric caeca and incorrectly placed liver and/or spleen (Fig. 10.43). In wild populations, fry with malformations typically die early, while they may survive for a while under the protected farming environment. The prevalence is generally low within a population, but may be highly variable between different parents. Most malformations are believed to be the result of genetic factors and aberrations, sometimes combined with unsuitable environmental factors during embryonic development, like hypoxia, stress factors and infections. It is difficult or impossible to determine the exact interrelationship between the different aetiologies and to pinpoint one particular cause. Embryogenesis is a complex process and abnormally developed individuals will seldom be visible before hatching. Variation in egg size, discolouration, pseudo-albinism, twins, and duplication of heads and tails typically become evident at hatching. Uni- or bilateral anopthalmia (Fig. 10.44) and abnormally small eyes (micropthalmia) are also considered to be of genetic origin. Several skeletal malformations, localized dysplasia and enlarged, duplicated or absent fins are examples of conditions with presumed genetic origin (Fig. 10.45).

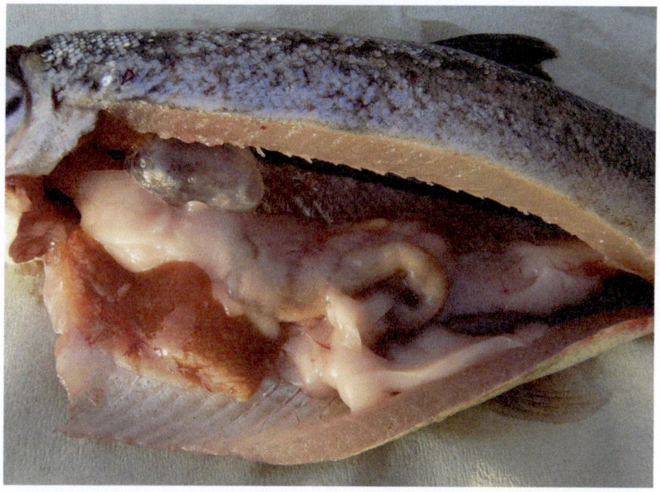

Fig. 10.43 Hypoplastic development of the swim bladder, aberrant location of the liver (situs inversus hepatis) and aplasia of pyloric caeca

Fig. 10.44 Anopthalmia in farmed Atlantic salmon

Fig. 10.45 Tail abnormality (double tail) in farmed Atlantic salmon

Fig. 10.46 Compressed and fused vertebrae in farmed rainbow trout

Fig. 10.47 Lateral compression and fusion of vertebrae from farmed rainbow trout

Abnormalities of the vertebral column, including kyphosis, lordosis, scoliosis, coiled vertebral column and 'corkscrew' fish are relevant examples (Figs. 10.46 and 10.47). Conjoined (Siamese) twins may be fused at the posterior or anterior end, or they may be fused via the yolk-sac in the pectoral region. It is rare for any of these twins to survive and typically die as the yolk sac is absorbed. However, conjoined twins (heteropagus) which survive to maturity have been described, and may occur when one of the twins is dominant (autosite) and outgrows the vestigial twin (Fig. 10.48). The two bodies fuse together along the flank, possibly with the duplication of the dorsal fin. Each fish may contain a complete set of internal organs or they are shared such as the intestinal tract, but often the organs in the vestigial twin are incomplete. It is considered that such fish arise from a single yolk and from a single blastoderm, at the margin of which two more or less separate centres of gastrulation and embryo-formation have appeared. A disturbance during the segmentation of the egg has caused this particular abnormality and therefore the result of natural causes.

10.10 Cardiac Abnormalities

A wide range of cardiac abnormalities or deviations from normal morphology may occur in salmonids. In particular, many different types have been recorded in intensively farmed Atlantic salmon and rainbow trout. There is a strong link between shape and function of the heart, and in general, a distinct triangular or pyramidal shape of the ventricle has been shown to give optimum performance. In particular, strains of salmon that undertake long migrations and negotiate high waterfalls are strongly dependant on optimum cardiac performance. The shape and size of the ventricle may be highly variable depending on species, gender, age and the fish immediate environment. For example, rainbow trout living in lakes and those living in rivers may have distinctly different cardiac shape and mass. Farmed fish are generally not exposed to severe physical challenges during normal production and handling and their requirement for a strong heart is therefore limited compared to wild fish. It is well

Fig. 10.48 Conjoined twin in an adult farmed Atlantic salmon. Notice heavy melanization and duplicated dorsal fin

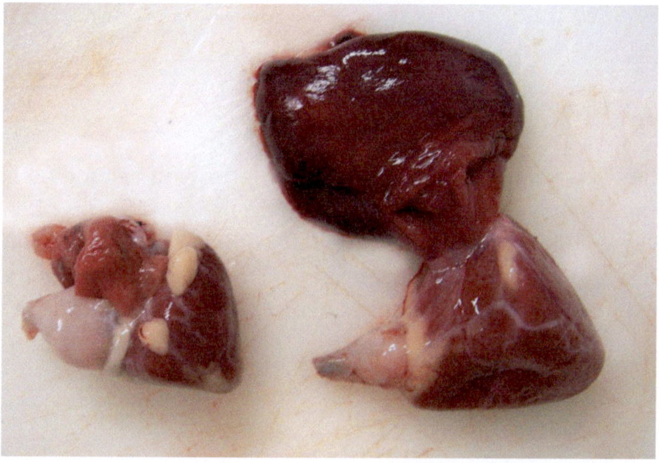

Fig. 10.49 Cardiomegaly (*right*) in farmed Atlantic salmon compared to a normal-sized heart from a fish of similar weight (*left*)

known among fish farmers that fish with different cardiac abnormalities suffer far higher mortality than normal fish during outbreaks of other diseases (e.g. pancreas disease). This clearly illustrates the compound effect of cardiac abnormalities for the general fish health. Specific defects in salmonid hearts are discussed below.

10.10.1 Cardiomegaly

Cardiomegaly of Atlantic salmon occurs in production fish in sea water and maybe linked to some aspect of the production process (Fig. 10.49). In some instances, atriomegaly may be recorded (Fig. 10.50). Such fish are very susceptible to stress and will often die during stressful operations like grading, transport or bath treatments.

Fig. 10.50 Severely dilated atrium (atriomegaly) in farmed Atlantic salmon with cardiac failure

One type of cardiomegaly in farmed rainbow trout has been linked to a spontaneous glycogen-storage disease. Clinical signs indicated heart failure with abnormal behaviour, exophthalmia, distended abdomen and ventral skin petechiae. Necropsy shows alterations in cardiac shape with distended atria and rounded ventricles. Microscopically, the compact wall of the ventricle is absent, or thinner than normal.

10.10.2 Absent Septum Transversum (Ectopia Cordis)

Absent septum transversum (ectopia cordis) of Atlantic salmon is characterized by total or partial absence of the septum transversum that separates the pericardial and the abdominal cavity. When the caudal wall of the pericardial cavity is absent, the heart will be found protruding into the abdominal cavity. The most common aberrant locations are ventral to the liver, or tilted dorsally cranial to the organ. In the latter, the apex of the heart is pushed caudo-dorsally or dorsally. In both cases, the shape of the heart is typically changed into a sac- or bean-shaped structure with resultant compromise of function (Fig. 10.51). The ventral aorta and the bulbus arteriosus will also be stretched and more or less angled in comparison to its normal straight alignment with the ventricle The aberrant heart typically makes an impression in the hepatic parenchyma and adhesions may occur between the heart and the liver. The condition results in restrictions on cardiac performance and reduced stress tolerance and swimming stamina. This condition is associated with high temperatures during the incubation of the eggs and during the yolk-sac period. Hyperthermia results in atrial natriuretic peptide (ANP) expression during critical

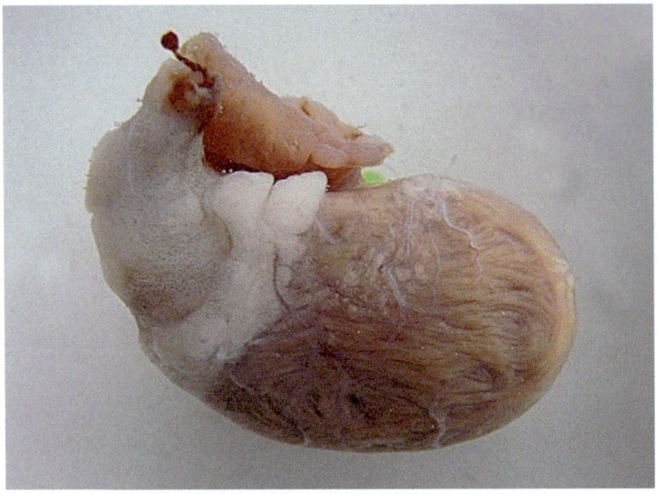

Fig. 10.51 Bean-shaped ventricle from farmed Atlantic salmon with aplastic septum transversum

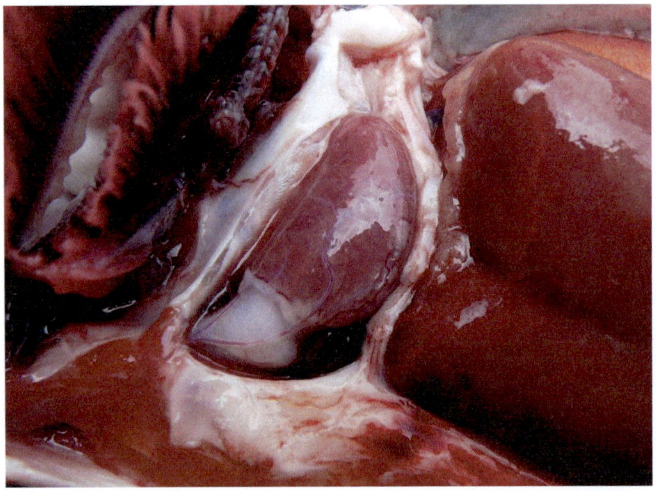

Fig. 10.52 Situs inversus of the heart in farmed Atlantic salmon. The apex of the ventricle is tilted upwards. Note the septum transversum is intact

periods of embryonic heart formation resulting in the abnormalities mentioned plus reduction in cardiosomatic index. Therefore optimal incubation temperature should be kept at <9 °C.

10.10.3 Situs Inversus Cordis

Situs inversus cordis (including 'bean hearts') is an aberrant location of the heart within an otherwise normal pericardial cavity. The shape of the ventricle will be altered and the bulbus misaligned as in fish with absent septum transversum, resulting in suboptimal cardiac function (Fig. 10.52).

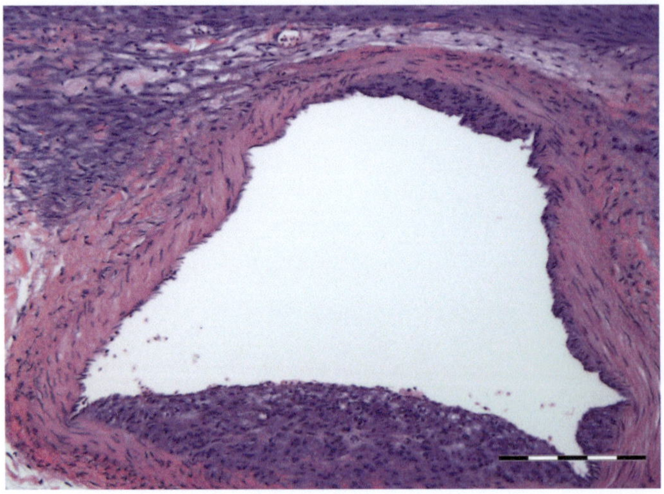

Fig. 10.53 Myointimal hyperplasia (arteriosclerosis) bulging into the vascular lumen in Atlantic salmon. Bar = 200μm

10.10.4 Skewed Ventricle

Skewed ventricle, while the ventricle in wild salmonids is fairly symmetrical when seen from the caudal side, it is often skewed in farmed fish, particularly the older individuals in the population. The significance of this is unknown, but a symmetrical heart is believed to be more effective than those with an aberrant shape.

10.10.5 Arteriosclerosis

Arteriosclerosis has been described as a 'fact of life', as this condition appears to occur in a majority of maturing and spawning fish, both wild and farmed, at a particular size and weight. Lesions are characterized by smooth muscle cells proliferating through the broken elastic lamina into the vascular lumen, therefore a proportion is obliterated. Lesions may occlude the lumen of the vessel and are typically confined to the part of the coronary artery close to the bifurcation on the ventral side of bulbus arteriosus (Fig. 10.53). Age and growth rate are probably the most important factors in the development of arteriosclerosis, but they can be modified both by sex hormones and diet composition. Farmed salmon grow faster than their wild counterparts, particularly during the freshwater phase, and therefore, accumulate lesions at a faster rate. The significance of partial or total obliteration of the coronary artery has been experimentally studied by ligation of the coronary artery. Fish with restricted blood flow cannot swim as fast as control fish and ventral aortic pressure is reduced. The most important effect(s) of coronary lesions are therefore probably during suboptimal environmental

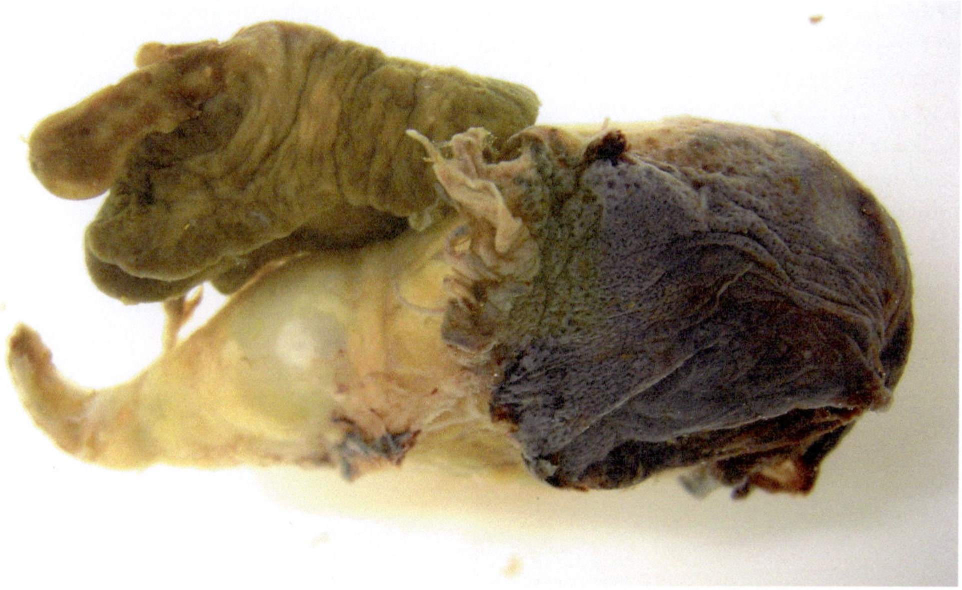

Fig. 10.54 Organized blood clot/haemopericardium in seawater-farmed rainbow trout

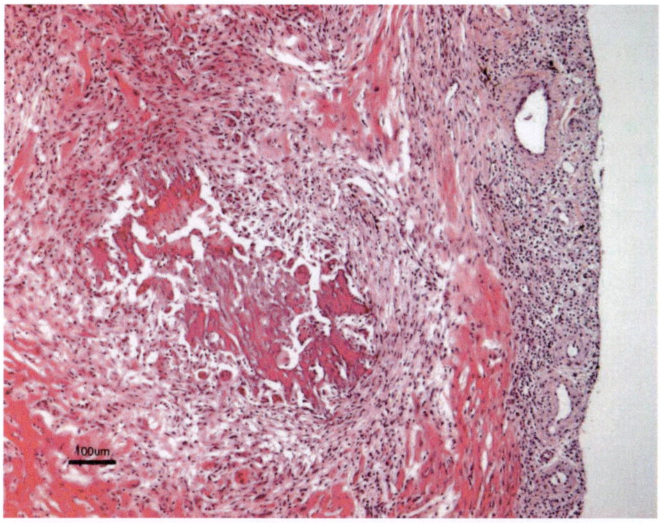

Fig. 10.55 Degeneration and dystrophic calcification of outer compact myocardium following arteriosclerosis in farmed Atlantic salmon broodstock. Low power

conditions and during upstream migration in wild fish, or handling, crowding and grading in farmed fish (Fig. 10.54). Fish recovering from full occlusion of the coronary artery may develop dystrophic calcification in parts of the outer compact myocardium and recanalization/neovascularization around the obliterated arteriosclerotic lesion (Fig. 10.55). Diagnosis of recent myocardial necrosis following arteriosclerosis is difficult as the fish usually die before obvious lesions develop.

10.10.6 Hypoplasia of the Outer Compact Myocardium

Hypoplasia of the outer compact myocardium has been diagnosed in both farmed rainbow trout and Atlantic salmon. The compact layer is either thin or absent, the ventricle is rounded and the atrium distended resulting in cardiac failure. There may be extensive fatty infiltration of the epicardium resulting in a condition known as 'white heart'. Affected fish may display severe exophthalmia, ascites and even protrusion of the heart through a hernia in the ventral part of the pericardium. The condition has occurred in several hatcheries and few, if any, fish survive to the sea water phase. The cause(s) of the condition is unknown.

10.10.7 Aneurysms

Aneurysms are defective, localized, blood-filled dilatation of a blood vessel which may be observed as 'balloon-like' protrusion (see Fig. 4.20). These are filled with blood or clots and may interfere with normal function due to their size. A few cases have anecdotally been reported which appear to be linked to certain fish strains.

10.10.8 Sub-endocardial Fibrosis

Sub-endocardial fibrosis or fibro-elastosis in farmed Atlantic salmon is characterized by elastic fibres between the endocardium and the cardiac myofibres and can be identified using Elastin-van Gieson staining. Lesions may be found in both atrium and ventricle (Fig. 10.56). Associated inflammatory response may be variable depending on the cause(s) of the condition. Some cases appear to be idiopathic with no other abnormalities recorded. However, several cases have been identified associated with other chronic cardiac diseases such as piscine myocarditis virus or cardiomyopathy syndrome, pancreas disease and piscine reovirus. In these cases, the lesions are believed to be attempts to repair damaged myocardium.

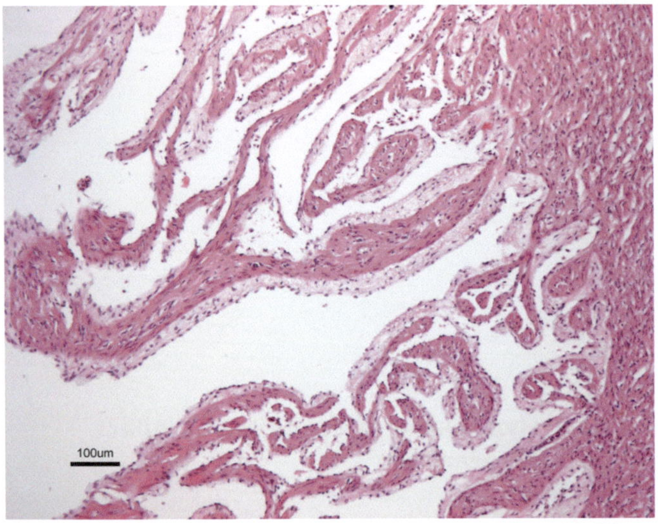

Fig. 10.56 Sub-endocardial fibrosis in the spongy ventricle of farmed Atlantic salmon

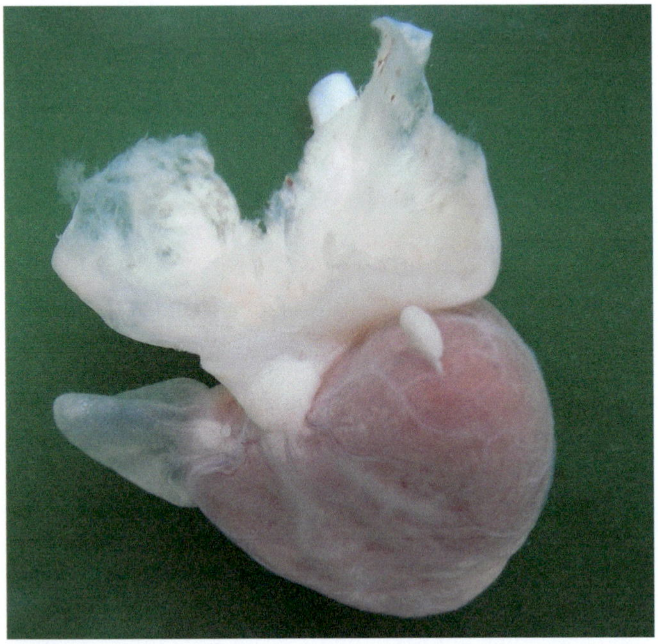

Fig. 10.57 Ventricular cyst in farmed Atlantic salmon. The cyst is filled with a clear, yellowish fluid

10.10.9 Ventricular Cysts

Ventricular cysts are epithelial lined, benign and observed in both farmed and wild Atlantic salmon (Fig. 10.57). The cysts usually originate from the edges of the ventricle and are filled with a clear fluid. At least for farmed stock the cause is considered to be related to high temperature during incubation of the eggs.

10.11 Predators

Both farmed and wild salmonids are subject to predator damage from birds, otters, mink, seals, bears, other fish and anthropogenic activity. Out-migrant wild smolts are vulnerable to predator attacks as are farmed smolts recently transferred to sea water. In many small rivers and creeks, up migrant fish are also frequently attacked by predators. Wounds inflicted by birds such as cormorants, herons, mergansers and gulls are typically more or less vertical streaks, often parallel lesions a few centimetres apart, and frequently with corresponding lesions on the opposite side reflecting the grip of a bird's beak (Fig. 10.58). Often birds try to take fish bigger than they can handle, the fish wiggles free and a wound of variable size and depth results. Beaks of cormorants and mergansers have a pointed 'tooth' that will often penetrate skin and muscle. Surviving fish may develop peritonitis and subsequent septicaemia where the wound

Fig. 10.58 Bird stab penetrating to the abdominal cavity in sea-run Arctic char

Fig. 10.59 Severe penetrating abdominal damage resulting from seal attack in Atlantic salmon

Fig. 10.60 Sea lamprey; (**a**) Characteristic circular lesion on body of sea trout. (**b**) Suction cup-like mouth used to attach to the skin of a fish, note keratinized teeth

penetrates to the abdominal cavity. Healing wounds are typically characterized by scale loss and melanization. Wounds inflicted by mammals can be variable, but in surviving fish punctures or raking lesions may occur. Seals often eat the head and/or anterior part of the abdomen (including the liver) and leave the remaining portion (Fig. 10.59).

A special type of lesion is caused by lampreys (Cyclostomata); ectoparasitic primitive jawless fish. They occur in both fresh- and sea water, the latter being anadromous. In particular, the sea lamprey may cause well-circumscribed lesions on the flank through the action of their circular suctorial mouth with horny teeth (Fig. 10.60). Considerable efforts are being made to control the

Fig. 10.61 Damage to the left maxilla from previous catch and release of wild Atlantic salmon

Fig. 10.62 Damage to both maxillae in grayling after previous catch and release; (*left*). Both maxillae have been tilted medially. The mandible has been removed. A grayling with normal maxillae is shown for comparison (*right*)

population of sea lampreys in the Great Lakes area where they have caused considerable damage to the fish population. Lampreys feed on muscular tissue, body fluids and blood and may eventually emaciate the host. Histopathological lesions include dermal penetration, muscle necrosis, oedema and haemorrhage. Circular skin ulcers (e.g. winter ulcers or pseudomonad infections) may pose differential diagnosis challenges.

10.12 Lesions Associated with Angling

Lesions associated with angling mainly result from physical hook damage to mouth and jaw, as well as scale loss and epidermal damage associated to handling and netting. In particular, these lesions may occur in a large proportion of populations when catch and release is the norm. Fish may be caught repeatedly in one season and reflected in different or accumulative lesions. Common hook-associated lesions include tearing, displacement or loss of maxillae, lesions in the eyes and hooks embedded in the oesophagus or stomach following deep hooking. Lesions on gill arches have also been reported. In the process of playing and landing the fish, damage to skin may result from contact with gravel, rocks, and landing net or contact with dry hands (Figs. 10.61 and 10.62). Depending on the extent and severity, lesions may progress into more severe secondary infections e.g. *Saprolegnia* and opportunistic bacterial infections e.g. *Pseudomonas* and *Aeromonas* spp. Lesions and mortality can be reduced by using barbless hooks, avoiding fishing at high water temperatures and correct handling of the fish.

Further Reading

Anon (2010) Risk assessment of catch and release. Opinion of the panel on animal health and welfare of the Norwegian Scientific Committee for Food Safety. Norwegian Scientific Committee for Food Safety (VKM), Oslo. ISBN 978 82 8082 396 0

Baxter EJ, Rodger HD, McAllen R, Doyle TK (2011) Gill disorders in marine-farmed salmon: investigating the role of hydrozoan jellyfish. Aquac Environ Interact 1:245–257

Berg A, Rødseth OM, Tangerås A, Hansen T (2006) Time of vaccination influences development of adhesions, growth and spinal deformities in Atlantic salmon *Salmo salar*. Dis Aquat Organ 69:239–248

Božidar S, Marko B, Zoran Z, Vesna D (2012) Histological methods in the assessment of different feed effects on liver and intestine of fish. J Agric Sci 56:87–100

Bruno D (1990a) Jaw deformity associated with farmed Atlantic salmon (*Salmo salar*). Vet Rec 126:402–403

Bruno DW (1990b) Occurrence of a conjoined twin among farmed Atlantic salmon, *Salmo salar* L. J Fish Biol 37:501–502

Bruno DW (1997) Cranial nodules in farmed Atlantic salmon, *Salmo salar* L., fry. J Appl Ichthyol 13:47–48

Bruno DW, Ellis AE (1985) Mortalities in Atlantic salmon associated with the jellyfish *Phialella quadrata*. Bull Eur Assoc Fish Pathol 5:64–65

Bruno DW, Ellis AE (1986) Multiple hepatic cysts in farmed Atlantic salmon, *Salmo salar* L. J Fish Dis 9:79–81

Bruno DW, Dear G, Seaton DD (1989) Mortality associated with phytoplankton blooms among farmed Atlantic salmon, *Salmo salar* L., in Scotland. Aquaculture 78:217–222

Bullock AM, Roberts RJ (1981) Sunburn lesions in salmonid fry: a clinical and histopathological report. J Fish Dis 4:271–275

Bylund G, Lerche O (1995) Thiamine therapy of M74 affected fry of Atlantic salmon *Salmo salar*. Bull Eur Assoc Fish Pathol 15:93–97

Clary JR, Clary SD (1978) Swim bladder stress syndrome. Salmonid (March/April):8–9

Colquhoun DJ, Skjerve E, Poppe TT (1988) *Pseudomonas fluorescens*, infectious pancreatic necrosis virus and environmental stress as potential factors in the development of vaccine related adhesions in Atlantic salmon, *Salmo salar* L. J Fish Dis 21:355–364

Farrell AP (2002) Coronary arteriosclerosis in salmon: growing old or growing fast? Comp Biochem Physiol A 132:723–735
Farrell AP, Steffensen JF (1987) Coronary ligation reduces maximum sustained swimming speed in chinook salmon, *Oncorhynchus tschawytscha*. Comp Biochem Physiol 87A:35–37
Farrell AP, Saunders RL, Freeman HC, Mommsen TP (1986) Arteriosclerosis in Atlantic salmon. Effects of dietary cholesterol and maturation. Arteriosclerosis 6:453–461
Farrell AP, Johansen JA, Saunders RL (1990) Coronary lesions in Pacific salmonids. J Fish Dis 13:97–100
Fisher JP, Fitzsimmons JD, Combs GF, Spitsbergen JM (1996) Naturally occurring thiamine deficiency causing reproductive failure in Finger Lakes Atlantic salmon and Great Lakes lake trout. Trans Am Fish Soc 125:67–178
Fitzsimmons JD, Brown SB, Hnath JG (1999) A review of early mortality syndrome in Great Lakes salmonids and its relationship with thiamine. Ambio 28:9–15
Fjelldal PG, Hansen T, Breck O, Ørnsrud R, Lock E-J, Waagbø R, Wargelius A, Witten PE (2012) Vertebral deformities in farmed Atlantic salmon (*Salmo salar* L.) – etiology and pathology. J Appl Icthol 28:433–440
Frischknecht R, Wahli T, Meier W (1994) Comparison of pathological changes due to deficiency of vitamin C, vitamin E and combinations of vitamins C and E in rainbow trout, *Oncorhynchus mykiss* (Walbaum). J Fish Dis 17:31–45
Gamperl AK, Farrell AP (2004) Cardiac plasticity in fishes: environmental influences and intraspecific differences. J Exp Biol 207:2539–2550
Graham MS, Farrell AP (1992) Environmental influences on cardiovascular variables in rainbow trout, *Oncorhynchus mykiss* (Walbaum). J Fish Biol 41:851–858
Grini A, Hansen T, Berg A, Wasgelius A, Fjelldal PG (2011) The effect of water temperature on vertebral deformities and vaccine-induced abdominal lesions in Atlantic salmon, *Salmo salar*. J Fish Dis 34:531–546
Karlsson L, Petterson E, Hedenskog M, Børjesson H, Eriksson R (1996) Biological factors affecting the incidence of M74. In: Report from the second workshop on reproduction disturbances in fish, Stockholm, Sweden, 20–23 November 1995. Swedish Environmental Protection Agency Report 4534 p 25
Koppang EO, Haugarvoll E, Hordvik I, Poppe TT, Bjerkås I (2004) Granulomatous uveitis associated with vaccination in the Atlantic salmon. Vet Pathol 41:122–130
Koppang EO, Haugarvoll E, Hordvik I, Aune L, Poppe TT (2005) Vaccine-associated granulomatous inflammation and melanin accumulation in Atlantic salmon, *Salmo salar* L., white muscle. J Fish Dis 28:13–22
Koppang EO, Bjerkås I, Haugarvoll E, Chan EKL, Szabo NJ, Ono N, Akikusa B, Jirillo E, Poppe TT, Sveier H, Tørud B, Satoh M (2008) Vaccination-induced systemic autoimmunity in farmed Atlantic salmon. J Immunol 181:4807–4814
Kuramoto T, Arima K, Kawakami S, Shimizu N, Nakawatari A, Hasegawa M, Hirama S, Moriyama K, Yotsugi K (1988) On the early development and the occurrence of twin malformation in chum salmon eggs and fry. Sci Rep Hakkaido Salm Hatch 42:59–73
Kvellestad A, Høie S, Thorud K, Thørud B, Lyngøy A (2000) Platyspondyly and shortness of vertebral column in farmed Atlantic salmon *Salmo salar* in Norway – description and interpretation of pathologic changes. Dis Aquat Organ 39:97–108
Lall SP, Lewis-McCrea LM (2007) Role of nutrients in skeletal metabolism and pathology in fish- an overview. Aquaculture 267:3–19
MacKenzie LA, Smith KF, Rhodes LL, Brown A, Langi V (2011) Mortalities of sea-cage salmon (*Oncorhynchus tshawytscha*) due to a bloom of *Pseudochattonella verruculosa* (Dictyochophyceae) in Queen Charlotte Sound, New Zealand. Harmful Algae 11:45–53
McArdle J, Bullock AM (1987) Solar ultraviolet radiation as a causal factor of "summer syndrome" in cage-reared Atlantic salmon, *Salmo salar* L.: a clinical and histopathological study. J Fish Dis 10:255–264
Meka JM (2004) The influence of hook type, angler experience, and fish size on injury rates and the duration of capture in an Alaskan catch-and-release rainbow trout fishery. N Am J Fish Manag 24:1309–1321
Müller R (1983) Coronary arteriosclerosis and thyroid hyperplasia in spawning coho salmon (*Oncorhynchus kisutch*) from Lake Ontario. Acta Zool Pathol Ant 77:3–12
Mutoloki S, Brudeseth B, Reite OB, Evensen Ø (2006) The contribution of *Aeromonas salmonicida* extracellular products to the induction of inflammation in Atlantic salmon (*Salmo salar* L.) following vaccination with oil-based vaccines. Fish Shellfish Immunol 20:1–11
Oliva-Teles A (2012) Nutrition and health of aquaculture fish. J Fish Dis 35:83–108
Poppe TT, Breck O (1997) Pathology of Atlantic salmon *Salmo salar* intraperitoneally immunized with oil-adjuvanted vaccine. A case report. Dis Aquat Organ 29:219–226
Poppe TT, Taksdal T (2000) Ventricular hypoplasia in farmed Atlantic salmon *Salmo salar*. Dis Aquat Organ 42:35–40
Poppe TT, Tørud B (2009) Intramyocardial dissecting haemorrhage in farmed rainbow trout *Oncorhynchus mykiss* Walbaum. J Fish Dis 32:1041–1043
Poppe TT, Midtlyng PJ, Sande RD (1988) Examination of abdominal organs and diagnosis of deficient septum transversum in Atlantic salmon, *Salmo salar* L., using diagnostic ultrasound imaging. J Fish Dis 21:67–72
Poppe TT, Helberg H, Griffiths D, Meldal H (1997) Swim bladder abnormality in farmed Atlantic salmon *Salmo salar*. Dis Aquat Organ 30:73–76
Poppe TT, Johansen R, Tørud B (2002) Cardiac abnormality with associated hernia in farmed rainbow trout *Oncorhynchus mykiss*. Dis Aquat Organ 50:153–155
Poppe TT, Johansen R, Gunnes G, Tørud B (2003) Heart morphology in wild and farmed Atlantic salmon *Salmo salar* and rainbow trout *Oncorhynchus mykiss*. Dis Aquat Organ 57:103–108
Poppe TT, Taksdal T, Bergtun PH (2007) Suspected myocardial necrosis in farmed Atlantic salmon, *Salmo salar* L.: a field case. J Fish Dis 30:615–620
Poppe TT, Bornø G, Iversen L, Myklebust E (2009) Idiopathic cardiac pathology in salt water-farmed rainbow trout, *Oncorhynchus mykiss* (Walbaum). J Fish Dis 32:807–810
Poynton SL (1987) Vertebral column abnormalities in brown trout, *Salmo trutta* L. J Fish Dis 10:53–57
Roberts RJ (1993) Ulcerative dermal necrosis (UDN) in wild salmonids. Fish Res 17:3–14
Roberts RJ, Shearer WM, Munro ALS, Elson KGR (1970) Studies on ulcerative dermal necrosis of salmonids. II: The sequential pathology of the lesions. J Fish Biol 2:373–378
Roberts RJ, Hardy RW, Sugiura SH (2001) Screamer disease in Atlantic salmon, *Salmo salar* L., in Chile. J Fish Dis 24:543–549
Salte R, Norberg K (1991) Disseminated intravascular coagulation in farmed Atlantic salmon, *Salmo salar* L.: evidence of consumptive coagulopathy. J Fish Dis 14:63–66
Sánchez RC, Obregón EB, Ruaco MJ (2011) Hypoxia is like an ethiological factor in vertebral column deformity of salmon (*Salmo salar*). Aquaculture 316:13–19
Saunders RL, Farrell AP, Knox DE (1992) Progression of coronary arterial lesions in Atlantic salmon (*Salmo salar*) as a function of growth rate. Can J Fish Aquat Sci 49:878–884
Segner H, Sundh H, Buchmann K, Douxfils J, Sundell KS, Mathieu C, Ruane N, Jutfelt F, Toften H, Vaughan L (2011) Health of farmed fish: its relation to fish welfare and its utility as welfare indicator. Fish Physiol Biochem 38:85–105

Seierstad SL, Poppe T, Larsen S (2005) Introduction and comparison of two methods of assessment of coronary lesions in Atlantic salmon, *Salmo salar* L. J Fish Dis 28:189–197

Silverstone AM, Hammell L (2002) Spinal deformities in farmed Atlantic salmon. Can Vet J 43:782–784

Skog Eriksen M, Espmark ÅM, Poppe T, Braastad BO, Salte R, Bakken M (2008) Fluctuating asymmetry in farmed Atlantic salmon (*Salmo salar*) juveniles: also a natural matter? Environ Biol Fish 81:87–99

Speare DJ (1990) Histopathology and ultrastructure of ocular lesions associated with gas bubble disease in salmonids. J Comp Pathol 103:421–432

Staurnes M, Andorsdottir G, Sundby A (1990) Distended, water-filled stomach in sea-farmed rainbow trout. Aquaculture 90:333–343

Steinum T, Kvellestad A, Colquhoun DJ, Heum M, Mohammad S, Grøntvedt RN, Falk K (2010) Microbial and pathological findings in farmed Atlantic salmon *Salmo salar* with proliferative gill inflammation. Dis Aquat Organ 91:201–211

Steucke EW Jr, Allison LH, Piper RG, Robertson R, Bowen JT (1968) Effects of light and diet on the incidence of cataract in hatchery-reared lake trout. Prog Fish Cult 30:220–226

Stewart LAE, Kadri S, Noble C, Kankainen M, Setälä J, Huntingford FA (2012) The bio-economic impact of improving fish welfare using demand feeders in Scottish Atlantic salmon smolt production. Aquac Econ Manag 16:394–398

Stradmeyer L (2008) Survival, growth and feeding of Atlantic salmon, *Salmo salar* L., smolts after transfer to salt water in relation to the failed smolt syndrome. Aquac Res 25:103–112

Takle H, Baeverfjord G, Helland S, Kjorsvik E, Andersen Ø (2006) Hyperthermia induced natriuretic peptide expression and deviant heart development in Atlantic salmon *Salmo salar* embryos. Gen Comp Endocrinol 147:118–125

Tørud B, Taksdal T, Dale OB, Kvellestad A, Poppe TT (2006) Myocardial glycogen storage disease in farmed rainbow trout, *Oncorhynchus mykiss* (Walbaum). J Fish Dis 29:535–540

Urán PA, Schrama JW, Rombout JHWM, Taverne-Thiele JJ, Obach A, Koppe W, Verreth JAJ (2009) Time-related changes of the intestinal morphology of Atlantic salmon, *Salmo salar* L., at two different soybean meal inclusion levels. J Fish Dis 32:733–744

Witten PE, Gil-Martens L, Huysseune A, Takle H, Hjelde K (2009) Towards a classification and an understanding of developmental relationships of vertebral body malformations in Atlantic salmon (*Salmo salar* L.). Aquaculture 295:6–14

Idiopathic Diseases

11

Abstract

Diseases with uncertain or compound aetiology occur in fish farming and defined as those where the aetiology is unknown, or idiopathic. However, this may be difficult to ascertain and furthermore many different conditions may occur concurrently with more well-defined diseases, such as those with an infectious aetiology. It is characteristic for many idiopathic diseases that they may not be lethal themselves, but when they affect an immunocompromised fish or occur with other agents, they may cause clinical disease and associated mortality. Similarly, the transition between production-related disease and infectious disease can be obscure.

Keywords

Idiopathic • Salmon • Trout

As with almost any farmed species, diseases with uncertain or compound aetiology also occur in fish farming and for the purpose of this book idiopathic diseases are those conditions where the aetiology is unknown. In some cases, these conditions are the result of different types of manipulations of the fish, the feed and the environmental factors in order to optimise the production and increase profitability. Due to the complexity, their aetiology may be difficult to ascertain and furthermore, many different conditions occur concurrently with more well-defined diseases, e.g. those of infectious aetiology. It is characteristic for many idiopathic diseases that they may not be lethal themselves, but when occurring together with other agents or in immunocompromised fish, they may cause clinical disease and mortality. For example, fish with shortened opercula, frayed fins or cardiac diseases, can cope with normal farming practices well, but are typically among the first to succumb during stressful events like crowding, treatments, suboptimal environment and outbreak of infectious diseases. The transition between production-related disease and infectious disease can also be obscure. Certain conditions such as skeletal malformations or metabolic disorders/dysfunctions are representatives of emerging or production disease and some examples of them (cardiac lesions, nutrition and deformities) are covered in other sections of this book. Selected idiopathic diseases are discussed below.

11.1 Ulcerative Dermal Necrosis

Losses in native Atlantic salmon and sea trout in Britain were the subject of investigation by an early fishery research project during the late 1800s and early 1900s. This work was carried out in Scotland into the so called 'Salmon disease', which is referred to as ulcerative dermal necrosis (UDN). Although many thousands of salmon died and the apparent spread between watersheds indicated an infectious cause, to date there has been no definitive agent identified. Furthermore, there are no records of UDN from farmed fish.

UDN is a chronic dermatological condition described from adult Atlantic salmon and brown trout returning to fresh water. Signs include small superficial grey-coloured head lesions above the eye, along the snout or adipose fin, often with erosion on the cranium, to deeper ulcers involving large areas principally on the top of the head (Figs. 11.1 and 11.2). In the latter stages, infection with *Saprolegnia parasitica* is frequently associated with the open wounds. In addition there

D.W. Bruno et al., *A Colour Atlas of Salmonid Diseases*,
DOI 10.1007/978-94-007-2010-7_11, © Springer Science+Business Media B.V. 2013

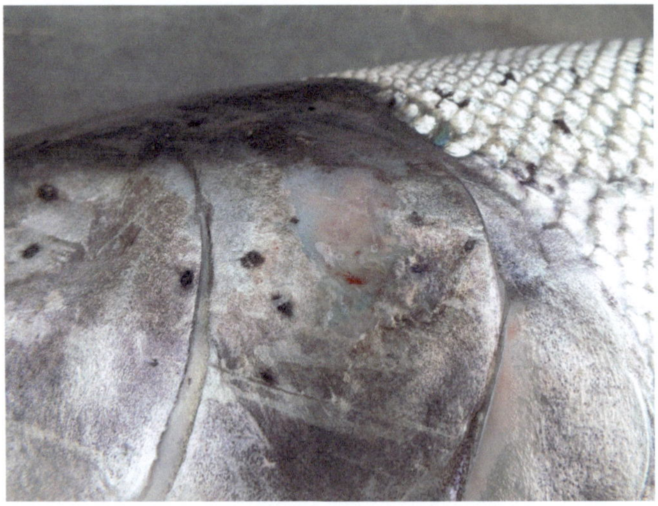

Fig. 11.1 Wild adult Atlantic salmon with early ulcerative dermal necrosis lesions on head

Fig. 11.3 Wild brown trout with ulcerative dermal necrosis lesion

Fig. 11.2 Wild adult Atlantic salmon with moderate ulcerative dermal necrosis

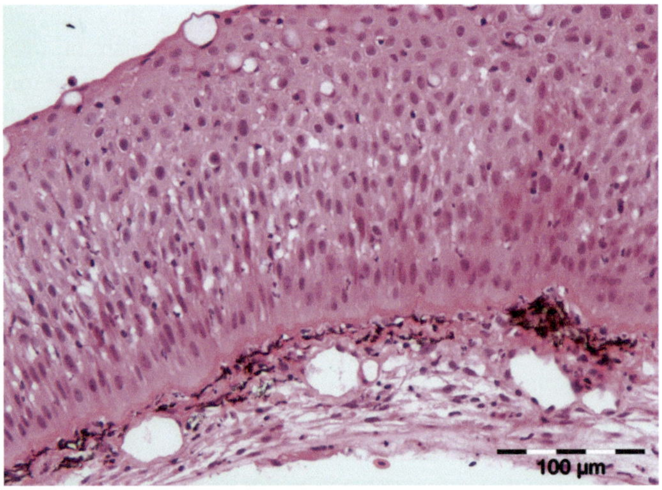

Fig. 11.4 Acantholysis and pemphigoid-like degeneration of the epidermis from Atlantic salmon with ulcerative dermal necrosis

are reports of UDN from wild brown trout (Fig. 11.3). Histologically the early, or pre-mycotic stage is limited to swelling and degeneration of melanophores below the basal layer, with acantholysis and pemphigoid-like degeneration mainly affecting the lower layers of the epidermis (Fig. 11.4). This is followed by a progressive cytolytic necrosis restricted to specific sites on the head, with foci of severe acantholysis. The pemphigoid bulla has been attributed to or linked to photosensitization. In more advanced lesions epidermal infiltration and/or necrosis, haemorrhage, dermal disarrangement, necrosis and infiltration can be observed (Fig. 11.5). The underlying skeletal muscle is not affected. In the final stage detachment or loss of the epidermal layer occurs and fungal hyphae are often detected but without any significant inflammatory response. Fish die from circulatory failure resulting from the osmotic haemodilution induced by wide areas of ulceration. The histological examination, together with the

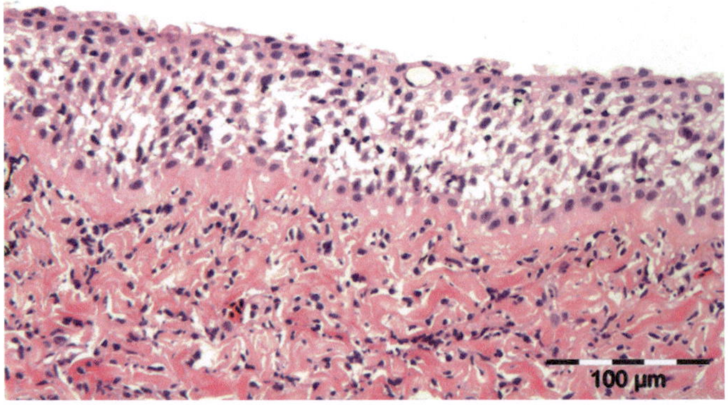

Fig. 11.5 Epidermal necrosis, dermal disarrangement and infiltration with inflammatory cells, in Atlantic with ulcerative dermal necrosis

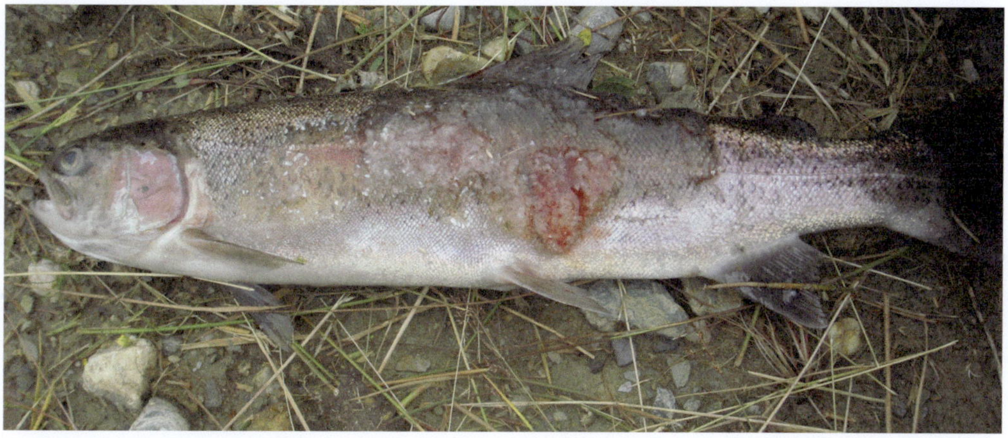

Fig. 11.6 Farmed rainbow trout with 'puffy skin'

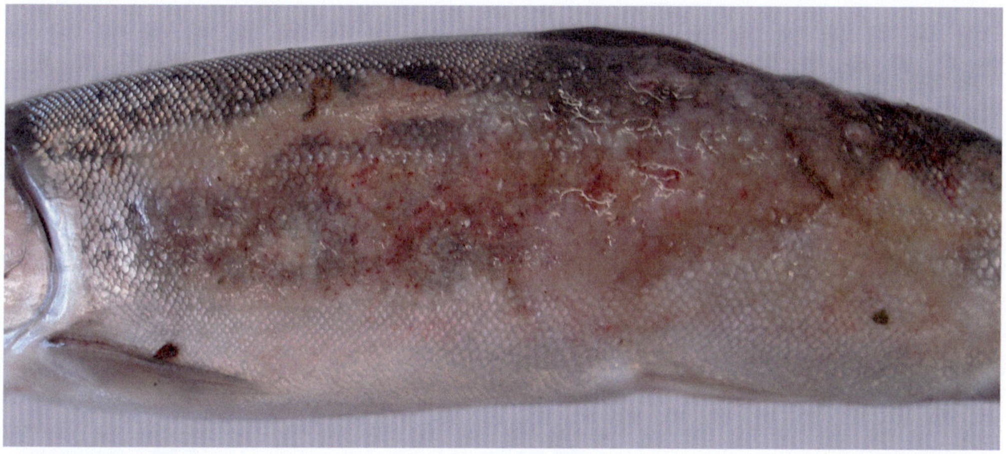

Fig. 11.7 Farmed rainbow trout showing close up of lesion associated with 'puffy skin'

macroscopic characteristics of early skin lesions is considered pathognomonic for UDN, however, the presence of hyphae reduces the likelihood of a definitive diagnosis. In the absence of a clear aetiology and an agreed case definition for this condition, it is fair to highlight that other skin conditions may also present some or all of the histological features described.

11.2 Puffy Skin

'Puffy skin' of farmed rainbow trout was first described in 1997 in Scotland and data suggests around 32 sites had reported this condition by 2011. The name is derived from the thick jelly-like mucus on the flanks of the fish, generally occurring in larger fish including triploids in freshwater (Figs. 11.6 and 11.7). Externally, fish show increase in deformities, loss of appetite and there is oedema, blistering and dark colouration, but internally the fish are within normal range. Antibiotics, brief salt dips (3 %), formalin and increased levels of vitamins in the diet, have been used to treat the condition, but overall it remains poorly researched. Histologically, there is an oedematous epidermal proliferative lesion, with and without capillary dilatation and diffuse haemorrhage (Fig. 11.8).

11.3 Polycystic Syndrome

Sporadic records of a polycystic liver or spleen have been reported from wild brown trout and farmed Atlantic salmon. Grossly affected fish show substantial abdominal distension. At necropsy, numerous non-pigmented, soft fluid-filled cysts are distributed throughout, and eventually obliterate the liver and spleen (Figs. 11.9 and 11.10). Histologically, these spaces are surrounded by a thin capsule of atrophic hepatocytes and loose connective tissue. The spaces vary considerably in diameter from microscopic to approximately 6 cm. Other

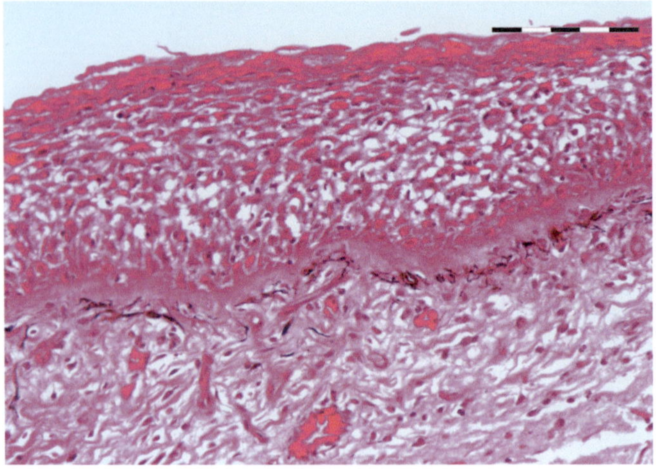

Fig. 11.8 Oedematous epidermal proliferative lesion from rainbow trout with 'puffy skin'. Medium power

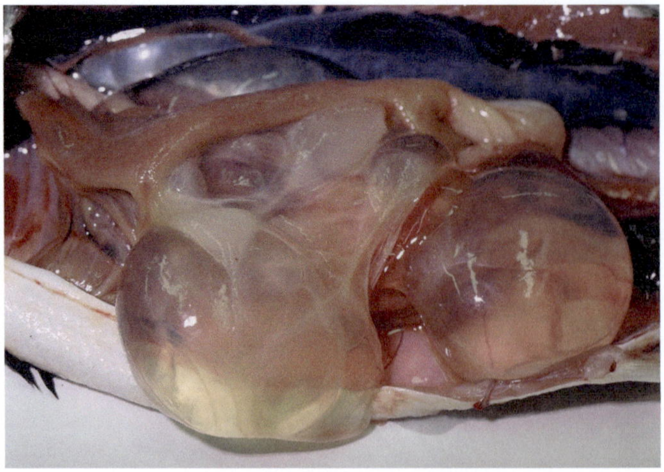

Fig. 11.9 Polycystic liver with cysts of variable size in farmed Atlantic salmon

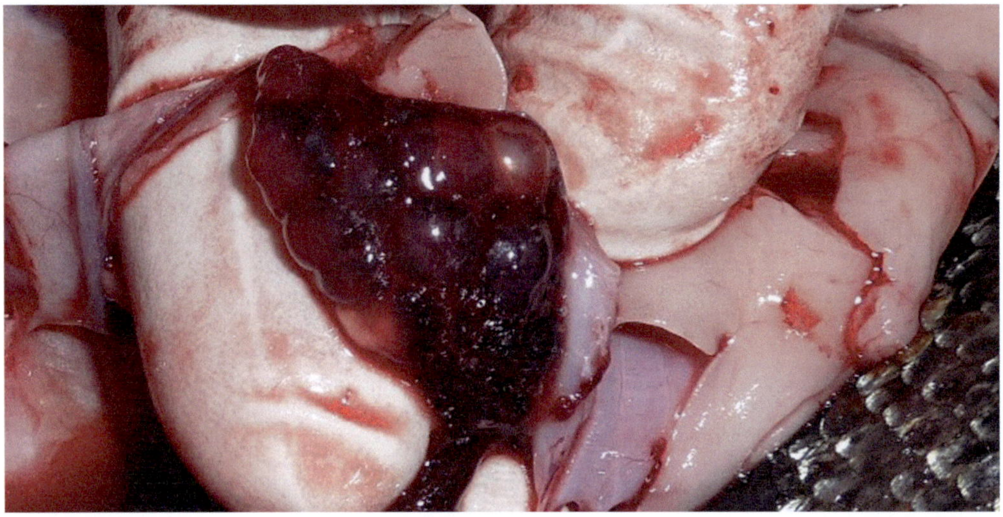

Fig. 11.10 Polycystic spleen in farmed Atlantic salmon

tissues in affected fish are generally within normal range. The cause of these conditions is unknown, but the apparent low prevalence in populations indicates the cysts might be of congenital origin. The diagnosis of polycystic condition and similar cystic conditions involving other tissues, is based on gross observations with additional information from histological assessment of stained sections.

11.4 White Eye Syndrome

White eye syndrome (or generalized soft tissue calcification) has occurred sporadically in hatcheries and smolt farms in Norway. Affected fish appear emaciated with exophthalmia and characteristic crescent-shaped white areas in front and behind the eye (Fig. 11.11). At necropsy, the body muscle is speckled with diffuse pale or white patches. Nodular white lesions may also be found in other organs including heart, kidney, liver and the gill arches (Fig. 11.12).

Histology reveals calcium deposits, extensive muscle degeneration and necrosis with calcification of body muscle, compact and spongy myocardium, cardiac valves and coronary artery. Heavy calcium deposits in the epicardium may also be found. Mineralized foci are also found in other organs such as liver, kidney, stomach wall and in the retrobulbar tissue (corpus choroidale). The aetiology is unknown, but is probably related to an alteration of the metabolism of the calcium-regulating mechanisms. Diagnosis is based upon characteristic gross and histological lesions.

Fig. 11.11 Farmed Atlantic salmon parr with white eye syndrome; characteristic crescent-shaped calcium deposits around the eyes

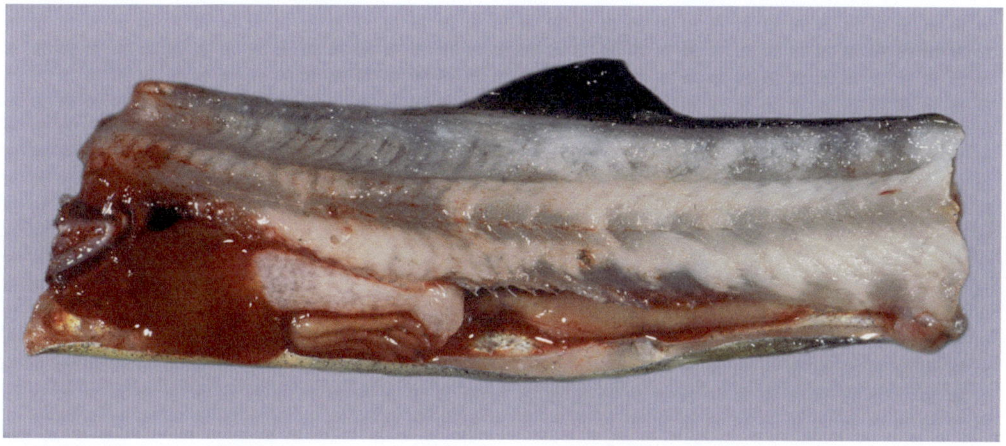

Fig. 11.12 Body muscles and intestine of farmed Atlantic salmon parr with white eye syndrome. Patchy calcium deposits are visible in white muscle and stomach wall

11.5 Haemorrhagic Smolt Syndrome

Farmed pre-smolt Atlantic salmon with clinical signs of haemorrhagic smolt syndrome (HSS) have been reported in Norway and Scotland. The seasonal condition is characterised by anaemia and extensive haemorrhage in most internal organs. Affected fish are lethargic and have pale gills, with extensive visceral and muscle petechiae and ecchymosis, but no major weight loss. Petechiae also occur on the gastrointestinal tract, swim bladder and peritoneum, heart and somatic musculature (Figs. 11.13 and 11.14). The liver can be bright yellow and mottled with petechiae. Ascites is reported in the visceral cavity.

Histological examination shows haemorrhage in most organs including pancreas, kidney and gut (Figs. 11.15, 11.16 and 11.17). The glomeruli appear degenerated and the renal tubules are full with erythrocytes. The condition is considered non-infectious as current evidence indicates there is no agent involved, however the aetiology requires further study.

11.6 Swim Bladder Stress Syndrome

Swim bladder stress syndrome may occur in several farmed species, but primarily, reported in rainbow trout (Fig. 11.18). The main feature is an over-inflation of the swim bladder and

Fig. 11.13 Farmed Atlantic salmon smolts with haemorrhagic smolt syndrome. Haemorrhage in white muscle and internal organs

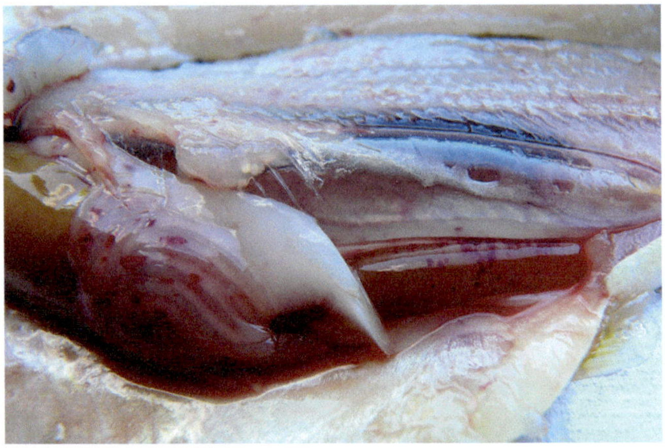

Fig. 11.14 Ascites and haemorrhage in the posterior intestine in farmed Atlantic salmon smolt with haemorrhagic smolt syndrome

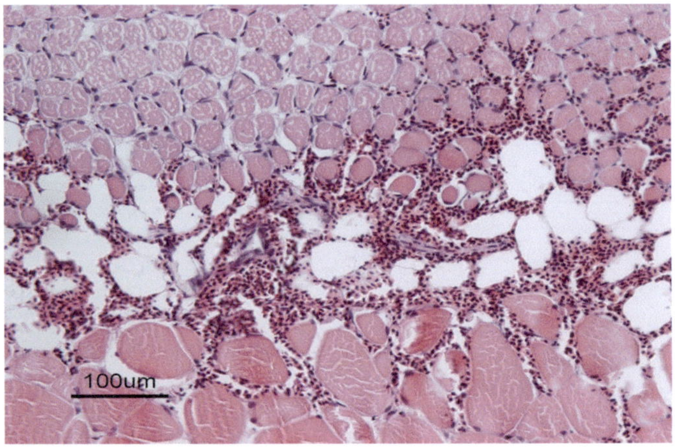

Fig. 11.15 Haemorrhage in the area between red and white muscle in farmed Atlantic salmon smolt with haemorrhagic smolt syndrome

inability to evacuate gas via the pneumatic duct. Affected fish have increased buoyancy and an altered centre of gravity causing them to swim erratically with the tail up, on their sides, or with their belly up. Fish with moderate over-inflated swim bladder may live for a long period, while severely affected fish can show a visibly enlarged abdomen and spend a lot of energy to stay in position, and may die from exhaustion. Affected fish are also more vulnerable to predators. At necropsy, the swim bladder is grossly over-inflated and may more or less displace other organs in the abdominal cavity. General stress or stress associated to inadequate water depth, is believed to be important for the development of this condition. Diagnosis is made from clinical observations and an enlarged swim bladder.

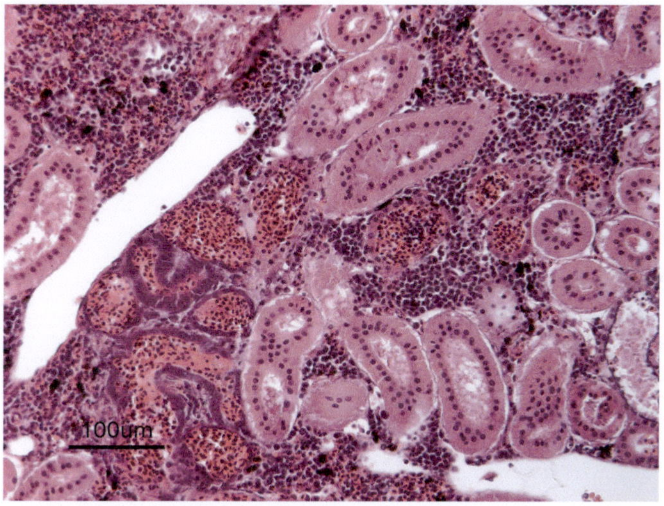

Fig. 11.16 Interstitial and tubular haemorrhage in the kidney of farmed Atlantic salmon smolt with haemorrhagic smolt syndrome

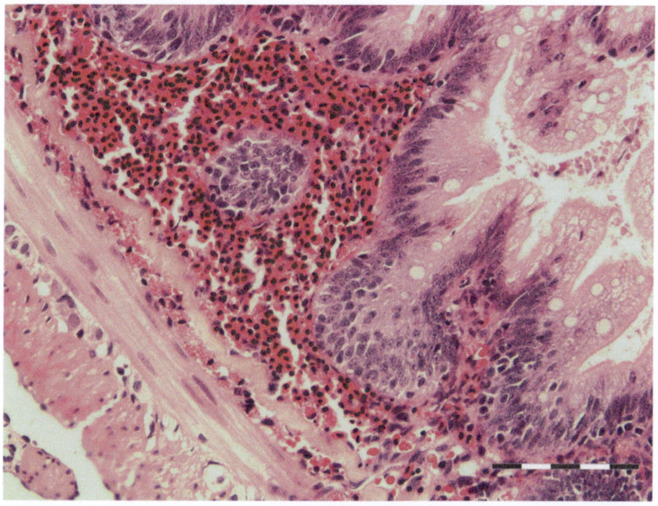

Fig. 11.17 Haemorrhage in stratum proprium of the posterior intestine in a farmed Atlantic salmon smolt with haemorrhagic smolt syndrome. Bar = 100 μm

Fig. 11.18 Distended swim bladder in farmed rainbow trout with swim bladder stress syndrome

Further Reading

Del-Pozo J, Crumlish M, Turnbull JF, Ferguson HW (2010) Histopathology and ultrastructure of segmented filamentous bacteria-associated rainbow trout gastroenteritis. Vet Pathol 47:220–230

Johansson N, Svensson KM, Fridberg G (1982) Studies on the pathology of ulcerative dermal necrosis (UDN) in Swedish salmon, *Salmo salar* L., and sea trout, *Salmo trutta* L., populations. J Fish Dis 5:293–308

Kolbeinshavn A, Wallace JC (1985) Observations on swim bladder stress syndrome in Arctic char (*Salvelinus alpinus*), induced by inadequate water depth. Aquaculture 46:259–261

Nylund A, Plarre H, Hodneland K, Devold M, Aspehaug V, Aarseth M, Koren C, Watanbe K (2003) Haemorrhagic smolt syndrome (HSS) in Norway: pathology and associated virus-like particles. Dis Aquat Organ 54:15–27

Roberts RJ (1993) Ulcerative dermal necrosis (UDN) in wild salmonids. Fish Res 17:3–14

Roberts RJ, Hill BJ (1976) Studies on ulcerative dermal necrosis of salmonids V. The histopathology of the condition in brown trout (*Salmo trutta* L.). J Fish Biol 8:89–92

Rodger HD, Richards RH (1998) Haemorrhagic smolt syndrome: a severe anaemic condition in farmed salmon in Scotland. Vet Rec 142:538–541

Neoplasia

12

Abstract

A neoplasm represents an altered process involving abnormal and uncontrolled growth of cells that is usually detrimental to the host. Neoplasm are classified as benign or malignant and named according to the features of differentiation recognised through histological examination, thus reflecting the tissue of origin. Neoplasia can arise following exposure to certain chemicals, heavy metals, ionizing radiation, chronic inflammation, ultraviolet radiation, certain viruses and pollution. The chapter covers a selection of reported neoplastic conditions affecting fish although overall, many of the cause(s) have not been determined.

Keywords

Neoplasia • Salmon • Trout

A neoplasm is an altered process involving abnormal and uncontrolled growth of cells that is usually detrimental to the animal. They are named according to the features of differentiation recognised through histological examination, thus reflecting the tissue of origin. In this context, neoplasm and tumour are interchangeable terms in current medical usage. The abnormal cell type may continue to increase even after the initiating mutagen is no longer present and generally more than one mutation is necessary for oncogenesis. Neoplasia can arise following exposure to certain chemicals, heavy metals, ionizing radiation, chronic inflammation, ultraviolet radiation, certain viruses and pollution. They are classified as benign or malignant (i.e. to include anaplasia, frequent mitotic figures and infiltration of neutrophils), but overall many of the cause(s) have not been determined. Examples of neoplasia representing a range of tissue origins are discussed.

12.1 Papilloma

Atlantic salmon papillomatosis is an epidermal benign neoplasia on the skin and scales of salmon parr in their second summer, but also occasionally of young adult fish (smolts and grilse) which have adapted to sea water.

The growths appear as single or multiple with a smooth to nodular texture, and range from white to brown or pink. Each papilloma may vary from a few millimetres to complex growths of up to 40 mm (Fig. 12.1). In heavily affected fish more than half of the body surface may be covered. Histologically, it is a plaque-like proliferation of the epidermis (Fig. 12.2). Each plaque is a stratified squamous epithelium with supporting stroma, hyperplastic epidermal cells containing prominent nucleoli with numerous atypical mitotic figures. There is usually little dermal involvement. The number of mucous cells is reduced and the basement membrane absent or indistinct. The papilloma is relatively harmless to the fish and eventually detaches allowing the skin to heal.

Electron microscopy studies of papillomas have indicated that virions with herpesvirus morphology are sometimes associated with the growth. Attempts to isolate this agent in culture have failed and therefore, the cause (or possibly causes) of papillomatosis remains unknown.

12.2 Chondroma

Single, firm, ovoid, white, smooth growths involving the branchial cartilage have been reported in salmonids and diagnosed as a benign cartilaginous neoplastic growth or

D.W. Bruno et al., *A Colour Atlas of Salmonid Diseases*,
DOI 10.1007/978-94-007-2010-7_12, © Springer Science+Business Media B.V. 2013

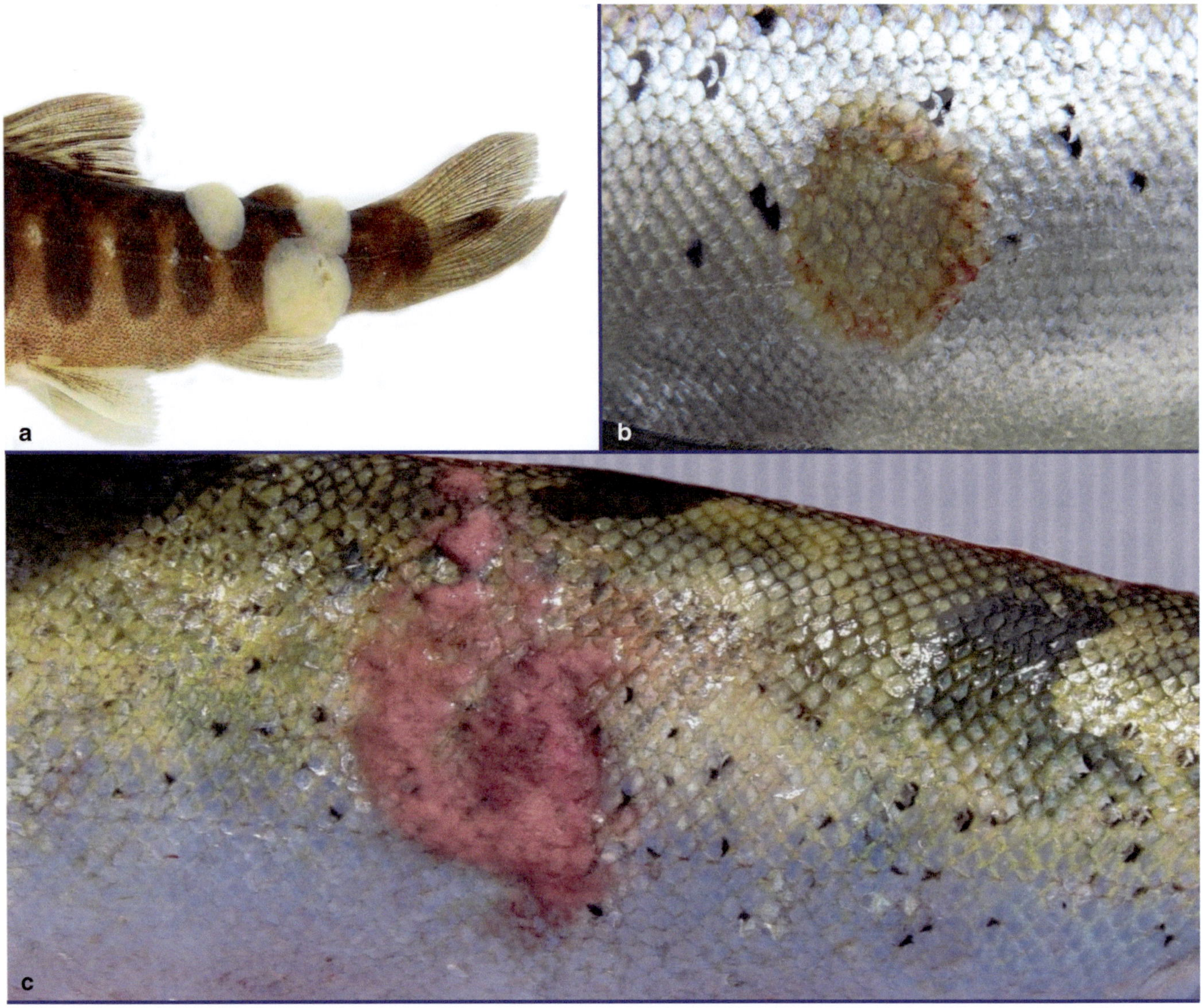

Fig. 12.1 (**a**) Papillomatosis in wild Atlantic salmon parr, alcohol-preserved specimen. (**b**) Healing papillomatosis in wild adult Atlantic salmon. (**c**) Papillomatosis in farmed adult Atlantic salmon. Note haemorrhage around margins

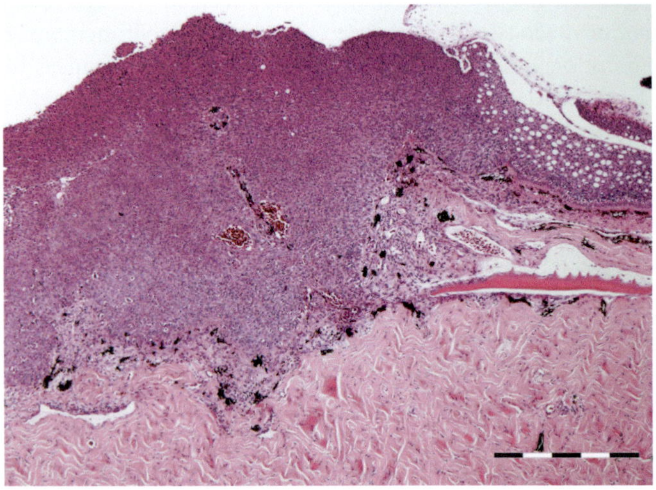

Fig. 12.2 Section showing papillomatosis in Atlantic salmon parr. Normal skin with mucous cells to the *right*. Bar = 500 μm

chondroma (Fig. 12.3). These are encapsulated with a lobular growing pattern that originates from the filament cartilage, possibly as an ingrowth of the surface covering. Normal squamous, or sometimes hyperplastic epithelium covers the surface of the mass with some occasional invaginations into the underlying stroma with increased mucous cell activity. Within the dermis, there is hypertrophy and some hyperplasia. Many of these growths contain a number of cystic spaces beneath the epidermis and are surrounded by a loose fibrous matrix of adipose tissue containing strands of immature cartilage (Fig. 12.4). Melanin granules are evenly spread throughout the dermis with no evidence of inflammation, infiltrative growth or distant metastasis. Chondrocytes are infrequent with no mitotic figures.

The absence of infectious agents coupled with the rarity of chondromas in populations indicates that these growths

Fig. 12.3 Chondroma in the gill of rainbow trout

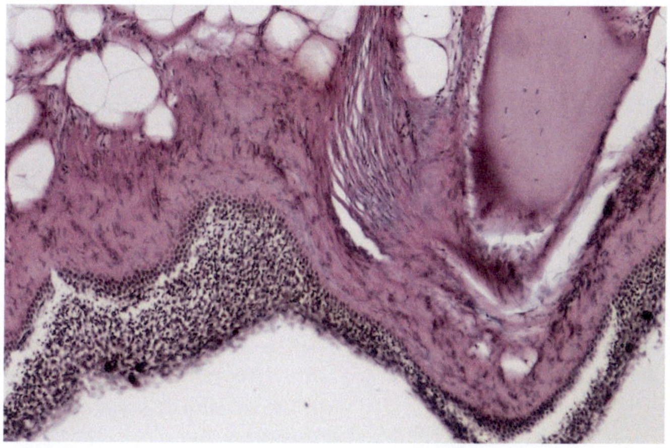

Fig. 12.4 Transverse section showing gill chondroma in rainbow trout. Lillie's allochrome stain. Low magnification

are natural occurrences, hence the cause or causes are considered to be spontaneous in nature. The topographic localization and histological features are used for diagnosis.

12.3 Pigment Cells

12.3.1 Melanoma

Pigment cell neoplasia is occasionally reported in salmonids, with melanomas being the most common type from this group. Both wild and farmed stock may be affected. The mature growths are generally raised, soft, black-pigmented areas which are visible on the body surface and underlying muscle (Fig. 12.5). The neoplasia shows invasion by melanomacrophages in varying degrees of differentiation, with fibrous deposition (Fig. 12.6). Metastasis has been recorded. Gross observations and the cellular detail described from stained histological sections are used to confirm the diagnosis.

Fig. 12.5 Well-defined melanoma in the fillet of a farmed Atlantic salmon

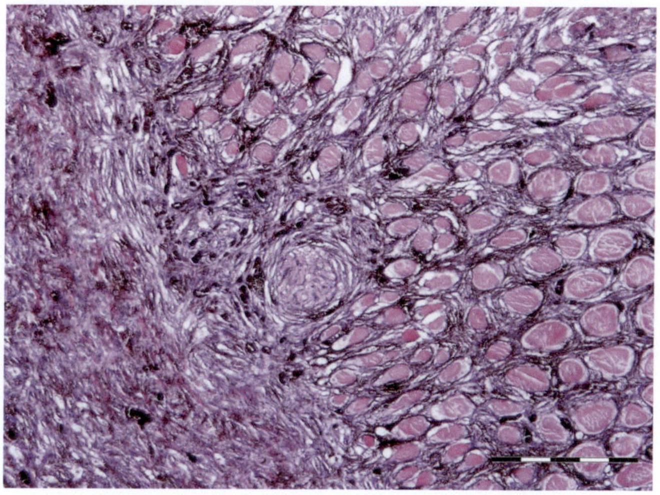

Fig. 12.6 Melanoma infiltrating white muscle of farmed Atlantic salmon. Bar = 200 μm

12.3.2 Iridophoroma

Iridophoromas consist of neoplastic iridophores, pigment fusiform cells which are arranged in bundles with moderate amount of cytoplasm containing olive to green pigment, crystalline and birefringent with polarized light. Their nuclei are round to ovoid with one to two nucleoli and mitotic figures are not reported. Macroscopically, iridophoromas comprise a well demarcated, whitish oval mass, raised above the surface of the skin. The overlying epidermis is slightly ulcerated with no additional pathological lesions or metastases reported. Histologically, the mass originates from the dermal pigment layer and normally encapsulated (Fig. 12.7). In a few areas the neoplastic cells infiltrate into the surrounding tissue. The epidermis is normally structured at the sides of the mass and completely eroded on the surface of the neoplasia. A severe oedema is reported in the dermis adjacent to the neoplasia.

12.4 Leiomyosarcoma

Swim bladder sarcoma (leiomyosarcoma) is infrequently reported, with a few cases described from maturing sea-reared Atlantic and wild sockeye salmon. Moribund fish are generally in poor condition and sluggish, but show no other external changes. At necropsy, multi-nodular masses of neoplastic cells occupying external and internal surfaces of the swim bladder, are randomly scattered. Single, hard nodular areas to a more extensive encrustment running the entire length of the organ protrude from the swim bladder surface into the abdominal cavity (Fig. 12.8). No evidence of infiltrative growth or distant metastasis has been reported.

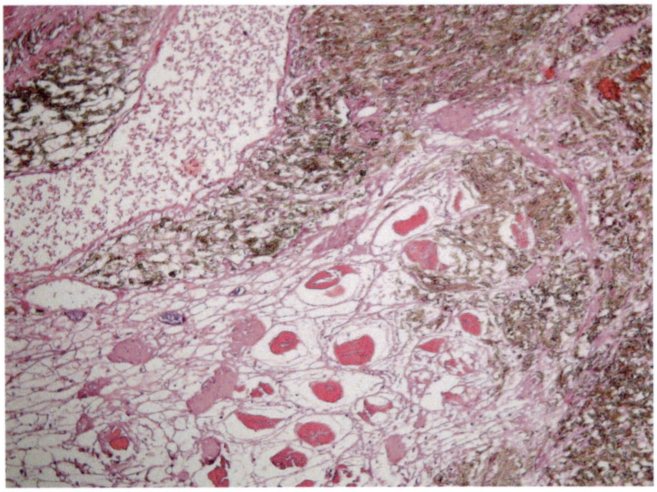

Fig. 12.7 Iridophoroma in the integument of rainbow trout. Low power

Histologically, the neoplasia is well differentiated and consist of interlacing bundles of spindle cells with round or elongated nuclei, which arise from the junction of the inner smooth muscle and the areolar zone of the swim bladder. These cells have round to oval nuclei with lightly stippled chromatin, and a single small nucleolus. Some nuclei may have slight margination of chromatin while other cells show enlarged irregularly shaped or clefted nuclei and form nests of anaplastic cells. A whirling arrangement of fibroblasts and smooth muscle makes up the mass. A novel piscine retrovirus (swim bladder sarcoma virus) has been identified in association with an occurrence in Atlantic salmon.

The diagnosis of leiomyosarcoma is based on the appearance of the well differentiated spindle-shaped cells with elongated cytoplasmic processes, low collagen and high mitotic index.

12.5 Fibrosarcoma

Fibromas and fibrosarcomas are masses of mesenchymal cell origin that are composed of benign and malignant fibroblasts and are usually found as nodular, well-defined lesions on or near the body surface. These masses are soft (myxomas) with a smooth, pale cut surface. Fibroblasts of variable differentiation and collagen, often arranged in sworls and whirls constitute the stroma. Central necrosis in the stroma may occur with occasional metastases in the kidney and swim bladder (Fig. 12.9). These neoplasms are readily distinguished histologically with the demonstration of elongated fibroblasts and dense collagenous fibres in characteristic whirling patterns (Fig. 12.10).

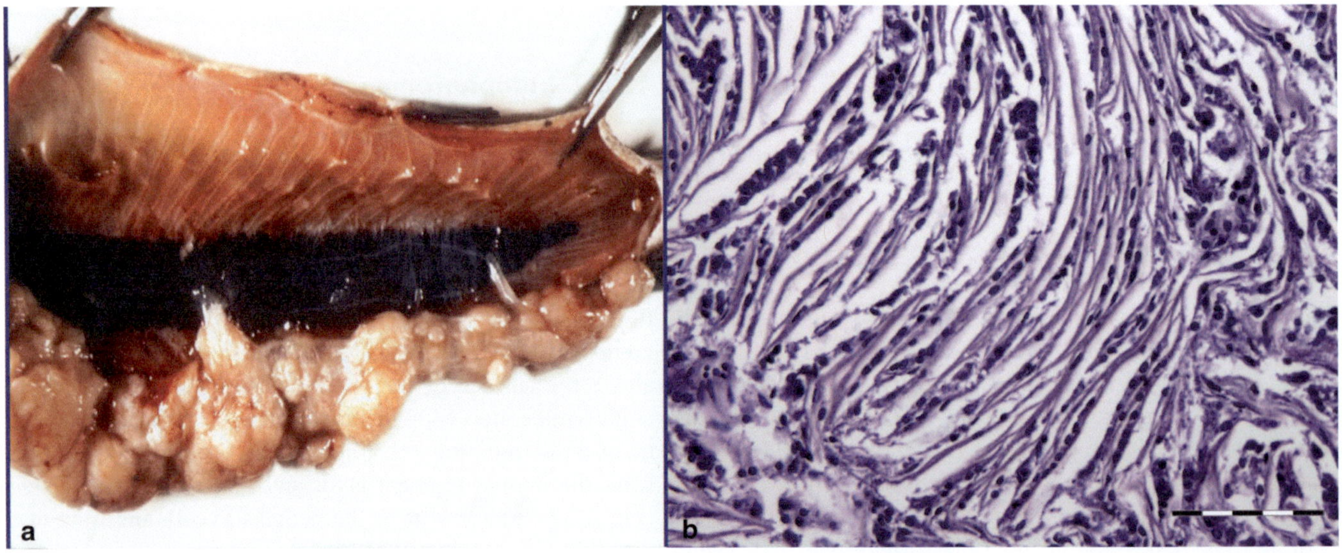

Fig. 12.8 (**a**) Leiomyosarcoma in Atlantic salmon. Scattered hard nodules protruding along the length of the swim bladder. (**b**) Leiomyosarcoma in wild sockeye salmon. Bar = 100 μm

Fig. 12.9 Fibrosarcoma in the swimbladder of farmed Atlantic salmon

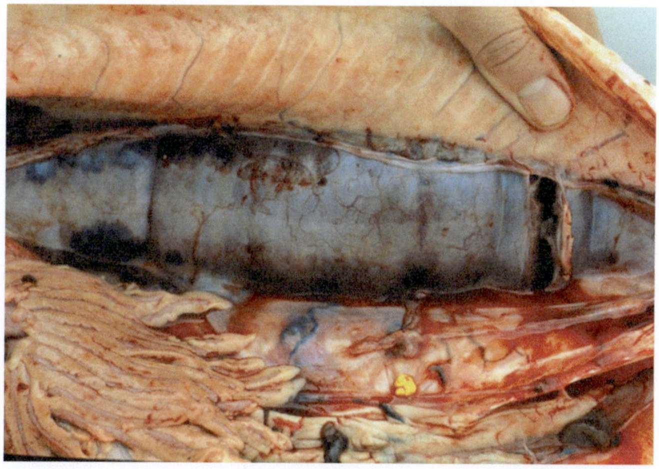

Fig. 12.11 Lymphosarcoma in the kidney of Atlantic salmon. Note pale colour and diffuse swelling of the organ

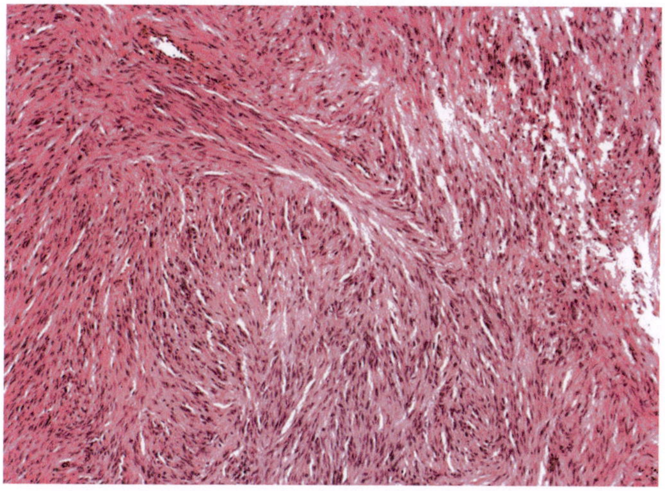

Fig. 12.10 Section of fibrosarcoma from rainbow trout. Low power

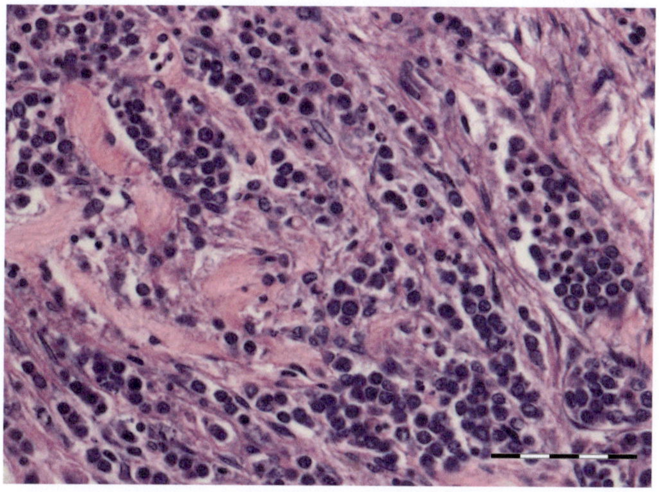

Fig. 12.12 Section showing lymphosarcoma with abundant lymphocytes from the kidney of farmed Atlantic salmon. Bar = 50 μm

12.6 Lymphoma and Lymphosarcoma

Lymphoid neoplasia has been described from many fish species, including several species of salmonids. In farmed rainbow trout this has been reported as grossly uni- and bilateral, oval masses that protrude beyond the operculum, resulting in a slight distortion that prevents complete closure of the gill cover. Each mass may extend into the gill chamber and conceivably interfere with respiratory movements. The mass is soft, smooth, oval and pinkish. Histologically, the mass is lightly encapsulated and contains darkly staining basophilic, uniform lymphoid-like cells. No evidence of metastasis or destruction of adjacent tissue is reported.

In cases where the growth is malignant this is referred to as a lymphosarcoma. Such masses vary in size and are most frequently found in the skin, subcutaneous tissue and trunk musculature, often with metastasis in the kidney, liver and spleen (Fig. 12.11). The cut surface of the mass is smooth, pale and homogenous. Superficial tumours may ulcerate. Histologically, lymphosarcoma growths consist of undifferentiated blast cells and immature lymphocytes infiltrating between normal cells (Fig. 12.12). Transmission has been successful with cell-free homogenates and a retrovirus has been reported to be associated with this condition. A diagnosis is based on demonstration of the characteristic lymphocyte-like cells in the mass.

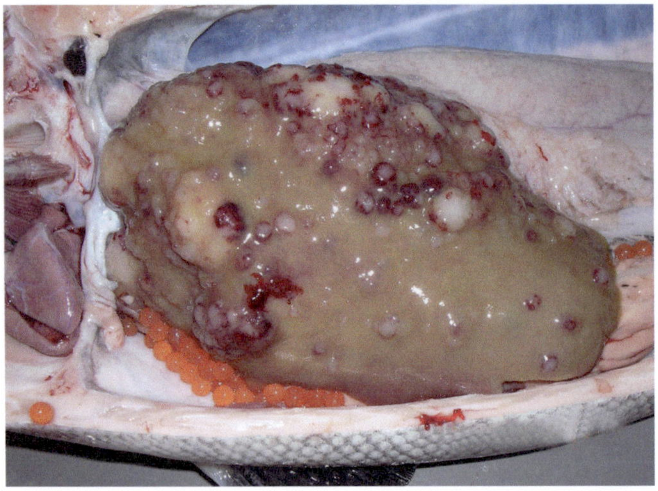

Fig. 12.13 Hepatocarcinoma in farmed Atlantic salmon broodstock

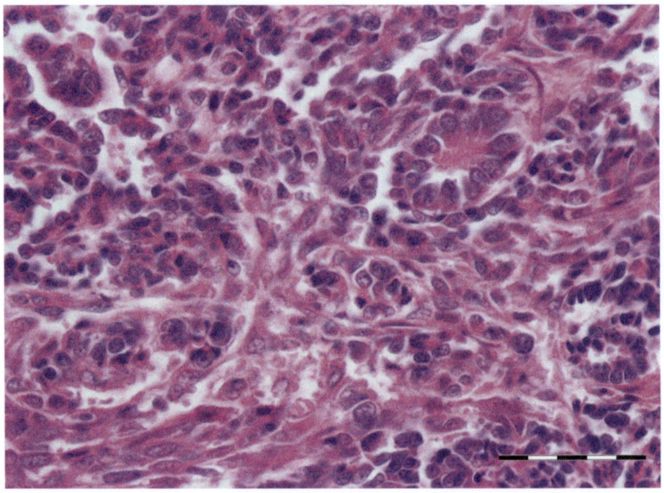

Fig. 12.14 Nephroblastoma from the kidney of rainbow trout. Bar = 50 μm

12.7 Hepatoma

Hepatomas and hepatic carcinomas were recorded in farmed rainbow trout in the 1960s as a result of *Aspergillus flavus* growing on oil seeds during warm and humid storage of the fish feed, but currently these are rare. Aflatoxins produced by *Aspergillus* are highly carcinogenic to rainbow trout and the affected fish show an enlarged abdomen, splenomegaly and massively swollen liver, with defined masses consisting of pale nodules protruding from the surface (Fig. 12.13). Occasionally there is widespread haemorrhage. Histologically, the growth consists of hypervascular masses and fibroblastic proliferation that frequently metastasize. Parenchyma cells display moderately enlarged and hyperchromic nuclei. Bile duct carcinomas are also frequently diagnosed in these fish. Hepatomas that are considered spontaneous in nature have also been reported in salmonids and similar to the above description. The diagnosis of a hepatoma is based on the characteristic gross and microscopical appearance of the mass.

12.8 Nephroblastoma

Nephroblastoma, also known as teratoma or embryonic nephroma, has been described in several fish species including rainbow trout. Grossly, affected fish may show an enlarged abdomen, skeletal deformity and compression of the swim bladder. The mass is often visible on the ventral surface where they are observed as dark/greyish rounded protrusions. The growth consists of multipotent tissue elements often comprising cartilage, connective tissue, nephrons and epithelial components with imperfect morphogenesis towards poorly differentiated tubules and glomeruli (Fig. 12.14). Metastasis is uncommon. The neoplasia can be diagnosed histologically.

12.9 Adenocarcinoma

Intestinal adenocarcinomas have been diagnosed in Atlantic salmon broodstock. Affected fish with primary neoplasia (i.e. only intestinal neoplasia) show no clinical signs of disease. At necropsy, focal lesions of variable size protrude into the anterior or posterior intestinal lumen. Early changes are characterized by irregular stratification of epithelial cells, hyperchromasia and mitotic figures. Intestinal folds are thickened due to influx of lymphocytes and eosinophilic granular cells. Subsequent stages may show growth of mucin-producing cells of epithelial origin, nuclear depolarization and basophilic cells with pleomorphic nuclei. In the liver, metastases are characterized by signet- ring nuclei and mucin-containing cells. The changes are believed to be the result of chronic intestinal inflammation associated with the use of particular diets containing high levels of plant ingredients.

12.10 Haemangioma

Haemangioma arises from endothelial cells that line blood vessels and can occur in any tissue where vasculature is present (Fig. 12.15). The tissue has very little or no collagen in the stroma, is often well differentiated having no microscopic features except for local invasion of surrounding normal tissues. A basophilic mass arising from the subcutis is composed of loosely organized to highly cellular areas, of spindle-shaped sarcoma-type cells in a whirled and palisade pattern, with numerous spaces or clefts of varying sizes containing erythrocytes (Fig. 12.16). The proliferating cell is well differentiated and contains an oval to elongate stippled nucleus, with pointed to rounded ends having one to two nucleoli and scant cytoplasm. Occasional areas of haemorrhage and necrosis can be present.

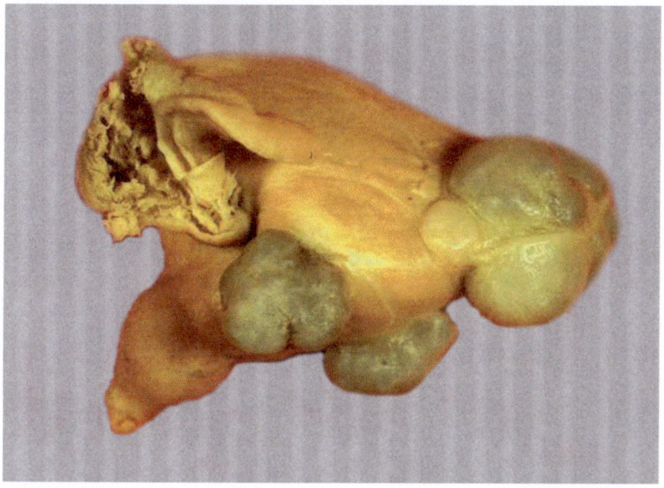

Fig. 12.15 Ventricular aneurisms (haemangiomas) in farmed Atlantic salmon. The endocardium is protruding through slits in the myocardial wall

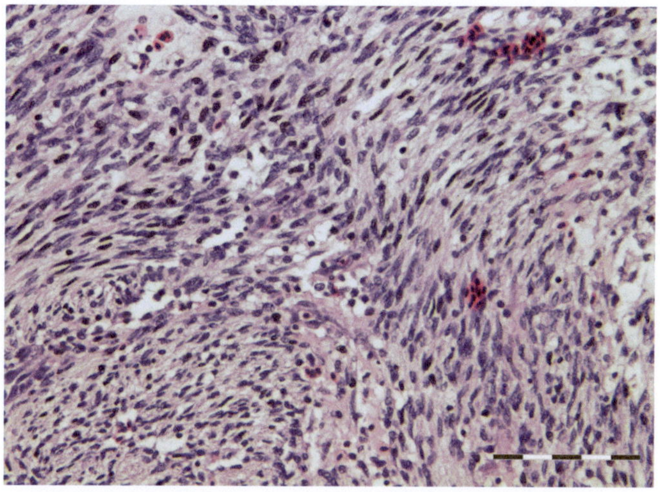

Fig. 12.16 Basophilic mass from the dorsal subcutis of the head of wild coho salmon classified as a haemangioma. Bar scale = 100 μm

Further Reading

Dale OB, Tørud B, Kvellestad A, Koppang HS, Koppang EO (2009) From chronic feed-induced intestinal inflammation to adenocarcinoma with metastases in salmonid fish. Cancer Res 69:4355–4362

Hoffmann RW, Fischer-Scherl T, Pfeil-Putzien C (1968) Lymphosarcoma in a wild grayling, *Thymallus thymallus*: a case report. J Fish Dis 11:267–270

McKnight IJ (1978) Sarcoma of the swimbladder of Atlantic salmon (*Salmo salar* L.). Aquaculture 13:55–60

Roald SO, Håstein T (1979) Lymphosarcoma in an Atlantic salmon *Salmo salar* L. J Fish Dis 2:249–251

Takashima F (1976) Hepatoma and cutaneous fibrosarcoma in hatchery-reared trout and salmon reared to gonadal maturation. Prog Exp Tumour Res 20:351–366

Glossary

Abrasion Superficial injury to skin or mucous membrane

Acantholysis Separation between adjacent Malpighian cells of the epidermis, following loss of function of the desmosomes

Acid-fast Bacteria not decolourised by weak acids e.g. mycobacteria

Acidophilic A substance within a cell or tissue that stains with an acid dye (e.g. eosin)

Acinus Any of the smallest lobules of a compound gland

Acute Characterised by a short and relatively severe course

Adenoma Benign neoplasia of glandular epithelium

Adhesion Joining of two structures by connective tissue which would normally be apart

Adjuvant An agent that may enhance the immune system and increase the response to a vaccine

Adipose Of a fatty nature

Adventitia The outermost connective tissue covering of any organ, vessel, or other structure

Aetiology Study of the causation of disease, both direct and predisposing

Afferent Leading or flowing into a named body e.g. glomerulus

Agglutination Clumping of bacteria or red blood cells in a fluid

Agranulocytosis Deficiency or absence of white blood cells

Amorphous Having no distinct form

Amyloid Glassy, homogenous substance appearing in the cytoplasm

Anaemia Deficiency of haemoglobin concentration in the blood or decreased number of red blood cells

Anamorphic Changing to a more complex form

Anaplasia Loss of the distinctive characteristics of a cell, associated with proliferative activity

Anastomosis Connection of the branches of two or more arteries or veins

Aneurysm Blood-filled dilatation or bulge of a vessel

Angioma Benign neoplasia of vascular tissue

Anisocytosis Unequal size of red blood cells

Ankylosis Stiffening of the joints between the vertebrae (and sometimes a fusion and shortening)

Anopthalmia Congenital absence of one or both eyes

Anorexia Loss or deficiency of appetite for food

Anoxia Inadequate supply of oxygen to the body tissues

Anterior The front end

Anterolateral Front and side

Anitschkow-like Large mononuclear cells found in the myocardium with an undulating, ribbon-like formation of nuclear chromatin

Aplasia Incomplete or defective development of tissue or organ

Aplastic Defective development or congenital absence of a tissue

Aplastic anaemia Defective, or a cessation of, regeneration of red blood cells e.g. drug induced

Apoptosis Programmed cell death, a normal component of the development and health of multicellular organisms

Artifact Artificial product or reaction resulting from physical or chemical process

Arteriosclerosis Chronic thickening and rigidity involving predominantly the middle coat of medium-sized arteries

Ascites Abnormal free serous fluid in the peritoneal cavity

Asphyxia Condition of suffocation, increased carbon dioxide tension in the blood

Asymptomatic An organism carrying a disease or an infectious agent but showing no overt signs of disease

Ataxia Defective muscular control resulting in irregular and jerky movements

Atheroma Deposits of lipid material in the inner wall of the arteries

Atherosclerosis Disease of the arteries in which lipid-like plaques develop on the inner wall lining

Atresia Abnormal closure, or congenital absence of a natural opening

Atriomegaly Increase in the inner size of the atrium

Atrium Thin-walled chamber of the heart

Atrophy Wasting, diminution in size and function, as a result of disuse, nutritional insufficiency

Atypical Not correlating with normal

Autoimmunity Failure of an organism to recognize its own constituent parts as self, which allows an immune response against its own cells

Autolysis Auto-digestion following release of digestive fluids or own enzymes

D.W. Bruno et al., *A Colour Atlas of Salmonid Diseases*,
DOI 10.1007/978-94-007-2010-7, © Springer Science+Business Media B.V. 2013

Autopsy Post-mortem examination by dissection to determine cause of death; see necropsy
Avulsion Forcible tearing away part of the body
Bacteria Small, simple prokaryotic microorganisms
Bacteraemia Presence of bacteria in the blood
Bactericidal Able to kill at least some types of bacteria
Bacteriostatic Ability to inhibit or retard bacterial growth
Basophilia Basophilic staining of cells
Benign Non-aggressive, innocuous (non-malignant) growth
Bifurcation The place where something divides into two branches or parts
Binucleate Two nuclei
Bipolar stain Particular staining pattern that colours only the two opposite poles of the microorganism
Birefringent Transmission of light unequally in different directions
Blood clot A soft, insoluble mass formed when fibrinogen is converted to Fibrin entrapping blood cells within coagulated plasma (i.e. thrombus)
Brachygnathia Abnormal shortness of the lower jaw
Brahycephaly Abnormal shortness of the head
Branchitis Inflammation of the gills
Bowman's capsule A double-walled, cup-shaped structure around the glomerulus of each nephron of the vertebrate kidney
Bryozoan Small aquatic animals of the phylum Bryozoa that reproduce by budding and form moss-like or branching colonies permanently attached to stones or seaweed
Cachexia Feeble state produced by serious disease, loss of weight, muscle mass, fatigue, weakness, loss of appetite
Calcareous Mostly or partly composed of calcium carbonate
Capillary Smallest thin-walled vessel
Carcinoma Malignant neoplasia whose parenchyma is composed of anaplastic epithelial cells
Cardiac failure Malfunction of the heart resulting in blood stasis, fluid accumulation, dilatation with or without hypertrophy
Cardiac tamponade An acute type of pericardial effusion in which fluid accumulates in the pericardium; see also haemopericardium
Cardiomegaly Enlargement of the heart
Cardiomyopathy Acute, subacute or chronic disturbance and enlargement of the myocardium
Cardiosomatic index Weight of cardiac ventricle*100/body weight
Carditis Inflammation Continued presence of an organism (bacteria, virus, or parasite) in the body that does not cause signs
Caseation Chronic process whereby a firm cheese-like mass is formed, then absorbed or converted into calcareous deposit
Caseous Development of a necrotic centre (cheesy appearance)
Cataract Partial or complete opacity of the crystalline lens or its capsule
Cavity An enclosed area
Cellulitis Localized or severe inflammation of the dermal and subcutaneous layers of the skin
Ceroid Golden-brown intracellular material, formed from indigestible remains
Ceroidosis A form of liver degeneration characterised by deposition of a pink/golden, fat material within cells. Associated with the use of rancid or vitamin E deficient feeds
Chalimus Developmental stage of parasitic lice (Copepoda) physically adhered to the skin of the host by a frontal filament
Chloride cell Acidophilic cells at the base of the gill lamellae which pump sodium and chloride ions out into sea water against a concentration gradient
Cholangiohepatitis Inflammation of bile ducts and associated liver parenchyma
Cholangitis Inflammation of the bile ducts
Cholecystitis Inflammation of the gall bladder
Cholestasis Bile accumulation within the liver
Chondritis Inflammation of the cartilage
Chondroma Benign neoplasia composed of cartilage
Choroid gland The vascular layer of the eye providing oxygen and nourishment to the eye
Chromaffin cells Neuroendocrine cells in the head kidney producing adrenalin/noradrenalin
Chromatolysis Disintegration or loss of cytoplasmic aggregates of basophilic material
Chronic A disease condition that is persistent or long-lasting
Cirrhosis Consequence of chronic liver disease characterized by replacement of liver tissue by fibrosis
Clinical Outward appearance of a disease in a living organism
Clitellum Swollen, glandular, saddle-like region in the epidermis of certain annelid worms
Cloudy swelling Degenerative change in cells, in which the cells swell due to injury to the membranes affecting ionic transfer
Coagulation Process whereby bleeding is normally arrested through clotting of the blood Proteins may also coagulate, e.g. in yolk-sac
Coalesce To grow together, to unite into a mass
Collagen Group of fibrous proteins that occur in vertebrate
Commensal An organism that benefits from another organism without affecting it negatively (see also parasite)
Confluent Becoming merged together, covering large area
Congenital Denotes the presence of an abnormality, condition or trait at birth, but does not imply that the defect is genetically related
Congestion Stagnant, abnormal accumulation of blood
Conidia Asexual, non-motile spores of a fungus

Conidiophore A structure that bears conidia
Constrictive Limiting
Copepodid Larval stage of parasitic lice (Copepoda) following the nauplius stage
Coracidium Free -swimming spherical, ciliated embryo of certain tapeworms
Corpuscles of Stannius Islands of endocrine eosinophilic cells found on the lateroventral surface of the kidney; regulates calcium metabolism
Cutaneous Belonging or pertaining to the skin
Cyst A closed, abnormal bladder-like sac or capsule
Cytolysis Breakdown of cells by destruction of their outer membrane
Cytomegaly Enlarged cells
Cytopathic Pertaining to disease of the living cell
Cytopenia A deficiency of cells, usually one or more of the various types of red blood cells
Cytotoxic Any substance which is destructive to cells
Debris Cellular fragments, remains of something destroyed
Definitive host Host in which an adult parasite with an indirect life-history lives and produces its eggs or off-spring (e.g. *Gyrodactylus*)
Deformity Distortion of any part or of the body in general
Degeneration Deterioration in quality of function, insufficient to cause necrosis
Depigmentation Loss of, or reduced pigment
Dermatomycosis Superficial infection of the skin by an oomycete
Desquamation Sloughing of cells from epithelial surfaces due to necrosis
Diagnosis Act of distinguishing one disease from another, but also the identification of the nature and cause of disease
Diapedesis Passage of red or white blood cells through the walls of the vessels that contain them without damage to the vessels
Diathesis (bleeding) Abnormal tendency towards bleeding
Differential diagnosis Process of weighing the probability of one disease versus that of other diseases
Dilatation Expansion of a cavity, may be part of a disease process or an adaptation to a disease
Diphtheritic Pertaining to features of the human disease diphtheria, the formation of a greyish membrane
Diplopagus Having one or more vital organs in common
Disease Condition in which the normal function or structure of part of the body or a bodily function is impaired
Disseminated Dispersed or spread throughout organ, tissue or body
Dysplasia Abnormality of development
Dystrophic Degeneration of tissue, in particular muscle
Ecdysis Sloughing of the epidermis
Ecchymosis Extravasation of blood from ruptured vessels into subcutaneous tissue under the skin or mucous membrane, bigger than petechiae
Ectasia Dilation or distention of a tubular structure
Ectoparasite A parasite that lives on or in the skin but not within the body
Ectozoic Living on the surface of an animal
Efferent Conveying away from a centre
Effusion Extravasation of fluid from vessels by rupture or exudation into body tissues or cavities
Ellipsoids Thick-walled capillary network
Emaciated Abnormally thin
Embolism Obstruction of a blood vessel by a solid body or gas bubble
Encapsulation Enclosure within a capsule
Encephalitis Inflammation of the brain
Encysted Enclosed within a bladder-like wall
Endarteritis obliterans Degeneration of the media of the larger vessels resulting in loss of potency
Endocarditis Inflammation of the inner membrane of the heart
Endocardium Lining membrane of the heart
Endoparasite Parasites that live inside the body of the host
Endophthalmitis Inflammation contained within the sclera
Endothelium Membrane lining various vessels and cavities
Enophthalmos Recession of the eyeball within the orbit
Endotheliotropic Having an affinity for endothelial cells
Endovasculitis Inflammation of the innermost layer of a blood or lymphatic vessel
Enteritis Inflammation of the intestine
Entomopathogenic fungus Fungus that can act as a parasite of insects which can kill or disable
Enzootic Localised disease, peculiar to, or constantly present in a given area, or location
Eosinophilia Increase in the number of eosinophilic staining cells
Epicarditis Inflammation of the epicardium
Epicardium The visceral layer of the pericardium
Epidemic A disease occurring more frequently than normal in a given population during a given time interval
Epidermis Outer, non-vascular layer of the skin
Epineurium Outermost layer of connective tissue surrounding a peripheral nerve
Epithelial Outer cell layer which is composed of stratified squamous epithelium
Epithelioma Abnormal growth of the epithelium
Epizootic Disease affecting many animals in a population, mostly over a large region
Erthropoiesis Production of red blood cells
Erythema Redness of the skin, occurring in patches of variable size
Erythroblast Cell from which red blood cells are derived
Erythrocyte Red blood cell
Erythrocythaemia Overproduction of red blood cells
Erythrocytopenia Deficiency in the number of red blood cells

Erythroderma Excessive redness of the skin
Erythrophagocytosis Ingestion of red blood cells by a macrophage or other phagocyte
Eutrophication Where an environment becomes enriched with nutrients
Extrasporogonic Sequence of a myxosporean developmental cycle
Exfoliation Scaling off of tissues in layers
Exophthalmia Abnormal protrusion of the eyeball from the orbit
Exotoxin Toxins released from bacteria
Extracellular Occurring outside the cell
Extrahepatic Outside the liver
Extravasation Force out from its proper vessel
Exudate Fluid with a high content of protein and cellular debris which has escaped from blood vessels usually as a result of inflammation
Exudation Oozing out of fluid through the capillary walls
Facultative anaerobe An organism that is able to grow under aerobic conditions but develops rapidly in an anaerobic environment
Fatty degeneration Accumulation of fatty droplets in the cytoplasm
Fatty necrosis Death of cells and tissues due to an imbalance of fat in cells and the rate of utilisation
Fenestrated Having one or more openings or pores
Fibrin Matrix on which a blood clot is formed
Fibrinogen Protein precursor from which the insoluble component of blood clots is made
Fibroblast Cell type common in developing or repairing tissues
Fibroma Benign neoplasia consisting of fibrous and muscle tissue
Fibroplasia Non-neoplastic increase in fibrous tissue
Fibrosarcoma Malignant neoplasia containing collagen fibres
Fibrosis Adequate or excessive fibrous tissue, which may replace other tissue as a repair response
Fixation The preservation of the structural organisation of a tissue
Flocculation Coalescence of colloidal particles in suspension
Fragmentation Separation into fragments
Furuncle Localised, subcutaneous haemorrhagic myositis
Fusion Joining together
Gastritis Inflammation of the stomach lining
Gastroschisis Cleft or fissure of abdomen with herniated viscera
Giant cell Multinucleated cell associated with inflammatory lesions formed by coalescence of epithelioid cells or by nuclear division without cytoplasmic division of monocytes
Glomerular Pertaining to the glomerulus
Glomerulonephritis Non-suppurative inflammation of the glomeruli
Glomerulopathy Disease of the renal glomeruli
Glomerulosclerosis Fibrosis of the glomeruli (result of inflammation)
Glomerulus Cluster of capillaries in kidney held together by an interstitium of mesangial cells
Glycogen Highly branched polysaccharide of glucose chains
Goblet cell Cells of the epithelial lining of the intestine which secrete mucous
Gram negative Bacteria which do not retain the primary violet but retain the counterstain in Gram's stain
Gram positive Bacteria which retain the primary violet in Gram's stain
Gram's stain Method for differentiating microorganisms, developed by Christian Gram in the nineteenth century
Granular Composed of granules or resembling granules
Granuloma Chronic inflammatory lesion or new growth made up macrophages
Granulomatous Having the characteristics of a granuloma
Grilse Atlantic salmon that have spent 1 year in sea water
Haemocoel Cavity between organs in invertebrates through which haemolymph circulates
Haematocrit The ratio of the volume occupied by packed red blood cells to the volume of the whole blood
Haematopoiesis The formation and development of blood cells
Haematopoietic Tissue forming red blood cells
Haematoma A swelling containing clotted blood under the skin, or deeper in the musculature following serious bruising
Haematoxylin Basic stain which gives a blue colour to cell nuclei
Haematogenous Disseminated through the blood stream
Haemoglobin Iron-containing oxygen-transporting metalloprotein of red blood cells
Haemolysis Disintegration of red blood cell membranes with liberation of haemoglobin
Haemolytic anaemia Condition resulting from reduced red blood cells survival time
Haemopericardium Blood in the pericardial cavity
Haemophthalmia Bleeding into the eyeball
Haemorrhage Escape of blood from a vessel
Haemorrhagic anaemia Loss of red blood cells due to bleeding
Haemosiderin Iron-containing substance resulting from the breakdown of red blood cells
Haemosiderosis Increased tissue iron stores
Haemostasis Arrest of bleeding through blood clotting
Halophilic Organisms surviving in environments with high salt concentrations

Hematophagous Animals that feed on blood
Hepatic Pertaining to the liver
Hepatitis Inflammation of the liver
Hepatocellular Pertaining to or affecting liver cells
Hepatocyte Main cell of the liver forming the parenchyma
Hepatoma Neoplasia whose parenchymal cells resemble those of the liver
Hepatomegaly Enlargement of the liver
Hepatotoxic Having an injurious effect on liver cells
Heteropagus Unequal conjoined twins where the imperfectly developed 'parasite' is attached to the ventral portion of the autosite
Histology Microscopic study of the structure of tissues
Histochemistry Specific and sometimes quantitative identification of chemical substances in tissues
Histopathology The microscopic study of diseased tissues
Histiocytic Animal cell that is part of the mononuclear phagocyte system
Histozoic Process of living within tissues but outside of the cell
Holocarpic Entire thallus developed into a fruiting body or sporangium
Holarctic Refers to habitats in the northern part of the northern hemisphere
Horizontal transmission Transfer of disease between individuals in a population
Hyaline degeneration Active accumulation (by pinocytosis) of protein in cytoplasmic organelles
Hyaloplasm Fluid portion of the cytoplasm as distinguished from the granular and netlike components
Hydatid Cyst formed by larvae of a tapeworm
Hydrocoele embryonalis Yolk sac dropsy
Hydropic swelling Intracellular oedema of keratinocytes
Hyperaemia Active, local congestion of blood in any part of the body
Hypercellularity Increase in number of cells
Hyperchromasia Increased colouring
Hyperpigmentation Darkening of an area of skin caused by increased melanin
Hyperplasia Increase in number of cells in a tissue with a corresponding increase in size of the tissue or organ
Hypertonic A solution with a higher salt concentration than normal cells
Hypertrophy Increase in size of individual cells and thereby also increase in size of the organ
Hypochromic Deficiency in colouring or pigmentation
Hypoplasia Excessive smallness of an organ or part, resulting from imperfect growth (reduced number of cells)
Hypoplastic anaemia Failure of the haematopoietic tissue to produce adequate numbers of cells
Hypoproteinaemia Decreased amounts of total protein in the circulating blood plasma
Hypotrophy Diminution in size, subnormal growth
Hypoxia A state of reduced oxygen supply to a tissue, organ or whole animal
Idiopathic Disease state of unknown or spontaneous origin
Incidence Number of cases developing per unit of population per unit of time
Inclusion bodies Round or oval bodies occurring in nuclei or cytoplasm especially at site of virus multiplication
Indurate The process of hardening or being hardened
Infarction Death of a section of tissue because the blood supply has been cut off
Infection Growth of pathogenic microorganisms in the body, whether or not body function is impaired
Infestation Incidence of a parasite which may lead to, or cause disease
Infiltration Penetration of the surrounding tissues, leaking of fluid into the tissues
Inflammation Complex reaction of living tissues to injury characterised clinically by heat, swelling and redness
Intercellular Between cells
Intermyotomal Between muscle blocks
Interrenal tissue Region where the major corticosteroid cortisol is produced
Interstitial Pertaining to the space between structures and tissue
Interstitium Space between (functional) parts of a tissue or organ
Intracellular Within cells
Intrahepatic Within the liver
Intraventricular Within a ventricle
Invagination Pushed forward forming a pouch
In vitro Tests or experiments conducted in an artificial environment
In vivo Tests or experiments on living organisms
Ischemia Deficient blood supply to part of the body or an organ
Karyolysis Necrotic change resulting in loss of nuclear morphology due to hydrolysis of chromatin following cell death
Karyomegaly Increase in the nuclear size of tissue cells
Karyorrhexis Nuclear membrane ruptures and the cell nucleus fragments following cell death
Kelt A salmon that has spawned and is in poor physical condition
Keratitis Inflammation with subsequent opacity of the cornea
Kypholordosis Coexistence of kyphosis and lordosis
Kyphosis An excessive upward curvature of the dorsal spine resulting in 'roundback' or 'hunchback'
Lacuna A space between cells
Lamellae Plate-like structures of the gills where gas exchange occurs
Leiomyoma Benign neoplasia of smooth muscle
Leiomyosarcoma Malignant neoplasia of smooth muscle

Leptomeningitis Inflammation of the pia or of the arachnoid membrane
Lesion Abnormality in the tissue of an organism
Lethargic Fatigue, exhaustion
Leucopenia Decreased number of white blood cells
Lipid Loose term, substance usually of fatty acids which are insoluble in water
Lipoid A group of compounds having a wide range of different lipid structures but which are insoluble in water
Lipoidosis Any disease of lipid metabolism within the cells of the body
Lipoma A benign neoplasia of fat cells
Lobule A small lobe or a subdivision of a lobe
Localise To limit the spread
Lordosis Concavity of the vertebral column (spine) which is directed downwards
Lunules Disc-like cups on the ventral surface of the anterior margin of the body (e.g. *Caligus* spp.)
Lymphocyte One type of white blood cell
Lymphocytopenia Reduced number of lymphocytes
Lymphocytosis An increase in lymphocytes in the blood
Lymphoid Pertaining to lymph
Lysis Dissolution and disintegration of cell membrane by the action of a lysin
Macrocephaly Abnormally large head
Macrognathia Refers to an overgrown lower jaw that juts out beyond the upper jaw
Macrophage A large mononuclear phagocyte white blood cell
Macroscopic Visible with the naked eye
Malignant Denoting aggressive, harmful growth which may spread to distant sites
Melanin Tyrosine derived black or brown pigment
Melanoma Neoplasia whose parenchyma is composed of pigment producing cells
Melanomacrophage Distinct group of phagocytic pigment cells
Melanophores Dark, pigment containing cells in the dermis
Melanosis Abnormal deposition of dark pigment of the skin
Meningitis Inflammation of the membranes of the brain and spinal cord
Mesangial cell Smooth muscle-like cell that occupies a central position in the renal glomerulus
Mesenchymal Part of the embryonic mesoderm, consisting of loosely packed, unspecialized cells set in a gelatinous ground substance
Metacercaria Encysted infective trematode larva in the intermediate host
Metachromasia Different colour staining of tissues than that of the original stain
Metaplasia Term applied to a change of one kind of tissue into another
Metastasis Process by which malignant disease spreads to other parts, of the body
Microcytic Undersized red blood cell
Micrognathia Condition where the jaw is undersized
Micropyle Pore in the membrane covering the ovum through which a spermatozoon can enter
Microstomia Abnormally small mouth
Microtome Instrument equipped with a steel blade which is used to cut thin sections of tissues embedded in paraffin wax
Mild Not severe
Milliary Term, expressive of size (about the size of millet seeds)
Miracidium Free-swimming ciliated larval form in the life cycle of a digenic trematode
Mitosis A type of cell division in which a single cell produces two identical daughter cells
Moderate Not excessive
Monocytes Partially differentiated end cells, leucocytes
Morbidity Frequency with which a disease appears in a population
Moribund Progressing towards death
Morphological Of form and structure
Mortality Fatal outcome
Mucosa A mucous membrane lining a cavity or organ lumen, composed of epithelial cells
Multicellular Many cells
Multinuclear Possessing many nuclei
Multivalent Vaccines that contain antigens from several different bacteria and viruses
Mural Pertaining to the wall of a cavity, organ or vessel
Mutagen An agent which produces a mutation or enhances the rate of mutation
Mycelium Mass of branching filaments of fungi or moulds
Mycetoma Chronic subcutaneous infection caused by actinomycetes or fungi, that can also appear in the brain, kidney, or other organs.
Mycosis Disease caused by the growth of any fungus
Myocarditis Inflammation of the myocardium of the heart
Myofibril Fibre, composed of a bundle of myofilaments that is found in striated muscle
Myophagia Invasion of degenerated muscle sarcoplasm by histiocytes
Myocardium Muscular tissue of the heart
Mycosis Any disease caused by fungi, or oomycete
Myodegeneration Muscle degeneration
Myolysis Disintegration or degeneration of muscle tissue
Myoma A neoplasia which consists almost totally of muscular tissue
Myotome A block of 'W' or 'V'-shaped segmental muscle
Myopathy Any abnormal condition or disease of muscular tissue

Myoseptum Connective tissue forming the boundary between successive myotomes
Myositis Inflammation of the muscle tissue
Myxoma Benign neoplasia of connective tissue origin
Necropsy Alternative name for autopsy, pertaining to animals; the same as post mortem or obduction
Necrosis Focal (limited) death of tissues and cells
Necrotic Death of circumscribed piece of tissue
Neoplasia A new growth, (usually abnormal)
Neovascularization Proliferation of blood vessels of a different kind than usual in tissue
Nephritis Inflammatory or inflammatory-like reaction of the kidney
Nephrocalcinosis Calcium deposits within kidney tubules
Nephromegaly Enlargement of the kidney
Nephron Basic structural and functional unit of the kidney
Nephrosis Degeneration of tubular epithelium of the kidney
Neuritis Inflammation/lesion of a nerve or their sheaths
Neuroma Neoplasia connected with a nerve
Neutrophil A leukocyte having no affinity for acid or basic dyes, but stainable by neutral dyes
Normal Anything which agrees with the regular or established type
Notifiable disease A serious infectious disease that must be reported to the appropriate authorities
Nucleus The inner essential part of a tissue cell
Obstruction Blockage of an organ when the normal passage is abnormally hindered
Occlusion The closure of an opening
Odontoblasts Cells forming the outer surface of dental pulp that produces the dentin of a tooth
Odontoma Neoplasia developing from or containing tooth structures
Oedema Abnormal infiltration of tissues with fluid
Oesophagus Muscular tube through which food passes from the pharynx to the stomach
Oncogenic Agent capable of neoplasia induction
Oncogenesis Formation of neoplasia
Operculum Moveable flap following the contours of the gill chamber downward and forward beneath the jaws
Optic chiasma Located at the bottom of the brain immediately below the hypothalamus
Oogenesis Cellular development that leads to the formation of a mature egg
Ossification Replacement of a tissue (usually cartilage) by bone
Osteitis Inflammation of bone
Osteoclast Large multinucleated cell responsible for the dissolution and absorption of bone
Overt disease Apparent disease
Pachymeningitis Inflammation of the dura, or external, fibrous, layer of the meninges
Pancreatitis Inflammation of the pancreas, may be acute or chronic
Pancytopenia Reduction in the number of red, white cells and platelets in the blood
Panophthalmitis Inflammation involving structures adjacent to the sclera
Pansporoblast Reproductive sporoblast that gives rise to more than one spore in the order Myxosporida
Pansteatis Generalised inflammatory infiltration of adipose tissues
Papillae Small projections
Papillitis Inflammation of the optic disc
Papilloma Benign neoplasia involving overgrowth of epithelial tissue
Papillomatosis Surface elevation caused by hyperplasia and enlargement of contiguous dermal papillae
Parasite An organism which obtains food or shelter from another host organism
Parenchyma All the soft tissue of internal organs except the muscular flesh. The essential cells of an organ (e.g. hepatocytes in the liver)
Perineurium The sheath of connective tissue that covers a bundle of nerve fibres
Patchy Irregular
Pathogen Any organism which by living on or within another organism causes disease in the host
Pathogenesis The origin and development of disease
Pathogenicity The potential to cause injury to the host
Pathogenic Producing disease or pathological changes
Pathological condition A deviation from normal to known or unknown origin
Pathognomonic Sign that is characteristic of a disease that it can be used to make a diagnosis
Pathology Science which deals with the causes of and the changes produced in the body by disease
Peliosis Extensive areas of red blood cells and dilated blood vessels within the parenchyma
Peduncle The narrow part of the body caudal to the vent to which the tail is attached
Pemphigoid-like Formation of 'vesicles' between epidermal cells
Pemphigoid Group of skin disorders similar to but clearly distinguishable from pemphigus
Pemphigus Distinctive group of skin and mucous membrane diseases
Pericarditis Acute or chronic inflammation of the pericardium
Pericardium The fibroserous sac surrounding the heart (this is closed in teleosts)
Periorbital Situated around the orbit of the eye
Perisplenitis Inflammation of the peritoneal capsule of the spleen

Peritonitis Inflammation of the peritoneum
Perivascular Surrounding a vessel (blood or lymph)
Petechiae Small, haemorrhagic spots
Phagocyte Cell capable of ingesting bacteria, foreign particles and other cells
Phagocytosis Ingestion of large foreign particles
Pernicious anaemia A progressive anaemia characterised by a decrease in numbers and variation of red blood cells
Phenotype Observable physical or biochemical characteristics of an organism
Photosensitivity Sensitivity of the skin to certain types of light
Physostomous Fish with a connecting tube between the swim bladder and a part of the alimentary canal
Physoclistous Having an swim bladder that is not connected to the alimentary canal
Pigmentation Deposit of pigment, especially when abnormal or excessive
Pillar cell Fine cytoplasmic filaments situated parallel to the collagen columns of the gills
PKX cell Earliest identifiable stage of *Tetracapsuloides bryosalmonae*
Platyspondyly Flattened vertebral body with reduced distance between the endplates
Pleomorphic The condition in which an individual assumes a number of different forms during its life
Plerocercoid Cestode larva that develops from a procercoid and the stage often found in fish tissues
Pneumatic duct The duct joining the swim bladder and alimentary canal of a physostomous fish
Podocytes Cells in the Bowman's capsule in the kidney that wrap around the capillaries of the glomerulus
Poikilocyte Irregular, malformed erythrocyte
Polychromasia Abnormal reaction of the red blood cells in severe anaemia, whereby they have a bluish tinge
Polycystic Composed of many cysts
Polycythaemia Increase in the number of circulating red blood cells
Posterior The hind end
Post-mortem After death, usually inferring dissection of the body; see also necropsy
Preopercle Flat membrane bone in the gill cover of most fish lying immediately in front of the opercula
Presporogonic The stage prior to the development of sporogonia
Prevalence Number of cases existing per unit of population at a given unit of time, generally represented as a percentage of the population. It is a static measure as compared with the dynamic measure incidence
Procercoid Cestode larva that usually infects the first intermediate host. In aquatic animals, this host is usually a crustacean
Proglottis Sexually mature segment of tapeworm
Prognathous Having a projecting lower jaw
Prognosis A forecast of the probable course and outcome of a disease
Prolapse Falling of a structure, descent
Proliferate Increase by cell division
Prognathous Projecting lower jaw
Prosector A person with the special task of preparing a dissection for demonstration
Proteinaceous Protein-like, resembling protein
Proteolysis The breaking down of proteins
Proximal Towards the centre
Pseudobranch The reduced first gill arch of a fish (on the inner surface of the operculum, near the junction of the preopercle)
Pseudomembrane False or new membrane (usually fibrinous)
Psychrophilic Organisms having an optimal temperature for growth at about 15 °C or lower
Punctate Dotted or spotted
Pycnidium Hollow fruiting body that produces pycnidiospores
Pyknosis Condensation of nuclear contents into a single, densely staining irregular-mass
Pyriform Pear-shaped
Regeneration Renewal of tissue
Regression Reversion to an earlier stage of development
Resolution Restoration from acute inflammation to normal
Resorption Disappearance of tissue by absorption into body fluids
Reticulocyte Immature red blood cell
Retrobulbar Pertaining to behind the eyeball
Rodlet cell Flask-shaped cell of uncertain identity found in many fish tissues, e.g. gills, intestine, renal tubules
Rhabdomyoma Benign neoplasia of striated muscle
Rudimentary An imperfectly developed or formed organ which is not functional and which may represent a normal or abnormal situation
Runt Non-feeding fish, with negligible body fat and very low condition factor
Rupture Burst, break
Saprophyte Organism that derives nourishment primarily from dead and decaying organic matter
Sarcoma Malignant neoplasia whose parenchyma is composed of anaplastic cells resembling those of the supportive tissues of the body
Sclerosis Hardening of tissues
Scolex Anterior, head-like segment of a tapeworm
Scoliosis Lateral curvature of the vertebral column (spine)
Scleritis Inflammation of the sclera (the white outer wall of the eye)
Secondary infection Infection in an animal which is already infected by another pathogenic organism
Septic Related to or caused by sepsis

Septicaemia Poisoning due to the presence and/or the multiplication of bacteria or viruses in the blood
Septum transversum Membrane that divides two cavities or soft masses of tissue in an organism e.g. the pericardial and abdominal cavity
Sequela Pathological consequences of a disease
Sequestrum Fragment of bone which is in the process of necrosis
Serosa Serous membrane
Siderosis Excess of iron (Perls' positive material) in the blood or tissues
Slight Superficial
Slough Tissue which becomes necrotic and separates from the healthy area
Smoltification The physiological process undergone by salmonid (salmon and trout) fish to allow them to migrate from freshwater to sea water as part of their lifecycle
Smooth muscle Involuntary, non-striated muscle found for example in the walls of blood vessels, urinary bladder and gastro-intestinal tract
Solitary Single, isolated
Splendore-Hoeppli reaction *In vivo* formation of intensely eosinophilic material (asteroid or club-shaped configurations) around microorganisms (e.g. fungi, bacteria and parasites) or biologically inert substances
Splenomegaly Abnormal enlargement of the spleen
Spongiosis Intracellular oedema of the epidermis
Sporoblast A cell of a sporozoan resulting from sexual reproduction and producing spores and sporozoites
Sporocyst Unicellular resting body from which asexual body from which asexual spores arise
Sporogony Multiple fusion of a cell to produce many dormant spores
Stasis Stagnation of the flow of blood
Steatitis Inflammation of adipose tissue resulting in a yellow discolouration of the fat
Steatosis Deposition of fat in the interstitial spaces of an organ
Stenosis Applied to a condition of unnatural narrowing in any passage of the body
Stomatitis Inflammation of the mucous lining of any of the structures in the mouth
Stratum A layer
Strobila Chain of proglottids constituting the bulk of the body of adult tapeworms
Stroma Matrix or supporting tissue of an organ
Stylet Hard and sharp anatomical structure, e.g. in *Argulus* spp.
Subacute Moderately severe, not acute
Subclinical Insufficient signs to cause classical identifiable disease
Subcutaneous Anything pertaining to the loose cellular tissue beneath the skin
Subendocardial Immediately beneath the endocardium
Sublethal Term applied to a dose which is not quite fatal
Superficial Peripheral, borderline
Suppurative Liquefactive necrosis with 'pus' formation
Swelling Protuberance
Synechiae Pathological adhesion between anatomical structures, e.g. the lens and iris, ventricular wall and pericardium
Systemic Widespread
Tachypnoea Increased respiration rate
Tamponade (cardiac) Rapid accumulation of fluid or blood in the pericardial sac
Telangiectasia Dilatation of small or terminal vessels
Teratogen Agent capable of causing malformation in embryos
Thrombosis Formation of a blood clot within the vessels or heart during life
Thrombus An intravascular blood clot which may impede circulation
Treatment Therapy used to remedy a health problem
Trophozoite The active, motile feeding stage of a sporozoan parasite
Torsion Twisting of a structure
Tubercle Solid, raised, round lump on the skin or surface of an organ or in an organ
Ulcer Non-healing breach of a body surface e.g. skin or intestine
Ulceration Discontinuity of the skin showing complete loss of the epidermis
Uraemia Raised level of urea in the blood but also defective renal function resulting in excessive nitrogenous compounds in the blood
Viviparous Giving birth to living offspring that develop within the animals body
Ureter Either of two ducts conveying urine from kidney
Uvulsion Forcible separation of a piece of tissue from the entire structure
Vaccine Suspension or extracts of dead or attenuated bacterial cells, viruses or parasites which retain the capacity to stimulate the immune system
Vascular Tissue consisting of, or containing a high proportion of blood vessels
Vasculitis Inflammation of the blood vessel wall
Vasodilation Widening of the lumen of blood vessels
Ventrum Pertaining to the underside or belly
Vertical transmission Parent to progeny transfer of disease agents
Vibriostatic agent 2,4-diamino-6,7-diisopropyl-pteridine (O/129); disk susceptibility testing to differentiate *Vibrio* spp.
Virulent Capable of producing disease
Vitellogenesis Yolk formation in the oocyte

Xenoma Superficial, hypertrophied host cell involving a parasitic infection caused by microsporidian protozoan parasites

Zoonoses Animal diseases which can be transmitted to man

Zoosanitary Pertaining to the health situation of animals or animal products

Zoonotic Infectious disease that is transmitted between species (may include a vector)

Zoosporangia A spore case in which zoospores are produced

Zymogen An inactive enzyme precursor (proenzyme), e.g. in the pancreatic acinar cells

Index

A
Absent septum transversum, 171, 172
Acanthocephalans, 142
Adenocarcinoma, 192
Adipose tissue, in abdominal cavity, 156, 157
Aeromonas hydrophila, 74–75
Aeromonas salmonicida subsp. *salmonicida*, 75–77
AGD. *See* Amoebic gill disease (AGD)
Algal blooms, 153
Aliivibrio salmonicida, 78
Ambiphyra, 112
Amoebic gill disease (AGD), 107–109
Anisakis simplex, 133–135
Anopthalmia, 169
Apatemon gracilis, 139, 140
Aquaculture, 1, 3, 4, 55, 59, 82, 107, 113, 145, 164
Argulus
  A. coregoni, 145–146
  A. foliaceus, 146
  A. japonicus, 146
Arteriosclerosis, 172–173. *See also* Myointimal hyperplasia
Ascites
  circulatory disorders, 46
  and haemorrhage, in posterior intestine, 183, 184
  and splenomegaly, BKD, 87, 88
Asymptomatic carriers, 73
Atlantic salmon
  absent septum transversum, 171, 172
  Anisakis simplex larvae, 33, 38, 39
  anopthalmia, 169
  arteriosclerosis, 172
  atriomegaly, 171
  bacterial diseases (*see* Bacterial diseases)
  BKD, 33, 37, 87–89
  blue sac disease, 164, 165
  cardiomegaly, 171
  cataracts, 162
  circulatory disorders, 44–46
  cold water aquaculture, 1
  commercial rearing, 1
  conjoined twin, 169, 171
  cranial nodules, 166
  double mouth deformity, 168
  EGCs, 43, 44
  epithelial necrosis, 47
  eye haemorrhage, 162, 163
  fibrosarcoma, 191
  functional anatomy (*see* Functional anatomy)
  gill hyperplasia, 42
  gill synechiae, 40, 41
  global production, 1
  glomerulonephritis, 44
  haemorrhage, *Listonella anguillarum*, 33, 38, 39
  with haemorrhagic smolt syndrome, 183–185
    clinical signs, 183
    interstitial/tubular haemorrhage, in kidney, 183, 185
    in islet of Langerhans, 183, 184
    posterior intestine, ascites and haemorrhage in, 183, 184
    in stratum proprium, 183, 185
    in white muscle and internal organs, 183, 184
  heart fat infiltration, 158
  hepatocarcinoma, 192
  hepatocyte erythrophagocytosis, 42
  hypoplasia of, outer compact myocardium, 173
  idiopathic highly cellular epicarditis, 43
  inverted operculum, 166
  ISA, 3
  leiomyosarcoma, 190
  lesions associated with angling, 176
  loser syndrome/failed smolt, 151, 152
  lymphosarcoma, in kidney, 191, 192
  mandibular ankylosis, 167
  melanomacrophages, 48
  metazoa parasites (*see* Metazoa parasites)
  microstomia, 167, 168
  M. truttae plasmodia, 33, 38
  multinucleated giant cell, 42
  M. viscosa infection, 33, 34, 36
  Mycobacterium infection, 33, 37
  myocardial haemorrhage, 35, 41
  non-specified kidney neoplasia, 33, 37
  nuclear hypertrophy, 36, 41
  oomycete infections (*see* Oomycetes infections)
  panopthalmitis, 162, 163
  papillomatosis, 33, 34, 36, 187–189
  pectoral fin erosion, 152, 153
  peribilary lymphoctic infiltration, 36, 41
  petechiae, 33, 38
  polycystic liver, 33, 38, 39
  polycystic syndrome, 181, 182
  protists (*see* Protists)
  P. salmonis infection, 33, 38
  pug-head condition, 167, 168
  punctured eyeball, 162, 163
  Saprolegnia infection, 33, 34, 36
  Saprolegnia spp., 99–102
  scale loss, 152, 153
  Screamer disease, 167
  seal attack in, 175
  situs inversus cordis, 172
  skin ulcers, 33, 34
  with skin wound, 152, 153
  soybean enteritis, 158, 159
  steatitis, of peripancreatic fat, 158

D.W. Bruno et al., *A Colour Atlas of Salmonid Diseases*,
DOI 10.1007/978-94-007-2010-7,

Atlantic salmon (*cont.*)
subendocardial fibrosis, 173, 174
swim bladder sarcoma virus, 190
tail abnormality, 169, 170
Tenacibaculum infection, 163
thiamine deficiency, 161
T. maritimum, 33–35
UDN, 179, 180
undigested pellets, 157
vaccination side-effects, 155–156
vaccine induced granulomatous peritonitis, 35, 41
ventral skin haemorrhage, 33, 34, 36
ventricular aneurisms, 193
ventricular cyst, 174
Vibrio infection, 33–35
viral diseases (*see* Viral diseases)
vitamin deficiency
acinar cell atrophy, 161
pseudobranch calcification, 159, 161
white muscle degeneration, 159, 161
well-defined melanoma, 189
with white eye syndrome, 182, 183
Atriomegaly, 171

B
Bacterial diseases
A. hydrophila, 74–75
A. salmonicida, 78
A. salmonicida subsp. salmonicida, 75–77
C. arthromitus, 94–95
and chlamydial diseases, 73, 74
Chryseobacterium spp., 86
C. maltaromaticum, 90
epitheliocystis, 93–94
facultative pathogens, 73
F. columnare, 83–85
F. noatunensis, 87
F. psychrophilum, 82–84
H. alvei, 85–86
L. anguillarum, 76–78
M. viscosa, 79–80
Mycobacterium spp., 91
nocardia, 91
P. fluorescens, 81, 82
primary pathogen, 73
P. salmonis, 91–93
P. skyensis, 86–87
RMS, 95–96
R. salmoninarum, 87–90
S. phocae, 90–91
T. maritimum, 84–86
Y. ruckeri, 80, 81
Bacterial kidney disease (BKD), 33, 37
ascites and splenomegaly, 87, 88
granulomas in kidney, 87, 88
necrotic focus, 88, 89
occurence, 87
Bivalvia, 146–148
BKD. *See* Bacterial kidney disease (BKD)
Blue sac disease, 164, 165
Branchial pallor, 44
Brook trout
M. cerebralis, 127, 128
nephrocalcinosis, 154
P. salvelini, 132
P. umblae, 139
Sanguinicola spp., 137
Brown trout
bacterial diseases (*see* Bacterial diseases)
polycystic syndrome, 181
UDN, 179, 180
B_6 vitamin deficiency, rainbow trout, 159, 160

C
Caligus
C. elongatus, 145
C. rogercresseyii, 145
Candidatus arthromitus, 94–95
Candidatus Paranucleospora theridon, 116–117
Capriniana piscium, 108, 110
Cardiac abnormalities, 170–171
absent septum transversum, 171–172
aneurysms, 173
arteriosclerosis, 172–173
cardiomegaly, 171
hypoplasia of, outer compact myocardium, 173
situs inversus cordis, 172
skewed ventricle, 172
subendocardial fibrosis, 173–174
ventricular cysts, 174
Cardiomegaly, 171
Cardiomyopathy syndrome (CMS)
cardiac tamponade, 67, 68
diffuse ventral scale-pocket oedema/haemorrhage, 67, 68
dilated atrium, 67, 68
epicarditis, 68, 69
myocardial degeneration and inflammation, 67, 68
spongy myocardium
necrosis, 67, 68
thrombus, 67, 69
Carnobacterium maltaromaticum, 90
Caseous necrosis, 46–47
Cataracts, 129, 138, 162–164
Cell death. *See* Necrosis
Ceratomyxa shasta
actinosporean stage, 123–124
characteristic elongate spores, 124, 125
diagnosis, 124
polymerase chain reaction, 124
Cestoda. *See* Tapeworms
Chilodonella piscicola, 110
Chinook salmon
C. maltaromaticum, 90
C. salmositica, 117
D. salmonis, 114
EIBS, 69
I. hoferi, 112, 113
ISA virus, 57
L. salmonae, 115
P. herbarum, 103, 104
plasmacytoid leukaemia, 60
R. salmoninarum, 89
S. destruens, 113
Chloromyxum spp., 124–125
Chondromas, 187–189
Chronic inflammation, 43
Chryseobacterium spp., 86

Ciliophora
Ambiphyra, 112
C. piscicola, 110
C. piscium, 108, 110
I. multifiliis, 110–111
Riboscyphidia, 112
Scyphidia, 112
T. truttae, 111–112
Circulatory disorders
anaemia, 44, 45
ascites, 46
branchial pallor, 44
haemolytic anaemia, 44, 46
haemorrhage, 44–45
hyperaemia and congestion, 45
hypoplastic anaemia, 44
myointimal hyperplasia, 45, 46
oedema, 46
Circulatory system
cardiovascular system, 8–11
secondary circulation, 11
Coagulative necrosis, 46
Coho salmon
C. rogercresseyii, 145
M. minteri, 126, 127
P. minibicornis, 130
Cold water strawberry disease, 96
Congenital malformations
anopthalmia, 169
compressed and fused vertebrae, 169, 170
conjoined twins, 169–171
hypoplasia, of swim bladder, 169
liver, aberrant location of, 169
pyloric caeca absence, 169
tail abnormality, 169, 170
Conjoined twin, 169, 171
Cotylurus spp., 140
Cryptobia salmositica, 117
Cryptocotyle lingua, 138
Cystidicola farionis, 133

D
Dactylogyrous spp., 141
Dermocystidium salmonis, 114, 115
Developmental disorders
blue sac disease, 164, 165
eggs and fry, deformities, 164
soft egg disease, 164
white spot disease, 164–166
yolk sac constrictions, 164
Dietary imbalance, 156–157
heart fat infiltration, 158
hepatic lipidosis, 157
mineral deficiencies, 162
soybean-induced enteritis, 158, 159
steatitis, 158
undigested pellets, 157
vitamin deficiencies, 158–162
Digestive system
intestine, 17–18
liver and gall bladder, 18–19
oesophagus, 17
oral cavity, 15–16
pancreas, 19
pharynx mucosa, 16–17
stomach, 17
swim/air bladder, 19, 20
vent, 18
visceral organs, 15
Dilated ureters, 154
Diphyllobothrium spp., 136–137
Diplostomum spathaceum, 138–139
Discocotyle sagittata, 142
Double mouth deformity, 168

E
Ectopia cordis. *See* Absent septum transversum
EGCs. *See* Eosinophilic granular cells (EGCs)
EIBS. *See* Erythrocytic inclusion body syndrome (EIBS)
Embryonic nephroma. *See* Nephroblastoma
Endocrine system, 22–23
Eosinophilic granular cells (EGCs), 43, 44
Epithelial necrosis, 47
Epitheliocystis, 93–94
Erythrocytic inclusion body syndrome (EIBS), 69, 70
Eubothrium
E. crassum, 136
E. salvelini, 136, 137
Eustrongylides sp., 133, 134
Excretory system, 14–15
Exophiala spp., 104–105
Exophthalmia, 43
Experimentally induced severe hyperplasia, 160, 161
Eye haemorrhage, 43, 162, 163

F
Fibro-elastosis. *See* Subendocardial fibrosis
Fibrosarcomas, 191
Fish health management, 3
Fish necropsy. *See* Necropsy
Flavobacterium
F. columnare, 83–85
F. psychrophilum, 82–84
Flexibacter
F. columnaris, 83–85
F. maritimus (*see Tenacibaculum maritimum*)
Francisella noatunensis, 87
Functional anatomy
brain, 21–22
circulatory system
cardiovascular system, 8–11
secondary circulation, 11
digestive system (*see* Digestive system)
endocrine system, 22–23
excretory system, 14–15
fins, 13
integument system, 11–12
kidney, 14–15
nervous system, 21–22
reproductive system
ovaries, 24
testes, 25
respiratory system (*see* Respiratory system)
sensory systems
mechanosensory lateral line system, 20
visual system, 20–21
skeletal system
body muscle, 12
myotomes, 12, 13
red aerobic muscle, 12–14
sympathetic and parasympathetic ganglions/nerves, 22

Fungal-like infections. *See* Oomycetes infections
Furunculosis, 75–77

G
Gas bubble disease (GBD), 153–154
Gill chondroma, 188, 189
Gill flukes, 141
Glochidiosis, 147–148
Glomerulonephritis, 44
Granulomatous inflammation, 43, 154
Gyrodactylus salaris, 141–142

H
Haemangiomas, 193
Haemorrhagic smolt syndrome (HSS), 183–185
Hafnia alvei, 85–86
Heart and skeletal muscle inflammation (HSMI). *See* Piscine reovirus (PRV)
Heart fat infiltration, 158
Henneguya
  H. cartilaginis, 126
  H. zschokkei, 126
Hepatic carcinomas, 191–192
Hepatocyte erythrophagocytosis, 42
Hepatocytes, vacuolation of, 157
Hepatomas, 191–192
Hirame rhabdovirus (HRV), 64
HSS. *See* Haemorrhagic smolt syndrome (HSS)
Hyaline droplet degeneration, 47
Hyperaemia and congestion, 45
Hypoplasia, of outer compact myocardium, 173

I
Ichthyobodo
  I. necator, 117, 118
  I. salmonis, 118
Ichthyocotylurus erraticus, 140
Ichthyophonus hoferi, 112–113
Ichthyophthirius multifiliis, 110–111
Idiopathic diseases
  HSS, 183–185
  polycystic syndrome, 181–182
  puffy skin, 181, 182
  swim bladder stress syndrome, 183, 184
  UDN (*see* Ulcerative dermal necrosis (UDN))
  white eye syndrome, 182, 183
IHNV. *See* Infectious haematopoietic necrosis virus (IHNV)
Infectious haematopoietic necrosis virus (IHNV)
  appearance of, 62
  diagnosis, 63, 64
  histopathological lesions, 63
  natural outbreaks, 63
Infectious pancreatic necrosis (IPN)
  clinical signs, 53–54
  diagnosis, 54–56
  feature of, 54
  mortality, 53
Infectious salmon anaemia (ISA), 3, 55–58
Integument system, 11–12
Intestinal adenocarcinomas, 192
Inverted operculum, 166
IPN. *See* Infectious pancreatic necrosis (IPN)
Iridophoromas, 189–190
Isaria farinosa, 101–103

J
Jaw and head deformities, 166–168
Jellyfish, 152–153

K
Kabatana takedai, 115, 116
Keratitis and exophthalmia, 162, 163
Kudoa thyrsites, 125–126

L
Leeches, 148
Leiomyosarcoma, 190
Lepeophtheirus salmonis, 143–145
Lesions associated with angling, 176
Lipidosis, 47, 48
Listonella anguillarum, 76–78
Loma salmonae, 115–116
Lymphoid neoplasia, 191
Lymphosarcoma, 191, 192

M
Magnesium deficiency, 162
Mandibular ankylosis, 167
Maxillopoda, 143
M74 condition, 161, 162
Melanoma, 188–189
Melanomacrophages, 48
Mesomycetozoea
  D. salmonis, 114, 115
  I. hoferi, 112–113
  S. destruens, 113–114
Metazoa parasites
  Acanthocephalus spp., 142–143
  A. coregoni, 145–146
  A. foliaceus, 146
  A. gracilis, 139, 140
  A. japonicus, 146
  A. simplex, 133–135
  C. elongatus, 145
  C. farionis, 133
  Chloromyxum spp., 124–125
  C. lingua, 138
  Cotylurus spp., 140
  C. rogercresseyii, 145
  C. shasta, 123–125
  Dactylogyrous spp., 141
  description, 123
  Diphyllobothrium spp., 136–137
  D. sagittata, 142
  D. spathaceum, 138–139
  E. crassum, 136
  E. salvelini, 136, 137
  Eustrongylides sp., 133, 134
  glochidiosis, 147–148
  G. salaris, 141–142
  H. cartilaginis, 126
  H. zschokkei, 126
  I. erraticus, 140
  K. thyrsites, 125–126
  leeches, 148
  L. salmonis, 143–145
  M. cerebralis, 126–127
  Myxidium spp., 126–127
  Nematodes, 131–135

Parvicapsula spp., 128–130
P. decipiens, 133
Pomphorhynchus spp., 143
P. oncorhynchi, 132–133
P. umblae, 139
Salmincola spp., 146–147
Sanguinicola spp., 137–138
S. tenue, 140
S. truttae, 130
T. bryosalmonae, 130–132
Microsporidia
Candidatus Paranucleospora theridon, 116–117
K. takedai, 115, 116
L. salmonae, 115–116
Microstomia, 167, 168
Milky flesh, 125
Mineralization, 48
Monogenea, 141
Moritella viscosa, 79–80
Motile aeromonad septicaemia, 74
Mycobacterium spp., 91
Myointimal hyperplasia, 45, 46
Myxobolus cerebralis, 126–127

N
Necropsy
euthanasia, 28
external examination
brain sample, 30
cool environment, 28
eye examination, 28, 30
gills, 28
histological examination, 28
mouth and oral cavity, 30
orientation planes, 28, 29
skin sample, 28
internal examination, 30–31
role in fish health, 27
sample size, 28
Necrosis, 46–47
Nematodes, 131–132
Neoparamoeba perurans, 107, 109
Neoplasia
adenocarcinoma, 192
basophilic mass, 193
benign and malignant, 48
chondromas, 187–189
fibrosarcomas, 191
haemangioma, 193
hepatic carcinomas, 191–192
hepatomas, 191–192
leiomyosarcoma, 190
lymphoid, 191
lymphosarcoma, 191, 192
nephroblastoma, 192
papillomas, 187, 188
pigment cell
iridophoromas, 189–190
melanoma, 188–189
Neoplasm, 187
Nephroblastoma, 192
Nephrocalcinosis, 139, 154–155
Nervous system, 21–22
Nocardiosis, 91
Nuclear hypertrophy, 36, 41

O
Oncorhynchus masou virus (OMV), 58
Oomycetes infections, 100
Exophiala spp., 104–105
I. farinosa, 101–103
molecular sequencing, 99
P. herbarum, 103, 104
Phialophora sp., 101
Saprolegnia spp., 99–102
Opercula, abnormalities of, 166
Organized blood clot/haemopericardium, 173

P
Paecilomyces farinosus. See Isaria farinosa
Pancreas disease (PD)
cardiomyocytes, focal necrosis/inflammation of, 65, 66
cellularity at junction, 65, 66
chronic, 64, 65
clinical signs, 64, 65
diagnosis, 67
endomysium, inflammation/fibrosis of, 66
histopathological changes, 64–65
myofibre hyaline degeneration, 65, 66
vs. SD, 66
Panopthalmitis, 85, 92, 162, 163
Pansteatitis. *See* Steatitis
Papillomas, 187, 188
Papillomatosis, 33, 34, 36
Parvicapsula spp., 128–130
Pasteurella skyensis, 86–87
Pathology and disease diagnosis
apoptosis, 47
artefacts, 49–50
circulatory disorders (*see* Circulatory disorders)
inflammation and proliferation, 33–44
lipidosis, 47, 48
melanomacrophages, 48
mineralization, 48
necrosis, 46–47
neoplasia, 48
pigments, 48
PD. *See* Pancreas disease (PD)
Peduncle disease, 33–35
Peribilary lymphoctic infiltration, 36, 41
Petechiae, 33, 38
Phialophora sp., 103
Philonema oncorhynchi, 132–133
Phoma herbarum, 103–104
Phyllodistomum umblae, 139
Pigment cell neoplasia
iridophoromas, 189–190
melanoma, 188–189
Pigments, 48
Piscine myocarditis virus (PMV), 67–69
Piscine reovirus (PRV), 59–60
Piscirickettsia salmonis, 91–93
Plasmacytoid leukaemia (PL), 60
Polycystic syndrome, 181–182
Pomphorhynchus spp., 143
Post-mortem examination. *See* Necropsy
Predators
bird stab, 174
sea lampreys, 175–176
seal attack, 175

Production related diseases/disorders
  associated with eye
    cataracts, 162
    eye haemorrhage, 162, 163
    keratitis and exophthalmia, 162, 163
    panopthalmitis, 162, 163
    parasitic infections, 164
    punctured eyeball, 162, 163
    superficial and physical damage, 164
  cardiac abnormalities (*see* Cardiac abnormalities)
  congenital malformations (*see* Congenital malformations)
  developmental disorders (*see* Developmental disorders)
  dietary imbalance (*see* Dietary imbalance)
  environment
    algal blooms, 153
    gas bubble disease, 153–154
    jellyfish, 152–153
    nephrocalcinosis, 154–155
    phytoplankton, 153
    water quality and husbandry, 151–152
  jaw and head deformities, 166–168
  lesions associated with angling, 176
  opercula, abnormalities of, 166
  predators
    bird stab, 174
    sea lampreys, 175–176
    seal attack, 175
  vaccination side-effects, 155–156
  vertebral column malformations, 168–169
Profitable fish farming, reduction of, 3
Proliferative gill inflammation (PGI), 94
Proliferative kidney disease (PKD)
  clinical signs, 130, 131
  diagnosis, 131
  histozoic and extrasporogonic stages, 130
  occurence, 130
Protists
  amoebozoa, 107–108
  ciliophora (*see* Ciliophora)
  description, 107
  mesomycetozoea
    D. salmonis, 114, 115
    I. hoferi, 112–113
    S. destruens, 113–114
  microsporidia
    Candidatus Paranucleospora theridon, 116–117
    K. takedai, 115, 116
    L. salmonae, 115–116
  sarcomastigophora (*see* Sarcomastigophora)
PRV. *See* Piscine reovirus (PRV)
Pseudomonas fluorescens, 81, 82
Pseudoterranova decipiens, 133
Puffy skin, 33–35, 181, 182
Pug-head condition, 167, 168

R

Rainbow trout
  adipose tissue, in abdominal cavity, 156, 157
  aflatoxins, 192
  anaemic gills, 44, 45
  bacterial diseases (*see* Bacterial diseases)
  B_6 vitamin deficiency, 159, 160
  cardiomegaly, 171
  cold water aquaculture, 1
  compressed and fused vertebrae, 170
  dilated ureters, 154
  early red mark syndrome, 33–35
  enteric red mouth, 33, 38
  experimentally induced severe hyperplasia, 160, 161
  fibrosarcoma, 191
  functional anatomy (*see* Functional anatomy)
  gill chondroma, 188, 189
  global production, 1
  granulomatous inflammation, 154
  haemosiderin deposits, 44, 46
  hepatocytes, vacuolation of, 157
  hepatomas, 191
  hyaline droplet degeneration, 47
  hypoplasia of, outer compact myocardium, 173
  iridophore, 190
  keratitis and exophthalmia, 162, 163
  lymphoid neoplasia, 191
  magnesium deficiency, 162
  metazoa parasites (*see* Metazoa parasites)
  mineralization, 48
  nephroblastoma, 192
  nephrocalcinosis, 154
  oedematous epidermal proliferative lesion, 181, 182
  oomycete infections (*see* Oomycetes infections)
  organized blood clot/haemopericardium, 173
  Peduncle disease, 33–35
  protists (*see* Protists)
  puffy skin, 33–35, 181
  with shortened operculum, 159, 161
  skin ulcer, 33
  swim bladder stress syndrome, 183, 185
  vacoluation in liver, 47, 48
  viral diseases (*see* Viral diseases)
  vitamin C deficiency, 158, 159
Rainbow trout gastroenteritis (RTGE), 94, 95
Red mark syndrome (RMS)
  antibiotic treatment, 95
  necropsy, 95
  occurence, 95
  rainbow trout, 33–35
  rickettsia-like organism, 95
  and strawberry disease, 96
Renibacterium salmoninarum, 87–90
Reproductive system
  ovaries, 24
  testes, 25
Respiratory system
  blood flow, in lamella capillaries, 7
  chloride cells, 6–7
  gill arches/filaments, 6, 7
  goblet cells, 6, 7
  intraepithelial lymphocytic cell accumulations, 7, 8
  pseudobranchial cells, 7, 8
Riboscyphidia, 112

S

Salmonid alphavirus (SAV)
  pancreas disease (*see* Pancreas disease (PD))
  SD
    clinical signs, 64, 65
    vs. PD, 66
Sanguinicola spp., 137–138
Saprolegnia spp., 99–102

Sarcomastigophora
C. salmositica, 117
I. necator, 117, 118
I. salmonis, 118
S. barkhanus, 119
S. salmonicida, 119, 120
S. salmonis, 119
Screamer disease, 167
Scyphidia, 112
SD. *See* Sleeping disease (SD)
Sensory systems
mechanosensory lateral line system, 20
visual system, 20–21
Situs inversus cordis, 172
Skeletal system
body muscle, 12
myotomes, 12, 13
red aerobic muscle, 12–14
Sleeping disease (SD)
clinical signs, 64, 65
vs. PD, 66
Soybean enteritis, 158, 159
Sphaerospora truttae, 130
Sphaerothecum destruens, 113–114
Spironucleus
S. barkhanus, 119
S. salmonicida, 119, 120
S. salmonis, 119
Steatitis, 158, 159
Steelhead trout, 124, 125
Stephanostomum tenue, 140
Streptococcus phocae, 90–91
Subendocardial fibrosis, 173, 174
Swim bladder mycosis, 102, 103
Swim bladder sarcoma. *See* Leiomyosarcoma
Swim bladder stress syndrome, 183–185
Sympathetic and parasympathetic ganglions/nerves, 22

T
Tail abnormality, 169, 170
Tapeworms, 135–136
Tenacibaculum infection, 163
Tenacibaculum maritimum, 84–86
Teratoma. *See* Nephroblastoma
Tetracapsuloides bryosalmonae, 130–132
Thiamine deficiency, 161, 162
Tissue abnormalities, recognition of. *See* Necropsy
Trematodes, 137
Trichodina truttae, 111–112

U
Ulcerative dermal necrosis (UDN)
acantholysis, of epidermis, 180
Atlantic salmon, 180
brown trout, 179, 180
dermal disarrangement, 180
epidermal necrosis, 180
histological examination, 180–181
infiltration, 180
pemphigoid-like degeneration, of epidermis, 180
signs, 179
Undigested pellets, 157

V
Vaccination side-effects, 155–156
Vaccine induced granulomatous peritonitis, 35, 41
Ventral skin haemorrhage, 33, 34, 36
Ventricular cyst, 174
Vertebral column malformations, 168–169
Viral diseases
EIBS, 69, 70
HRV, 64
IHNV, 62–64
IPN, 53–56
ISA, 55–58
OMV, 58
plasmacytoid leukaemia, 60
PMV, 67–69
PRV, 59–60
salmonid alphavirus, 64, 66
VHS, 60–62
Viral haemorrhagic septicaemia (VHS)
description, 60
diagnosis, 62
emergence of, 61
endurance, 62
signs, 61
transmission, 61–62
Vitamin deficiency
Atlantic salmon
acinar cell atrophy, 161
pseudobranch calcification, 159, 161
white muscle degeneration, 159, 161
rainbow trout, 158, 159

W
Water quality and husbandry, 151–152
Whirling disease (WD), 127, 128
White eye syndrome, 182, 183

Y
Yersinia ruckeri, 80, 81

Z
Ziehl-Nielsen stain, 91
Zona pellucida, 24

MIX
Papier aus verantwortungsvollen Quellen
Paper from responsible sources
FSC® C105338

If you have any concerns about our products,
you can contact us on
ProductSafety@springernature.com

In case Publisher is established outside the EU,
the EU authorized representative is:
Springer Nature Customer Service Center GmbH
Europaplatz 3, 69115 Heidelberg, Germany

Printed by Libri Plureos GmbH
in Hamburg, Germany